RADIATION DAMAGE PROCESSES IN MATERIALS

NATO ADVANCED STUDY INSTITUTES SERIES

Proceedings of the Advanced Study Institute Programme, which aims at the dissemination of advanced knowledge and the formation of contacts among scientists from different countries.

The series is published by an international board of publishers in conjunction with NATO Scientific Affairs Division.

A	Life Sciences	Plenum Publishing Corporation
B	Physics	London and New York
C	Mathematical and Physical Sciences	D. Reidel Publishing Company Dordrecht and Boston
D	Behavioural and Social Sciences	Sijthoff International Publishing Company Leiden
E	Applied Sciences	Noordhoff International Publishing Leiden

Series E: Applied Sciences – No. 8

RADIATION DAMAGE PROCESSES IN MATERIALS

edited by

C. H. S. DUPUY

Professor of Physical Sciences
Claude Bernard University, Lyons, France

NOORDHOFF – LEYDEN – 1975

Proceedings of the NATO Advanced Study Institute
on Radiation Damage Processes in Materials,
held on Corsica, France, August 27-September 9, 1973

ISBN-13: 978-94-010-1916-3   e-ISBN-13: 978-94-010-1914-9
DOI: 10.1007/978-94-010-1914-9

Copyright © 1975 by Noordhoff International Publishing, a division of A. W. Sijthoff International Publishing Company BV
Softcover reprint of the hardcover 1st edition 1975

V

PREFACE

Thirty years ago, the sharp development of the nuclear phy-sics has given scope to some connected areas such as radiochemistry, radiobiology, radioprotection, radiation damages.

In this last subject - damages induced by radiations in materials, the earlier studies are essentially connected to the mechanism of defect creation. Several workers, for instance, SEITZ (1949), DIENES and VINEYARD (1957), BILLINGTON and CRAWFORD (1961) have developed the first approach in the damage processes theories.

In the 65th years a "saturation effect" occurs in the studies of the mechanisms and correlatively a strong development appears in the physics of the defect itself. If it is possible in many cases to study defects without a good knowledge of their origin many researchs, in particular in the field of defects induced by energetic heavy ions, needs a better understanding of the damage processes.

The track phenomena for instance is of special interest in heavy ions problems, cosmic ray tracks in lunar and meteorite crystals or glasses are a good indicator of the solar activity. On the other hand, color centers, induced by energetic heavy ions in alkali-halides crystals, shown a quite different behaviour than those created by light particles, it is necessary to assume that the ionic bombardement creates centers in a well located region : a core around the path of the incident particle.

Due to these remarks, JH.CRAWFORD and myself have concluded to the necessity of organizing a Summer School in the field of "Radiation Damage Processes in Materials". It was thus possible to make a summary on this subject and, by pannel discussions, to impulse new ideas to find new ways for research.

To increase the efficiency of this Summer School it was decided to class the lectures not in term of materials but in terms of mechanisms : ionizing or elastic collisions processes for instance. This was a sort of wager because specialists used to consider their materials as the "good one" and are often less interested in general processes. However, it seems that the participants think now that this wager has been laid.

In particular the study of damages connected to the energetic particles-solids interactions seems to have raised up a great interest and this year in Clermont-Ferrand the french physicists met at a "Table Ronde" on this subject. I hope in some years another international meeting will perhaps permit to summarised again the knowledge in the field of Radiation Damage Processes .

It is with the aid of many persons that this Summer-School could have performed and I want to acknowledge everybody who have givent support.

In particular, I must emphasize activity of JH.CRAWFORD who gave me the initial support and I want to acknowledge Professor H.CURIEN and Professor W.DEKEYSER for their aid.

Doctor T.KESTER from N.A.T.O. and S.BEDERE from C.E.A. have allowed us to obtain the necessary financial support. I thank them.

The lecturers and Doctor Y.QUERE, Doctor Y.FARGE and Professor P.BARUCH, members of the Scientific Committee have given up the scientific potential to this School. I give them my best acknowledgement.

I am indubted to Doctor JP. de ROCCA-SERRA, Porto-Vecchio's Mayor and J. de ROCCA-SERRA, Chef de la Mission Régionale de Corse, for their very precious aid and I thank them.

Finally, I want to acknowledge my coworkers of the "Département de Physique des Matériaux" de Lyon for their material aid. Specially I thank Doctor A.CACHARD, J.PIVOT and G.CHASSAGNE. I must mention specially Mrs AM.PIVOT, M.PEREZ and N.MARTINOD.

This Summer School was possible because of the financial support of the N.A.T.O. and of the C.E.A. France. The participants and myself are very indubted to these organisms.

TABLE OF CONTENTS

VIII

Part 1

General Principles

ENERGY LOSS OF CHARGED PARTICLES IN SOLIDS

P.SIGMUND

H.C. Ørsted Institute
DK-2100 COPENHAGEN

1. INTRODUCTION

Let a monoenergetic beam of fast charged particles (α-particles, electrons, fission fragments, etc.) bombard a thin foil of solid material. If the foil is thin enough, you will be able to verify -by means of a particle detector- that most of the impinging particles have penetrated the foil. However, even for the thinnest foils that you might be able to produce, you will be forced to appreciate that the beam has not just penetrated without interacting: pratically all detected beam particles have lost part of their energy, most of them have undergone a (possibly slight) change of direction, an α-particle may have captured one or more electrons, and a large variety of ionized and excited states may be observed when the beam consists of heavier atomic particles. In addition, a vast number of secondary processes may be observed such as emitted electrons and photons, products from chemical and nuclear reactions, and sputtered target atoms. Moreover, the target has changed temperature, and may have undergone a change in its physical and chemical properties. The energy required for these processes to go on is largely taken from the incoming beam. Thus, in order to investigate radiation effects caused by energetic charged particles, it is convenient to start by considering the process of energy loss (or stopping).

The stopping of charged particles in matter has played a significant role in the investigation of ato-

mic and nuclear properties from the discovery of cathode and canal rays in the last century, and in particular of α- and β-rays around the beginning of this century. It became evident at a rather early stage that a
detailed knowledge of stopping was necessary in order
to determine the energies of the products of nuclear
reactions in cloud chambers, emulsion, etc. Reversely,
the understanding of energy loss followed perhaps more
closely in the footsteps of atomic theory than most
other applications of quantum theory. Although the study of radiation effects dates back to at least half a
century before the discovery of the atomic nucleus, it
is mainly the nuclear applications, and consequently
the high-energy aspects, that motivated the development
of stopping theory during the first half of this century. Gradually, during the study of radiation effects
in materials, and because of the increasing attention
to ion implantation and sputtering, the low-energy aspects of stopping theory received considerable interest.
A more recent development is the need for accurate energy -loss data for the purpose of surface analysis by
particle beams.

A charged particle interacting with matter loses
energy to :
    i - Excitation and ionization
   ii - Nuclear motion
  iii - Photons
   iv - Nuclear reactions.

The most prominent energy-loss mechanism is usually excitation and ionization, i). This dominates the
slowing-down of α- and β-particles as well as fission
fragments from natural sources. However, energy loss
by photon emission iii) (bremsstrahlung, Cerenkow radiation) is important and may dominate at relativistic
velocities, while energy loss to nuclear motion is most
important at low velocities, in particular for high-
mass ions. Whether or not nuclear reactions iv) influence the analysis of energy loss, depends on the projectile energy, the desired accuracy, and on the projectile-target combination considered.

With a view at the scope of this school it appears
fair to concentrate on energy loss mechanisms i) and
ii) i.e. electronic and nuclear stopping, that dominate
the energy loss of heavy particles at MeV energies and
below, and even light particles (electrons) up to well
into the MeV region. This restriction will not preclude occasional discussion of high-energy phenomena. How-

ever, the basic aim is to provide the necessary background to understand the energy loss of primary bombarding particles in the energy range in question and to provide you with some of the tools you need if you want to work in this field theoretically or experimentally. I shall discuss stopping of ions more than stopping of electrons, mainly because the former phenomenon is closer to my own field of interest. I also concentrate on electronic stopping rather than nuclear stopping, mainly because I have treated the latter aspect in two previous reviews[191]. A major part of these lectures deals with the elementary process of energy loss in a single collision. Obviously, such events are studied experimentally most conveniently under single collision conditions, i.e. in gas targets under low pressure. In many applications of stopping theory, a solid can be considered as being a gas under high pressure, and we shall follow this approach whenever possible. For both reasons, these lectures will deal as much with energy loss in gases as in solids. When considering the effects of many collisions, we normally avoid the situation of complete slowing down, i.e. restrict our attention to reasonably thin bombarded foils. For a review of phenomena in thick targets you are kindly referred to the two previous reviews quoted above.

The literature on penetration phenomena is extensive, and I have made no attempt of giving a complete list of references, nor even of review articles and books. I have tried to give you some of the key references in the beginning of each chapter. If you want to enter this field, it will pay off to get your private copy of Bohr[36] and Fano[78], and have them neatly bound. However, you will find a considerable number of excellent reviews to select your own, favored one, dependent on your scope of interest. By the way, some of the best papers in this field are rather old.

2. ELEMENTARY STOPPING THEORY

General references : Bethe and Ashkin[21]
Bohr[36]
Jackson[108]

It is instructive to briefly outline the classical theory for the stopping of α- or β- particles, which was pioneered by Bohr[37,38]. You can here study a number of useful concepts, and get a qualitative orientation about characteristic energies and lengths. If modified the right way, the theory can be made quantitative to a surprisingly high degree.

2.1. Cross section

There are several, roughly equivalent definitions of the cross section σ_A for some event A. It is useful to be aware of some of them.

i) Classically σ_A is an effective area per target atom "seen" by a beam particle at large distance ; i. e., an event A occurs if and only if the trajectory of a projectile particle is aiming at a point within the area σ_A. Most often, σ_A is the area of a circular disk of width Δp surrounding the target particle at a distance p, the impact parameter (fig.1), so

$$\sigma_A = 2 \pi p \Delta p \qquad (2.1)$$

ii) Considering a thin target with area S and thickness Δx, containing N' target particles (e.g. atoms, electrons, impurity atoms, etc.) per unit volume, the probability for a projectile hitting the target area to give rise to an event A is given by

$$P_A = \frac{\text{"opaque" area}}{\text{total area}} = \frac{n\sigma_A}{S} = N'\Delta x \sigma_A \qquad (2.2)$$

where $n = N'S \Delta x$ is the total number of target particles. Eq.(2.2) is a probability statement and as such does not depend on the applicability of a classical orbital picture, as does eq.(2.1).

iii) With a current density J in the incoming beam, the average number of events A per unit time is given by

$$JSP_A = J n\sigma_A \qquad (2.3)$$

Eq.(2.3) is a prescription for how to measure a cross section experimentally.

iv) Let $n = 1$ in (2.3). Then, σ_A is given by the number of events A per unit time, devided by the incoming beam flux J. This definition provides a convenient basis for quantum mechanical calculations of cross sections.

In the following, all four definitions will be used without (normally) being specified. We note, however, the slight differences between them ; first, i) is the only one making use of a classical orbital picture ; second, ii) and iii) require a massive, but "thin" target, i.e. Δx or n small enough so that $P_A \ll 1$; finally, the assumption of randomness enters in diffe-

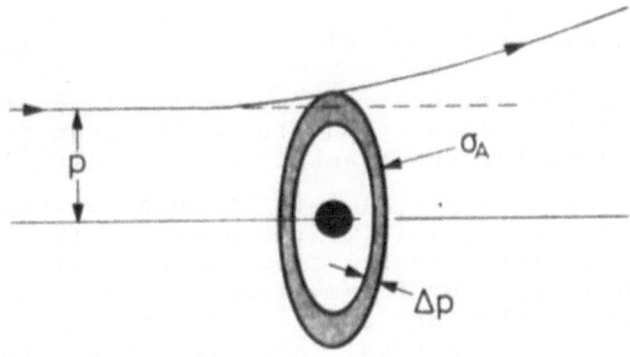

Fig.1 : Classical definition of the cross section σ_A for an event A, the event takes place if and only if the projectile aims at the area σ_A, characterized here by an impact parameter p and a width Δp.

rent ways ; in ii) the target atoms must be distributed at random locations, in iv) it is the beam that
must be homogeneous, and in iii) both assumptions are
made, although only the former is necessary.

2.2. Characterization of Energy Loss

In stopping theory, it is first of all the energy
loss cross section that is of interest. The event A is,
then, taken to be a collision leading to an energy loss
within the interval (T,T + dT). We write

$$\sigma_A = d\sigma\ (E,T) = \frac{d\sigma}{dT}\ dT \qquad (2.4)$$

where E is the particle energy, and call $d\sigma(E,T)$ the
differential cross section. You will often find this
notation used for the quantity $d\sigma/dT = K(E,T)$.

The two main problems in stopping theory are I)
the calculation of $d\sigma(E,T)$ for a given situation from
first principles, and II) from there to arrive at an
energy loss spectrum for a given experimental geometry.
The latter problem may also include the calculation of
the distribution in total range in a thick target.

Let a beam pass a thin layer of thickness Δx with
N' target particles per unit volume. A number of collisions may happen, each giving rise to some energy loss
(T_1, dT_1), (T_2, dT_2), etc., the probability P_i for
each of these being small. Then, the average total
energy loss is given, according to (2.2), by

$$<\Delta E> = \sum_i T_i\ P_i = N'\Delta x \int T\ d\sigma \qquad (2.5)$$

where the integral extends over all possible energy
losses in individual collisions. The quantity

$$\int T \cdot d\sigma = S(E) \qquad (2.6)$$

is called the stopping cross section, and N'S(E) the
stopping power. Let, for the moment, n_i be the number
of collisions with energy loss (T_i, dT_i) in an individual passage, so that $\Delta E = \sum T_i n_i$ and $<n_i> = P_i << 1$.
Then, if subsequent collisions are independent events,
we have :

$$<(\Delta E - <\Delta E>)^2> = <(\sum_i T_i(n_i - P_i))^2> =$$
$$= \sum_i T_i^2 < (n_i - P_i)^2 > \approx$$

$$\approx \sum_i T_i^2 P_i = N' \Delta x \int T^2 d\sigma \qquad (2.7)$$

In the next to last step, the assumption was made that the n_i follow a Poisson distribution around P_i. In order that this be true, many collisions must take place, i.e. Δx may not be <u>too</u> small. We shall come back to this point in sect.5.

Eqs.(2.5) and (2.7) give a simple way to characterize the energy distribution of a beam after passage of a foil, once $d\sigma$ has been found. But remind that (2.7) is less well justified than (2.5)

2.3. <u>Rutherford Scattering</u>

Consider a projectile, mass m_1, charge e_1, penetrating a medium consisting of electrons and nuclei and assume, somewhat uncritically, that you can treat the interaction between the projectile and the target particles by classical mechanics assuming Coulomb's law. If the target particles are free, the scattering is governed by the equations of Kepler motion, where

$$\tan \frac{\theta}{2} = \frac{b}{2p} \qquad (2.8)$$

and θ is the center-of-mass scattering angle, p the impact parameter, and b the collision diameter defined by :

$$\frac{e_1 e_2}{b} = \frac{m_0}{2} v^2 \quad ; \quad m_0 = \frac{m_1 m_2}{m_1 + m_2} \qquad (2.8a)$$

e_2 and m_2 are the charge and mass of a target particle (electron or nucleus). If the target is assumed to be at rest initially, the energy loss follows by energy-momentum conservation,

$$T = T_{max} \sin^2 \frac{\theta}{2} \quad ; \quad T_{max} = \frac{4 m_1 m_2}{(m_1 + m_2)^2} E \qquad (2.9)$$

and, inserting (2.8),

$$T = \frac{T_{max}}{1 + (\frac{2p}{b})^2} \qquad (2.10)$$

Rutherford's cross section reads

$$d\sigma = \pi(\frac{b}{2})^2 \frac{\cos \theta/2}{\sin^3 \theta/2} \, d\theta = \pi(\frac{b}{2})^2 T_{max} \frac{dT}{T^2} \qquad (2.11)$$

in this notation. The latter form is the more convenient one in stopping theory.

From (2.11) you obtain the average energy loss (2.5)

$$< \Delta E > = N'\Delta x \frac{2\pi \, e_1^2 e_2^2}{m_2 \, v^2} \int_{T_{min}}^{T_{max}} \frac{dT}{T} \qquad (2.12a)$$

and the straggling

$$<(\Delta E - <\Delta E>)^2 > = N'\Delta x \frac{2\pi \, e_1^2 e_2^2}{m_2 \, v^2} \int_{T_{min}}^{T_{max}} dT \qquad (2.12b)$$

Alternatively, you can use (2.1) and (2.10) directly to obtain

$$< \Delta E > = N'\Delta x \frac{2\pi \, e_1^2 e_2^2}{m_2 \, v^2} \int_{P_{min}}^{P_{max}} \frac{2p\,dp}{(\frac{b}{2})^2 + p^2} \qquad (2.12a')$$

and a similar expression for straggling. The limits T_{min} and P_{max} have been introduced arbitrarily since the integrals (2.12a) and (2.12a') diverge for $T_{min} = 0$ and $P_{max} = \infty$. In (2.12a'), we have $P_{min} = 0$.

Disregarding the divergences for a moment, we first investigate the relative contributions of electrons and nuclei to (2.12). Take a monatomic target (atomic number Z_2 ; nuclear mass M_2) with N nuclei per unit volume, then,

$$m_2 = M_2 \quad ; \quad e_2 = Z_2 e \quad ; \quad N' = N \text{ for nuclei} \qquad (2.13a)$$

$$m_2 = m \quad ; \quad e_2 = -e \quad ; \quad N' = Z_2 N \text{ for electrons} \qquad (2.13b)$$

where m and -e are the electron mass and charge, respectively. From (2.12) and (2.13), we obtain

$$<\Delta E>_n = N\Delta x \frac{2\pi e_1^2 Z_2^2 e^2}{M_2 v^2} \log \frac{T_{max}}{T_{min}} \qquad (2.14a)$$

$$<\Delta E>_e = N\Delta x \frac{2\pi e_1^2 Z_2 e^2}{m v^2} \log \frac{T_{max}}{T_{min}} \qquad (2.14b)$$

$$<(\Delta E - <\Delta E>)^2>_n = N\Delta x \frac{4\pi e_1^2 Z_2^2 e^2 m_1^2}{(m_1+M_2)^2}(1-\frac{T_{min}}{T_{max}}) \quad (2.14c)$$

$$<(\Delta E - <\Delta E>)^2>_e = N\Delta x \frac{4\pi e_1^2 Z_2 e^2 m_1^2}{(m_1+m)^2}(1-\frac{T_{min}}{T_{max}}) \quad (2.14d)$$

T_{min} is normally different for nuclei (index n) and e-
lectrons (index e).

The following qualitative conclusions can be drawn
from eqs. (2.14). First, within the range of validity
of eqs.(2.14), the electronic energy loss is much greater than the nuclear one. Disregarding the logarithmic
terms, we obtain :

$$\frac{<\Delta E>_n}{<\Delta E>_e} \sim Z_2 \frac{m}{M_2} \ll 1 \quad (\text{normally} \sim \frac{1}{4000}) \qquad (2.15)$$

Second, the average energy loss depends primarily on
the projectile velocity rather than its energy, i.e.
an electron and a proton with the same velocity have
roughly the same energy loss per path length ; if they
have the same energy, the electron suffers much less
energy loss than the proton.

Third, the energy loss decreases with increasing
energy except when T_{max} approaches T_{min}.

For the straggling the relation corresponding to
(2.15) reads

$$\frac{<(\Delta E - <\Delta E>)^2>_n}{<(\Delta E - <\Delta E>)^2>_e} \sim \frac{(1+m/m_1)^2}{(1+M_2/m_1)^2} Z_2 \qquad (2.16)$$

If the incident particle is an electron ($m_1=m$), this
quantity is extremely small, while for heavier projectiles the nuclear contribution may become sizable.

You may wonder how to explain qualitatively the enormeous difference in electronic and nuclear stopping that is evident from (2.15). Take, as an example, the case of hydrogen gas, where the interaction force between the projectile and the nuclei is the same as the one between the projectile and the electrons except for the sign. Since the distribution in space is random, the only essential difference between electrons and nuclei within the present picture is their mass. Integrating, for a given impact parameter, the interaction force over time, you obtain the transferred momentum, which is roughly the same for electrons and nuclei except for the sign. However, in order to obtain the transferred energy, you divide the square of the momentum by twice the mass, and this causes the large difference between electrons and nuclei as target particles.

We conclude that, in the velocity range where the present picture is at least qualitatively correct, the average energy loss of charged particles is determined predominantly by the target electrons rather than the nuclei. This is the reason why we shall devote considerable space in the following to a more sophisticated treatment of electronic stopping. However, the nuclear effects should not be forgotten, not only do they give rise to all those radiation effects that originate in atomic displacement, but they also play an important role in the scattering and hence the trajectory of the bombarding particle. Moreover, the stopping due to nuclear collisions may dominate at lower velocities when the present picture does not apply.

2.4. Adiabaticity

Reference : Bohr[37]

One necessary (and obvious) condition for the description of electronic stopping given in the previous section to be valid, is that the projectile velocity is large compared to the velocity of the most loosely bound target electrons, i.e.

$$v \gg v_0 = e^2/\hbar \qquad (2.17)$$

In the opposite case, i.e., $v \ll v_0$, all electrons are capable of adjusting their orbital motion in accordance with the instantaneous position of the projectile and, therefore, are expected to absorb comparatively little energy from the latter during a scattering event. This is the first part of Bohr's adiabaticity argument.

Even when (2.17) is fulfilled, a collision event may be adiabatic when the collision time τ is large compared to the orbiting period of the electrons. In fact, for an impact parameter p, the collision time τ is of the order of

$$\tau \sim 2p/v \qquad (2.18)$$

hence collisions with

$$p \gg p_0 = v/\omega_0 \; ; \; \omega_0 = \text{revolution frequency} \quad (2.19)$$

will be adiabatic, i.e. <u>classically</u> lead to a much smaller energy loss than anticipated in the previous section that dealt with free target electrons. Hence, p_0 is a rough estimate for a maximum impact parameter p_{max} for collisions leading to electronic energy loss.

For outer-shell electrons, you may apply the order-of-magnitude estimate $\hbar \, \omega_0 \sim e^2/a_0$, where a_0 is the Bohr radius. Then, (2.19) reads

$$p_0 \sim \frac{v}{v_0} \, a_0 \qquad (2.21)$$

This is a rather significant result. Since (2.21) characterizes an effective range of the Coulomb interaction, it does not only define an upper integration limit for the integral (2.12a') in case of electronic stopping but, even more important, it shows that electronic stopping acts over a fairly large distance, certainly, for $v \gtrsim v_0$, over the dimensions of an atom, and much longer for $v \gg v_0$. This has the following consequences :

i) The two-particle Rutherford cross section (2.8) is not a sufficient basis for electronic-stopping theory.

ii) At high velocities, in dense media like solids or liquids, a moving projectile interacts with the electrons of many atoms at a time. Collective effects (polarization) must become significant.

The first problem is avoided in most treatments by the use of perturbation theory from the beginning. The second one was the main motivation to treat electronic stopping within the framework of electromagnetic theory84. The latter approach in particular turned out to be extremely fruitful .

2.5. Perturbation Approach

References : Bohr[37]

Fermi[84]

Let us first go back to the case of a projectile interacting with free electrons. The perturbation approach is based on the assumption that the projectile-target interaction is "weak". Then as a zero'th approximation, the motion of the collision partners is the same as with no interaction, i.e. uniform motion of the projectile, and the target electron at rest.

By use of a well-known argument, called momentum approximation, you can find the momentum transferred to the target by integrating the interaction force over the straight-line trajectory. Hence, the energy loss becomes

$$T = \frac{1}{2m_2} \left(\int_{-\infty}^{\infty} \frac{e_1 e_2 p}{(p^2 + v^2 t^2)^{3/2}} \, dt \right)^2 = \frac{2e_1^2 e_2^2}{m_2 v^2} \cdot \frac{1}{p^2} \qquad (2.22)$$

Although this is just the lowest order of an expansion of the two-particle expression (2.10) in terms of $(e_1 e_2)^2$, the advantage of (2.22) is that it applies equally well to a situation where the projectile interacts simultaneously with many target electrons, provided that the mutual interaction between the target electrons is neglected.

In calculating the stopping cross section $\int 2\pi p \, dp \, T(p)$ from (2.22), a new divergence occurs at $p = 0$. One way to remove this divergence is to split up the integral into two parts, one applying to "single-particle" interactions, $p = 0 \ldots p^*$ where (2.10) is used, the other applying to "collective" interactions, $p = p^* \ldots p_0$. From (2.10) and (2.22), a rough estimate for p^* is $p^* \sim b/2$. In practice this implies using (2.10) for $p = 0 \ldots p_0$, hence

$$\langle \Delta E \rangle = NZ_2 \, \Delta x \, \frac{4\pi \, e_1^2 \, e^2}{m \, v^2} \log \frac{2p_0}{b} \qquad (2.23)$$

with $\frac{2p_0}{b} \sim \frac{m_0 v^3}{e_1 e_2 \omega_0}$. This is similar to Bohr's[37] stopping formula, except for a constant in the logarithm that arises from the specific assumption of classical harmonic binding forces of the electrons, and a more careful treatment by Bohr of the electrons in different shells.

Fermi[84] applies (2.22) for $p = p_{min} \cdots p_0$, and sets $p_{min} = \lambda = \hbar/mv$ for $\lambda > b/2$, by the argument that the classical picture breaks down for impact parameters smaller than the De Broglie wavelength. Using this, the logarithm in (2.23) reads

$$\log \frac{mv^2}{\hbar\omega_o} \qquad\qquad (2.23')$$

which is formally very similar to Bethe's formula discussed below.

2.6 Simple Dielectric Description

Reference : Landau and Lifshitz[121]

The perturbation treatment sketched in the foregoing paragraph has a somewhat too narrow range of applicability. However, the one assumption that the projectile moves essentially uniformly over a distance of many atomic diameters appears to be well justified, even when the momentum approximation breaks down. Hence, at least for the distant interactions, a field description of the stopping appears appropriate. This description, that was first suggested by Fermi[84a], and developed later by Fermi[84b], Kramers[119], Lindhard[129] and many others, can be developed at several levels of sophistication. It allows very conveniently to include the magnetic interaction between the projectile and the medium ; more important in the present context, the restriction to high velocities is, although necessary, much less stringent than (2.17).

The formulation presented in this section goes just as far as necessary to show some of the advantages of the dielectric description, and makes use of the conventional concept of a dielectric constant. The description will be refined in sect.3.

We start with a projectile (e_1, m_1) moving uniformly, i.e., $\vec{r} = \vec{v} t$. The electrostatic field in vacuum is determined by Poisson's equation

$$\nabla^2 \phi(\vec{r}, t) = -4\pi \, e_1 \delta(\vec{r} - \vec{v}t) \qquad\qquad (2.24)$$

where ϕ is the scalar potential. By Fourier transform, the solution of (2.24) becomes

$$\phi(\vec{r},t) = \frac{1}{(2\pi)^3} \int d^3k \, \frac{4\pi e_1}{k^2} \, e^{i\vec{k}(\vec{r}-\vec{v}t)} \qquad (2.25)$$

Since (2.25) represents an analysis in terms of the frequency components $\omega = \vec{k}.\vec{v}$, the field in a medium that is characterized macroscopically by a frequency -dependent dielectric constant $\varepsilon(\omega)$ becomes

$$\vec{E}(\vec{r},t) = \frac{1}{(2\pi)^3} \cdot \int d^3k \; \frac{-4\pi i e_1 \vec{k}}{k^2 \varepsilon(\vec{k}.\vec{v})} \; e^{i\vec{k}.(\vec{r}-\vec{v}t)} \qquad (2.26)$$

The force acting on the projectile is then given by

$$\vec{F} = e_1\left(\vec{E}(\vec{v}t,t) - \vec{E}_{vac}(\vec{v}t,t)\right)$$

i.e. the field is taken at the position $\vec{r}=\vec{v}t$, and $\vec{E}_{vac}$ is the field for $\varepsilon = 1$, with no medium present. Hence,

$$\vec{F} = -\frac{i e_1^2}{2\pi^2} \int d^3k \; \frac{\vec{k}}{k} \; (\frac{1}{\varepsilon(\vec{k}.\vec{v})} - 1) \qquad (2.27)$$

For reasons of symmetry, $\vec{F}$ must be directed opposite to $\vec{v}$, further it must be real, and its numerical value is equal to the stopping power dE/dx. Transforming (2.27) to spherical coordinates, and substituting $\vec{k}.\vec{v} = \omega$ yields the stopping power

$$\frac{dE}{dx} = -\frac{e_1^2}{\pi v^2} \; J_m \int_0^\infty \frac{dk}{k} \int_{-kv}^{kv} \omega d\omega (\frac{1}{\varepsilon(\omega)} - 1) \qquad (2.28)$$

This expression has first been derived by Lindhard[129] under slightly more general assumptions (see sect.3) Landau and Lifshitz[121] offer a slightly different, but equivalent, integral expression.

As it stands, eq.(2.28) is a convenient tool for the calculation of the long-range part of the stopping power. It is based on perturbation theory in the sense that the reaction of the projectile on the disturbance it creates is neglected (no slowing-down, no scattering) but it goes beyond the perturbation approach in that the reaction of the medium upon the disturbing field can be characterized much more adequately than by the perturbation approach presented in the foregoing chapter. Thus long-range screening of the Coulomb-field of the projectile charge is included except for retardation effects.

The major shortcoming of the present formulation is that it does not allow for rapid spatial variations of the electric field, i.e., close collisions are not described properly. This difficulty has been overcome by Lindhard[129] by introducing a wavenumber dependence in the dielectric constant. We shall make use of this concept later, in connection with the quantum theory of stopping.

2.6.1. Application : Individual Oscillators

We consider a gas at low pressure, e.g. hydrogen gas. To a good approximation, the electrons may be considered classical oscillators with the dielectric constant

$$\varepsilon(\omega) = 1 + \frac{4\pi N Z_2 e^2}{m} \sum_\nu \frac{f_\nu}{\omega_\nu^2 - \omega^2 - i\omega\Gamma_\nu} \qquad (2.29)$$

where ω_ν are the eigenfrequencies of the electrons, Γ_ν the corresponding damping constants, and f_ν the oscillator strengths satisfying the sum rule

$$\sum_\nu f_\nu = 1 \qquad (2.30)$$

(strictly speaking, in classical theory we have $f_\nu = Z_2^{-1}$ for $\nu = 1 \dots Z_2$ and $f_\nu = 0$ for $\nu > Z_2$; however, introducing oscillator strengths at this point does not complicate the formalism, but brings the results much closer to reality). Within the range of validity of (2.29), $\varepsilon(\omega)$ is close to 1 and hence, up to first order in N,

$$\frac{1}{\varepsilon(\omega)} - 1 \approx - \frac{4\pi N Z_2 e^2}{m} \sum_\nu \frac{f_\nu}{\omega_\nu^2 - \omega^2 - i\omega\Gamma_\nu} \qquad (2.31)$$

Inserting this into (2.28) and assuming Γ_ν small, you obtain after integration

$$\frac{dE}{dx} = \frac{4\pi N Z_2 e_1^2 e^2}{mv^2} \sum_\nu f_\nu \log \frac{k_{max}v}{\omega_\nu} \qquad (2.32)$$

where k_{max} has been introduced as a cut-off in the wave number spectrum and can be determined similarly as p_{min} in ch.2.5. At present, the key point is that k_{max} hinges on close collisions, and thus does not depend on ν. Making use of (2.30), and introducing the "average ionisation potential" I by the definition

$$\sum_\nu f_\nu \log \hbar\omega_\nu = \log I \qquad (2.33)$$

18

(2.32) reads

$$\frac{dE}{dx} = \frac{4\pi N e_1^2\, e^2}{mv^2}\, Z_2\, \log \frac{\hbar k_{max} v}{I} \qquad (2.34)$$

Setting, as a crude estimate, $\hbar k_{max}$ equal to the maximum momentum transfer,

$$\hbar k_{max} = 2m_0 v \qquad (2.35)$$

you obtain

$$\frac{dE}{dx} = \frac{4\pi N Z_2 e_1^2\, e^2}{mv^2}\, \log \frac{2m_0 v^2}{I} \qquad (2.36)$$

which is identical with the nonrelativistic form of Bethe's formula. Since the assumptions entering (2.36), apart from the use of oscillator strengths, are those of Bohr's classical treatment[37], and since (2.29) can be derived under identical assumptions as in the conventional derivation of Bethe's formula (see sect.3), the intimate connection between classical and quantal stopping theory becomes evident already at the present stage, but will be followed up more closely.

2.6.2. Application : Free Electron Gas

In the Drude theory of the free electron gas (Jackson[108]), the dielectric constant takes the simple form

$$\varepsilon(\omega) = 1 - \frac{\omega_p^2}{\omega^2} \qquad (2.29')$$

where ω_p is the plasma frequency. If (2.29') is inserted into (2.28), the integral reduces to a similar form as the one in the previous paragraph, and the energy loss becomes

$$\frac{dE}{dx} = \frac{e_1^2 \omega_p^2}{v^2}\, \log \frac{2m_0 v^2}{\hbar \omega_p} \qquad (2.36')$$

where again an upper integration limit (2.35) was introduced. This formula was first derived by Kramers[119]. Because of the long range nature of the present description, its validity is confined to the velocity range $v \gg v_0$; it will be shown later how proper inclusion of short-range interactions modifies eq.(2.36') considerably.

2.7. Relativistic Effects

References : Bohr[39]
Jackson[108]
Landau and Lifshitz[121]

The treatment of relativistic effects will be cursory since they are of minor importance within the scope of this school. We can distinguish between the following effects :

   i) Modifications of the electric interaction at high velocities
  ii) Collective effects
 iii) Magnetic interaction
  iv) Bremsstrahlung.

The first effect is the most important one at moderate velocities. To estimate its significance, we note that, according to (2.22), the dominating contribution to the stopping are distant collisions with a momentum transfer roughly perpendicular to the velocity of the projectile. The electric field component in this direction, according to the Lorentz transformation for the fields, increases by a factor of $\gamma = (1-v^2/c^2)^{-1/2}$. However, this increase is compensated by a corresponding decrease of the length of the interaction region, and hence the expression for the momentum transfer at large impact parameters does not change. However, the change in collision time does affect the adiabatic limit (2.19). This, together with another term that arises via the field component parallel to the velocity of the projectile* causes a correction to the various stopping formulas derived so far. In case of eq.2.36, the correction can be incorporated by replacing the logarithmic term by (Bohr[38])

$$\log \frac{2m_o v^2}{I} - \log \left(1-\frac{v^2}{c^2}\right) - \frac{v^2}{c^2} \qquad (2.37)$$

---

* This component, which has been treated in a consistent manner by Bohr[37], has largely been neglected in this presentation. In the perturbation approach (ch. 2.5) it gives a vanishing contribution, in Bohr's treatment it yields a factor of 1.261 under the logarithm. In order to recover this factor in the dielectric treatment (sect.2.6), one would need a more carefull consideration of k_{max}.

Obviously, for v → c, the variation of the stopping power as a function of velocity is determined predominantly by the term $-\ln(1-v^2/c^2)$, hence the stopping
power <u>increases</u> with increasing v in this limit, whereas
it <u>decreases</u> with increasing v in the nonrelativistic
limit (2.36).

In the extreme relativistic region, strong polari-
sation effects (ii) occur that limit the size of the
interaction region and hence tend to cause the stopping
power to saturate ;(this is often called density effect). At the same time, part of the energy loss goes
into Cerenkov radiation. Proper calculation of the e-
nergy loss in this region is a nontrivial task and has
attracted the attention of many theoreticians since
Fermi[84], e.g. Aa.Bohr[39], Halpern et al[196], Sternheimer[195].
A most recent treatment of this complex (i)-iii) together) has been given by Brynjolfsson[58].

Bremsstrahlung (iv) is an effect connected predominantly with nuclear rather than electronic collisions
and thus somewhat unrelated with the processes discussed up to now. The main reason why it is mentioned here is its significant contribution to the stopping power of high-energy particles. The classical, nonrelativistic and relativistic theory has been summarized
by Jackson[108]. Results based on quantum mechanical perturbation theory (Sauter[182], Bethe and Heitler[22]) have
been reviewed by Bethe and Ashkin[21]. The latter give
the following estimate for the ratio between the energy losses due to radiation (Bremsstrahlung) and (electronic) collisions

$$\frac{(dE/dx)_{rad}}{(dE/dx)_{coll}} \sim \frac{E_{tot}\, Z_2}{1600\ mc^2} \qquad (2.38)$$

for relativistic electrons (E_{tot} is the total energy,
including rest energy). For heavy projectiles, this
expression would have to be multiplied by $(Z_1 m/M_1)^2$;
at nonrelativistic velocities, Jackson[108] obtains

$$(\frac{dE}{dx})_{rad} = \frac{16}{3}\, NZ_2\, \frac{Z_2 e^2}{\hbar c}\, \frac{e_1^4}{m_1 c^2} \qquad (2.39)$$

which is extremely small for heavy projectiles, while
some caution may be required for electrons.

2.8. Nuclear Collisions

References : Bohr[39]
Lindhard et al[131]
Sigmund[191] a,d

Although it was mentioned in ch.2.3 that nuclear stopping is a relatively small effect in the Rutherford collision region, some attention needs to be given to this effect, in particular with a view at its significance in the theory of radiation effects.

As in the case of electronic stopping, the divergence of the stopping power at small energy transfers (2.12a) or large impact parameters (2.12a') needs some consideration. It is easily seen that an adiabatic cut-off is not relevant in case of solids (the limiting impact parameter would be about 1000 times as great as the one given in (2.21) and not existent in gases). A natural cut-off is given through the screening of the nuclear charge by the electrons of the target nucleus and, in case of heavy projectiles, of the projectile. In cases where the target atoms are ionized (plasmas) the screening is provided more collectively. This latter case will be left out of consideration here.

Assume a screening parameter a for the atomic interaction that is of the order of a_o (the Bohr radius); then the logarithmic term in the nuclear stopping power (2.12a') becomes, roughly, $\ln(a/b)$, and this becomes small when $b \sim a$, or when the dimensionless energy measure

$$\varepsilon = \frac{a}{b} = \frac{m_o v^2/2}{e_1 e_2/a} \qquad (2.40)$$

is of the order of 1. Above this energy, the nuclear stopping power behaves roughly as E^{-1}, but below this value it is expected to decrease rapidly. While the reduced mass m_o in (2.40) is of the order of nuclear masses, the corresponding quantity in the electronic stopping power (2.36) is of the order of electronic masses. Hence, the "cut-off" or, more realistically, the maximum of the electronic stopping power occurs at much higher velocities than the maximum of nuclear stopping. Consequently, below the adiabatic limit where the electronic stopping power decreases with decreasing energy, the nuclear stopping power still increases for a considerable while and becomes competitive, and often dominating for heavy ions, at low velocities.

At energies where $\varepsilon < 1$, the description of nucle-

ar collisions by truncated -Coulomb interactions bra-
kes down, and a proper treatment of the scattering on
screened Coulomb potentials becomes essential. Bohr[39]
used the term "excessive screening" for the region
$\varepsilon \lesssim 1$. This is now conveniently called the Thomas-Fermi
region, since atomic interactions in the range of dis-
tances up to a few screening radii are described with
reasonable accuracy by the Thomas-Fermi model of the
atom. The term "excessive screening" can then be reser-
ved for distant interactions involving only the outermost atomic shells.

Scattering cross sections for heavy projectiles in
the Thomas-Fermi and excessive-screening region have
been reviewed previously (Sigmund[191 a,d]). We briefly
sum up the main results.

i) In the Thomas-Fermi region, Lindhard et al[131]
   and Firsov[86 a,b] suggested the following form
   of the interatomic potential,

$$V(r) = \frac{Z_1 Z_2 e^2}{r}\, \phi_0(r/a) \qquad (2.41)$$

with

$$a = 0.8853\ a_0 (Z_1^{2/3} + Z_2^{2/3})^{-1/2} \qquad (2.42a)$$

or

$$a = 0.8853\ a_0 (Z_1^{1/2} + Z_2^{1/2})^{-2/3} \qquad (2.42a)$$

and ϕ_0 a universal screening function, taken to be
the screening function of a neutral atom, either
according to the straight Thomas-Fermi model or
the Lenz-Jensen version.

ii) Accurate computer codes exist to calculate scat-
    tering angles and cross sections from a potential
    like (2.41) by means of classical scattering theo-
    ry. One of several extensive tabulations has been
    published by Robinson[176].

iii) An approximate expression for the scattering
     angle and the cross section that depends only on
     one variable instead of two, and that is accurate
     within a few percent in the Thomas-Fermi region,
     has been derived by Lindhard et al[131]. For the
     scattering angle, they write

$$\frac{\theta^2}{4} = - \frac{3}{16E_r^2} p^{1/3} \frac{d}{dp} \{v^2(p)p^{2/3}\} \tag{2.43}$$

where

$$E_r = \frac{m_0}{2} v^2 \tag{2.43a}$$

This yields the cross section

$$d\sigma = \pi a^2 \frac{d\, t^{1/2}}{t} f(t^{1/2}) \tag{2.44}$$

where

$$t^{1/2} = \epsilon \sin \theta/2 \tag{2.44a}$$

and f is a function determined by ϕ_0 and drawn up in fig.2a for two representative cases. Figs. 2b and 2c show the stopping power and straggling.

iv) Measurements of the angular distribution of scattered light and heavy ions (Andersen et al[2], Hvelplund et al[106], Loftager et al[135], v.Wijnga-arden et al[210]) tend to confirm the qualitative features of the theory, while some disparity e-xists about the detailed shape of the screening function (fig.2a). It appears desirable to have experimental results available on a greater varie-ty of ion-target combinations.

v) Measurements of multiple scattering of ions are only partially in accordance with the theore-tical expectations (Meyer[146]), but it is not cle-ar at present whether the existing disparities (e.g.Andersen et al[2]) originate in the basic scat-tering cross section or another effect.

vi) Calculations of interatomic potentials in the region of exclusive outer-shell interaction (e.g. Gaydaenko & Nikulin[89], Gilbert & Wahl[91]) indicate a tendency towards exponential shape of the poten-tial (Born-Mayer potential) in the medium and lo-wer eV region,

$$V(r) \approx A_e^{-\alpha r} \tag{2.45}$$

24

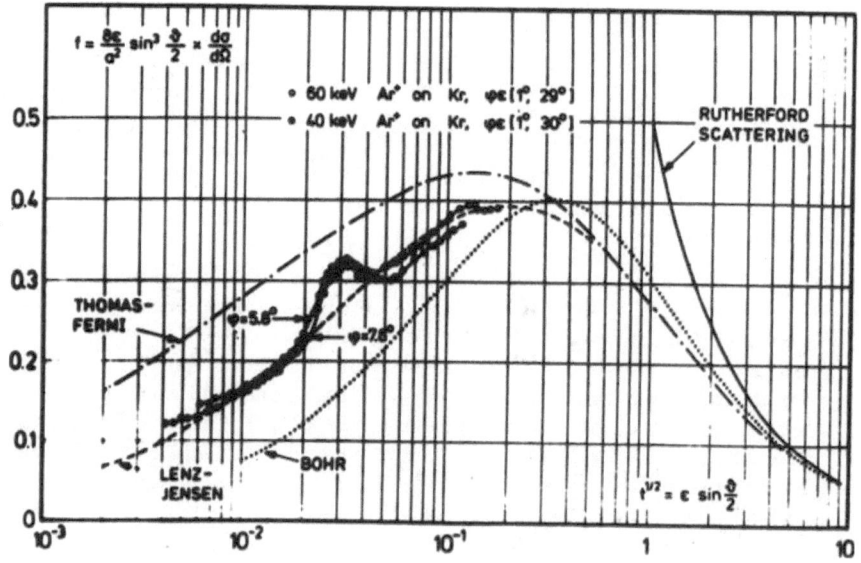

Fig.2a : Calculated and measured differential cross sections for ion-atom scattering. Calculated curves refer to eq.(2.44) and are based on three different screening functions in eq.(2.41). Measured cross sections are from the ion-gas collision work of Loftager & Claussen[135]. Courtesy of P.Loftager (unpublished).

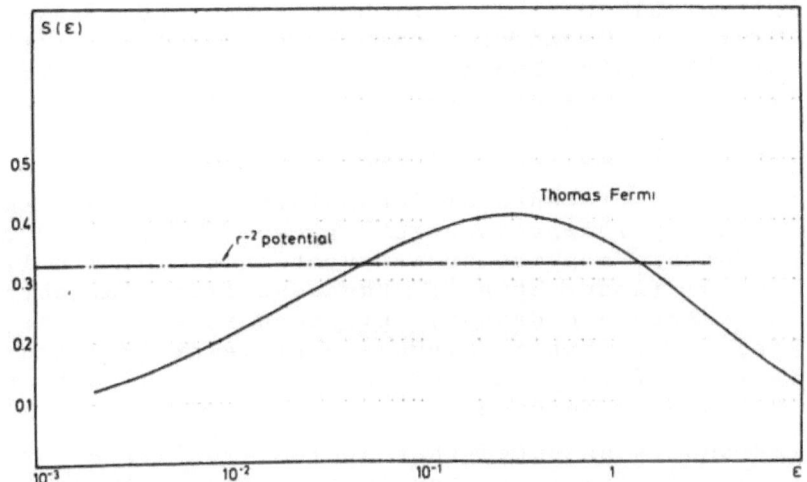

Fig.2b : Nuclear stopping power based on Thomas-Fermi interaction. In dimensionless units (2.40) and $\int Td\sigma = s(\varepsilon)4\pi Z_1 Z_2 e^2 aM_1/(M_1+M_2)$. From Lindhard et al[131].

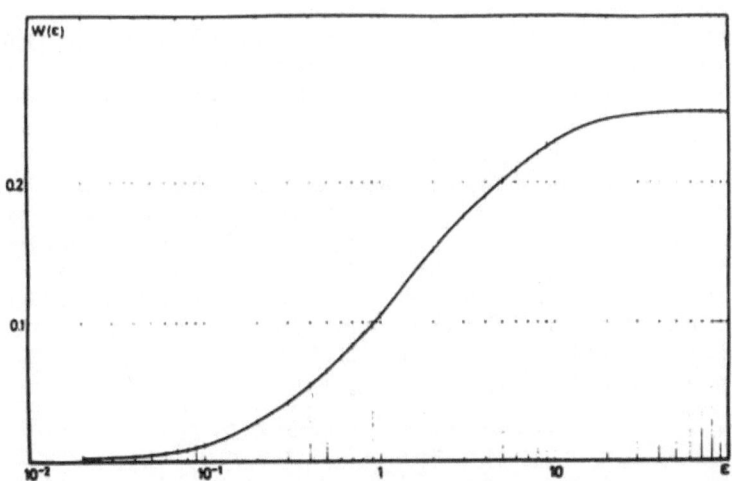

Fig.2c : Energy loss straggling due to nuclear collisions based on Thomas-Fermi interaction. In dimensionless units (2.40) and $\int T^2 d\sigma = w(\varepsilon)\pi(4M_1 Z_1 Z_2 e^2/(M_1+M_2))^2$. From Lindhard et al[131].

Potentials of this form have been used for many years on a semiempirical (sometimes "semi-tentative") basis. The screening radius $1/\alpha$ occuring in (2.45), being determined by outer-shell interactions, is not related in an obvious way to the Thomas-Fermi radius (2.42).

vii) Measurements of incomplete total cross sections by ion-gas scattering experiments lend considerable support to interaction potentials like (2.45) (Amdur et al[1]) ; the conditions for obtaining accurate data by this technique have been redefined recently (Sigmund & Lillemark[192]) and a straightforward method to invert cross sections has been developed.

It should be noted that nuclear collisions are not a phenomenon of mere energy loss, but are always accompanied by significant scattering, except when very heavy ions, say gold ions, collide with a very light target, say hydrogen. Therefore, the direct measurement of integrated quantities like stopping power or straggling is connected with significant difficulties. Since this is a rather central problem in connection with current work in the field of ion penetration, we shall go into more detail in sect.5.

Finally, a word may be said about the "primary recoil spectrum". This is a well-known concept in radiation damage theory, especially when the source of radiation delivers neutrons or electromagnetic radiation. When a beam of charged particles bombards a thin foil, i.e. of thickness small compared to the particle range, energy loss in the foil is negligible as a first approximation, and the primary recoil spectrum given by (2.2) $N\Delta x\ d\sigma(E,T)$, where T is now the recoil energy and Δx the foil thickness. For Rutherford scattering in particular, one obtains the characteristic T^{-2} spectrum (2.11) with a sharp cutoff at $T = T_{max} = \gamma E$, where $\gamma = 4M_1M_2/(M_1+M_2)^2$.

If you irradiate with charged particles you will often find, contrary to the situation with neutrons, that your target is so thick that considerable energy loss takes place. You should be aware of the fact that the primary recoil spectrum then changes shape. In the extreme case of complete slowing down, you may write :

$$F(T)\ dT \cong NdT \int_0^R dx\ \frac{d\sigma(E',T)}{dT} \qquad (2.46)$$

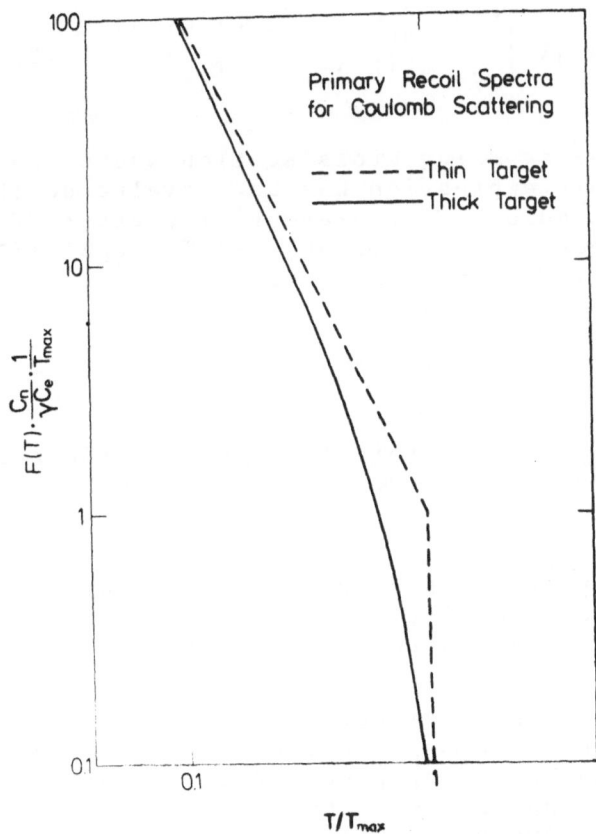

Fig.3 : Primary recoil spectra from Rutherford collisions. Full-drawn line : Eq (2.48), complete slowing-down of the primary ion. Dashed line : Rutherford spectrum neglecting energy loss.

where $E' = E'(x)$ is the projectile energy as a function of travelled path length x and R the total path length down to zero energy. Replacing $dx = dE'/(dE'/dx)$ and recalling that $d\sigma/dT = 0$ for $T > E'$, we can rewrite (2.46) in the form

$$F(T)dT \cong dT \int_{T/\gamma}^{E} \frac{dE'}{S(E')} \frac{d\sigma(E', T)}{dT} \qquad (2.47)$$

where S is the stopping cross section and E the initial energy. This expression has been evaluated (Sigmund[191]) for a number of representative cases. For example, for Rutherford scattering, $d\sigma/dT = C_n E'^{-1} T^{-2}$, and $S(E') \cong S_e(E') \cong C_e/E'$, you obtain

$$F(T)dT = \frac{C_n}{\gamma C_e} \frac{dT}{T^2} (T_{max} - T) \qquad (2.48)$$

The difference between this spectrum and the thin-foil estimate is most pronounced near the cutoff (fig.3).

2.9. Summary

i) Energy loss has been treated in this section, on the basis of classical mechanics, as a sequence of scattering processes between a projectile and individual electrons as well as individual nuclei.

ii) On the basis of Rutherford's cross section, electronic energy loss gives the dominating contribution to the total stopping power. Both the electronic and nuclear stopping power show a logarithmic divergence at small energy transfers.

iii) Within the classical treatment, the divergence of the electronic stopping power is removed by taking into account the harmonic coupling of the electrons either to individual atoms or, via the plasma frequency, to an electron gas.

iv) The adiabatic limit occurs for $v \gg v_0$ at impact parameters corresponding to several atomic diameters. Therefore, electronic energy loss is of collective nature to a considerable extent in this velocity region. It is appropriate, therefore, to use a dielectric treatment of electronic stopping.

v) Both the dielectric treatment and the individual-particle one assume, in the form discussed up to now, that the projectile velocity is large compared to v_o. For light ions and electrons, this condition is also sufficient to ensure that the theory is at least qualitatively correct. For heavy ions, the additional requirement that the projectile be essentially fully ionized moves the lower limit up to higher velocities.

vi) There is a close connection between the theory of electronic stopping and that of optical dispersion ; this is revealed in particular by the dielectric treatment.

vii) The treatment concentrated on the total stopping power rather than straggling, inner-shell excitations, etc. The latter fall much more often -at moderate energies- outside the adiabatic limit and thus need a more sophisticated treatment. The former effect is influenced predominantly by individual-particle collisions rather than distant interactions. As a first approximation, eq.(2.12b) with $T_{min} = 0$ is quite satisfactory for both electronic and nuclear interactions at sufficiently high velocities.

viii) The divergence in the nuclear stopping power is removed by taking into account the effect of screening on the Coulomb force of the nucleus. The nuclear stopping power approaches a maximum at a much lower energy than the electronic one, and therefore can become considerable at low energies, in particular for heavy ions. For ions, nuclear scattering effects can largely be treated on the basis of classical mechanics (see below).

ix) A number of relativistic effects have been outlined.

3. QUANTUM THEORY OF STOPPING

General references : Bethe[20]
Fano[78]
Inokuti[107]

Most of the considerations of the previous section were hinging on the existence of a well-defined impact parameter. As was mentioned briefly in sect.2.5, that quantity is no longer defined when the collision diameter b becomes of the order of, or smaller than the de Broglie wavelength $\lambdabar$. Thus one is lead to the criterion

$$x = \frac{|b|}{\lambdabar} = \frac{2|e_1 e_2|}{\hbar v} \gg 1 \qquad (3.1)$$

for a classical impact parameter treatment to be valid. Bohr[39] showed, by a wave-packet argument, that $x^{-1/2}$ is indeed a rough measure of the relative uncertainty in scattering angle θ due to the uncertainty in impact parameter, and Lindhard[129] generalized the argument to other than pure Coulomb interactions.

For an incident proton hitting an electron, (3.1) reads $v << 2v_0$, which is complementary to (2.17). Thus, a proper electronic stopping theory in the high-velocity region needs to be based on quantum mechanics. Conversely, classical calculations are feasible (from the present point of view) mainly in the region where the projectile moves more slowly than the electrons. Note however, that other quantum effects may be important : the bound states of electrons have entered, so far, in a rather rudimentary way, and also the Pauli principle will play an important role at a later stage. Conversely, classical calculations need not actually lead to incorrect results : they may just have to be verified by a quantal calculation.

3.1. Elements of the Bethe Theory

Bethe's quantum theory[20] of electronic stopping is based on a plane-wave Born approximation, and is thus the quantum analogue of the perturbation approach ch.2.5. We first sketch the derivation of the stopping formula -following closely the conventional treatment- and then mention a number of extensions and modifications.

Let a projectile with a momentum $\vec{p}$ (charge e_1, mass m_1) hit an atom (atomic number Z_2). If we consider only interactions between the projectile and the target electrons, the cross sections for scattering of the projectile into the momentum interval ($\vec{p}'$, d^3p'), and excitation of the electronic system from the ground state E_0 into the state E_n is given by Fermi's golden rule,

$$d\sigma = \frac{2\pi}{\hbar v} \mid <\vec{p}',n\mid V\mid \vec{p}, o>\mid^2 .$$

$$\cdot \delta(E'-E+E_n-E_0) \frac{d^3p'}{h^3} \tag{3.2}$$

where V is the interaction potential,

$$V = \sum_{i=1}^{Z_2} \frac{-e_1 e}{|\vec{r}_i - \vec{r}|} \tag{3.3}$$

and the states $|\vec{p}, n >$ are products of projectile and atomic wave functions, i.e.

$$|\vec{p}, n > = e^{i\vec{p}\cdot\vec{r}/\hbar}|n> \tag{3.4}$$

E and E' are the kinetic energies of the incoming and scattered projectile. Eq.(3.2) is exact to the lowest order in V, i.e. e_1^2, and a derivation has been sketched in appendix A.

Applying Fourier transform on the potential V similar to (2.25), the matrix element in (3.2) can be written in the form

$$<\vec{p}', n|V|\vec{p}, 0 > = - \frac{4\pi e_1 e}{(\vec{p}-\vec{p}')^2/\hbar^2} < |\sum_i e^{i(\vec{p}-\vec{p}')\cdot\vec{r}_i/\hbar}|0> \tag{3.5}$$

Inserting (3.5) into (3.2), and introducing the transferred momentum,

$$\vec{q} = \vec{p}' - \vec{p} \tag{3.6}$$

the cross section reads

$$d\sigma = \frac{d^3q}{(2\pi)^2 v} (\frac{4\pi e_1 e}{q^2})^2$$

$$|<n|\sum_i e^{i\vec{q}\cdot\vec{r}_i/\hbar}|0>|^2 \delta(\frac{\vec{p}\cdot\vec{q}+q^2/2}{m_1} + E_n - E_0) \tag{3.7}$$

In order to obtain the stopping power from (3.7), we sum over the excited states n and integrate over all momentum transfers $\vec{q}$,

$$S_e(E) = \sum_n (E_n - E_0) \int d\sigma \tag{3.8}$$

If the atoms of the medium have no preferred orientation, any dependence of the matrix elements in (3.7) upon the direction of $\vec{q}$ will disappear after summation over those values of n that belong to a given value of E_n. Hence, the angular integrations in (3.7) and (3.8) are performed easily, and, introducing the quantity

$$Q = q^2/2m \qquad (3.9)$$

we obtain

$$S_e(E) = \frac{2\pi e_1^2 e^2}{mv^2} \sum_n (E_n - E_o)$$

$$\int \frac{dQ}{Q^2} \; |<n| \sum_i e^{i\vec{q}\cdot\vec{r}_i/\hbar} |0>|^2 \qquad (3.10)$$

From the integration over the polar angle in (3.7) one has to keep the bound

$$E_n - E_o < \frac{q}{m1} \; (p - \frac{q}{2}) \qquad (3.11)$$

If the incident particle is an ion, we have always q << p, hence, using (3.9), this condition reads

$$Q > \frac{(E_n - E_o)^2}{2mv^2} \qquad (3.12)$$

Considerably more care is required in case of incident electrons (see below).

Before going on, some remarks will be made on the physical content of (3.10), which we now write in the form,

$$S_e(E) = \frac{2\pi e_1^2 Z_2 e^2}{mv^2} \sum_n (E_n - E_o) \int \frac{dQ}{Q^2} |F_n(q)|^2 \qquad (3.13)$$

$$Q > (E_n - E_o)^2/2mv^2$$

introducing the generalized oscillator strengths

$$F_n(q) = Z_2^{-1/2} \sum_{i=1}^{Z_2} (e^{i\vec{q}\cdot\vec{r}_i/\hbar})_{no} \qquad (3.14)$$

If we would deal with free instead of bound electrons, the oscillator strengths $F_n(q)$ would have to be such as to give nonvanishing contributions only for $Q = E_n-E_O$, and therefore, (3.13) would go over into (2.12a) and (2.14b). Note that the integration condition then reads

$$Q < 2mv^2 \qquad\qquad (3.15)$$

This defines the classically allowed maximum energy transfer. With the binding taken into account, q is the momentum transfer, Q is the corresponding energy transfer <u>if no binding where there,</u> and E_n-E_O is the actual energy transfer, corresponding to T in sect.2. $|F_n(q)|^2$ is a measure for the probability density of various energy losses (E_n-E_O) for a given momentum transfer q.

Bethe's further evaluation in the general case hinges heavily on the condition $v \gg v_O$ that needs to be fulfilled for the Born approximation to be applicable (Mott & Massey[154]). You remember from the classical case, eq. (2.21) that the range of impact parameters that contribute to the stopping power, extended over several atomic diameters, $p \lesssim p_{max}$, $p_{max} \gg a$. Since there is a rough correspondance of impact parameter and momentum transfer, i.e., $p \sim \hbar/q$, the above approximate inequality can also be written as $q \gtrsim q_{min}$, $q_{min} \ll \hbar/a$. Hence there is an essential contribution to the integral (3.13) from q-values $\hbar^{-1}|\vec{q}.\vec{r}_i| < \hbar^{-1}qa \ll 1$, i.e. where the exponentials in (3.14) can be expanded up to first order (dipole approximation).

$$|F_n(q)|^2 \simeq \frac{q^2}{Z_2\hbar^2} |(\textstyle\sum x_i)_{no}|^2 \equiv \frac{Q}{E_n-E_O} f_n \qquad (3.16)$$

where the (arbitrary) direction of $\vec{q}$ has been chosen to be the x-axis ; here the usual dipole oscillator strengths

$$f_n = \frac{1}{Z_2} \frac{2m}{\hbar^2} (E_n-E_O)|(\textstyle\sum x_i)_{no}|^2 \qquad\qquad (3.17)$$

have been introduced. Assuming that the above approximation is valid up to some value $q_1 \sim \hbar/a$, the integral (3.13) can be split up into a low-Q and a high-Q portion with $Q \lesssim Q_1 = q_1^2/2m$; for low Q, inserting (3.16) we obtain

$$S_{e,low\ Q} = \frac{2\pi e_1^2 Z_2 e^2}{mv^2} \sum_n f_n \log \frac{2mv^2 Q_1}{(E_n - E_o)^2} \qquad (3.18)$$

The second region, where $Q > Q_1$, can now be assumed to be far enough away from the adiabatic limit so that the influence of atomic binding can be ignored. Consequently, we have $Q \approx E_n - E_o$, and (3.12) goes over into (3.15)'. Using Bethe's sum rule (a proof of which is given in appendix B),

$$\sum_n (E_n - E_o)|F_n(q)|^2 = Q \qquad (3.19)$$

we obtain

$$S_{e,high\ Q} = \frac{2\pi e_1^2 Z_2 e^2}{mv^2} \log \frac{2mv^2}{Q_1} \qquad (3.20)$$

The sum rule for the dipole oscillator strengths reads in the present notation

$$\sum_n f_n = 1 \qquad (3.21)$$

(this follows from (3.19) for $q \to 0$) ; adding (3.18) and (3.20) we obtain Bethe's nonrelativistic stopping formula (Bethe[20]),

$$S_e(E) = \frac{4\pi e_1^2 Z_2 e^2}{mv^2} \sum_n f_n \log \frac{2mv^2}{E_n - E_o} = \frac{4\pi e_1^2 Z_2 e^2}{mv^2} \log \frac{2mv^2}{I}$$

$$(3.22)$$

where I is the mean ionization potential defined by

$$\log I = \sum_n f_n \log(E_n - E_o) \qquad (3.23)$$

Note that the (somewhat arbitrary) boundary value Q_1 between the high and low-Q regions drops out. An exceptionally clear discussion of the limitations of this division can be found in Fano's article[78].

You may have noted that the integration limit at high Q, or small impact parameters, is just the clas-

sical one, and not the one suggested by Fermi that lead
to (2.23'). On the other hand, a very striking difference between the classical and the present treatment
is the occurence of equally large terms containing
$\ln(2mv^2)$ in both the high-Q and the low-Q contributions. For $2mv^2 \gg I$ this leads to an equipartition of
the energy loss from close and distant collisions which is to be discussed below.

You may note that, as it must be in a perturbation
treatment, the primary interaction, characterized by
e_1^2, enters only into the factor in front of (3.22)
and not, as in (2.23), also into the logarithm. A possible disadvantage of this kind of a consistent scheme
of approximation is that an essential improvement of
the theory may have to involve higher-order Born approximations ; this is rather cumbersome. However, alternative approximation schemes are available, just as in
the classical theory.

3.2. Extensions and Modifications

Bloch[31] reproduced the essential parts of the
Bethe theory by use of an impact parameter treatment,
i.e. considering the excitation of an atom by the field
of a point charge passing by at a distance p from the
nucleus with a constant velocity and integrating subsequently over p. This treatment is formally very similar
to Bohr's classical theory[37,38]. By neglecting, for
distant interactions, the variation of the field across
the diameter of the target atom, the response becomes
linear in the field, and hence Bloch's treatment is
equivalent to Bethe's perturbation approach. Such a
treatment has a number of obvious advantages.

i) The splitting-up into distant and close collisions is done in real rather than Fourier space. The
connection with the classical theory becomes much more
obvious, and a stopping theory becomes feasible for
crystalline media where the atoms are not distributed
at random. An attempt in this latter direction has been
made recently by Dettmann & Robinson[70].

ii) In cases where the first Born approximation is
not accurate enough -either because of high experimental accuracy or because of low velocity or high charge-
and a higher-order perturbation treatment is desirable
it is mostly the motion of the target electrons rather
than the projectile that needs to be described more accurately. Bloch's treatment readily allows for extension, while a full second-order Born approximation requires corrections both on the target and projectile

motion. Calculations using Bloch's formalism have been performed recently by Hill & Merzbacher[100].

iii) In case of inner-shell ionization, where the first Born approximation breaks down because of too strong interaction, an effective improvement can be achieved by use of an impact parameter treatment involving actual (non-straight) Coulomb trajectories (Bang & Hansteen[15]).

iv) Finally, the weaker assumptions necessary in order to derive Bethe's formula by Bloch's procedure extend the range of validity of this formula (Inokuti[107]).

The theory has been extended to relativistic velocities by Bethe[20] and Møller[149]. The relativistic stopping formula reads

$$S_e(E) = \frac{4\pi e_1^2 Z_2 e^2}{mv^2} \left(\log \frac{2mv^2}{I} + \log \frac{1}{1-\frac{v^2}{c^2}} - \frac{v^2}{c^2} \right) \quad (3.24)$$

for heavy projectiles ; it is slightly more complicated for incident electrons. (See also Gould[95]). The relativistic extension is obviously the same as in the classical case, eq.(2.37) ; however, a number of complications arise for close collisions. When the projectile is an electron, the indistinguishability of projectile and target electron in close collisions causes a modification of the Rutherford cross section already at nonrelativistic velocities. At high velocities, deviations from Rutherford scattering occur for both heavy and light projectiles (Mott[153], Mc Kinley & Feshbach[144]). Jackson & Mc Carthy[109], referring to unpublished work by Fermi, noted that this gives rise to a substantial deviation from the Born-approximation result, causing the stopping power to depend on the sign of the projectile charge. (See also Morgan & Eby[150]).

The straggling of energy loss has been treated along similar lines as the stopping power (Sternheimer[195]). According to Fano[78], the basic formula reads in the nonrelativistic limit

$$\langle \Delta E^2 \rangle = N \sum_n (E_n - E_0)^2 \int d\sigma =$$

$$4\pi e_1^2 e^2 N Z_2 \Delta x \{1 + \frac{2}{3} \frac{<0|\sum_i \vec{v}_i|^2|0>}{Z_2 v^2} \log \frac{2mv^2}{I_1}\}$$ (3.25)

for incident ions, where a new ionization potential I_1 is introduced, and defined by

$$\log I_1 = [\sum_n (E_n - E_0) f_n \log(E_n - E_0)] / \sum_n f_n (E_n - E_0)$$

This formula is very similar to (2.14d).

3.3. Dielectric Theory

References Lindhard [129]

In principle, the dielectric theory is just a re-formulation of Bloch's stopping calculation[31]. Used in connection with the Thomas-Fermi principle, i.e. replacing an actual electronic charge distribution by a Fermi gas of varying density, the dielectric theory is a powerful tool in the evaluation of ionization potentials. Moreover, inner-shell corrections -deviations from the Bethe stopping formula arising at low (although higher than v_0) velocities- can be calculated. Finally, stopping formulae for $v < v_0$ can be derived.

It was shown in ch.2.6 how distant interactions can be described quantitatively by a field description ; in order to include properly the close collisions in such a description, allowance must be made for a rapid spatial variation of the electric field. This, in turn, requires retardation effects to be included, and hence the dielectric constant needs to be wave-number dependent. Due to this, (2.28) is modified to

$$\frac{dE}{dx} = -\frac{e_1^2}{\pi v^2} Jm \int_0^\infty \frac{dk}{k} \int_{-kv}^{kv} \omega d\omega (\frac{1}{\varepsilon(k,\omega)} - 1)$$ (3.27)

where $\varepsilon(k,\omega)$ is a longitudinal dielectric constant or response function. Magnetic interactions can be described similarly, by means of a transverse dielectric constant, when necessary. You should note that, as it stands (3.27) is no more or less classical a relation than (2.28). In either case, you have the choice of inserting a quantal or classical expression for the dielectric constant of your medium. However, in practical applications (2.28) has been evaluated so far mostly by applying classical oscillator models instead of the quantal expression (2.29), while the main domain of application of (3.27) has been the degenerate Fermi gas.

There is a very close similarity between (3.27) and (3.13). You may try to derive the correspondence by starting from the very beginning, just as in ch. 2.6.1. This involves calculation of the (time-dependent) perturbation up to first order of the electrons of an atom by a field $\vec{E}(\vec{k},\omega)\exp i(\vec{k}.\vec{r}-\omega t)$, and from the induced charge to arrive at the dielectric constant. Fano[78] gives a short review of such calculations. At present we note that $\hbar k$ in (3.27) is identical with the q in ch. (3.1). We rewrite (3.27)

$$\frac{dE}{dx} = \frac{2\pi e_1^2 Z_2 e^2}{mv^2} N \int \frac{dQ}{Q^2} \frac{i}{\pi \omega_0^2} Q \int \omega d\omega \left(\frac{1}{\varepsilon(k,\omega)} - 1\right) \qquad (3.27')$$

introducing the plama frequency

$$\omega_0^2 = \frac{4\pi Z_2 Ne^2}{m} \qquad (3.28)$$

Then, the relation between (3.27') and (3.13) reads

$$\sum_n (E_n - E_0) |F_n(\vec{q})|^2 \rightarrow \frac{iQ}{\pi \omega_0^2} \int \omega d\omega \left(\frac{1}{\varepsilon(k,\omega)} - 1\right) \qquad (3.29)$$

and the sum rule (3.19) would be expected to have the form

$$\frac{i}{\pi \omega_0^2} \int_{-\infty}^{+\infty} \omega d\omega \left(\frac{1}{\varepsilon(k,\omega)} - 1\right) = 1 \qquad (3.30)$$

Eq.(3.30) can be proved directly (Nozières & Pines[161]; Lindhard & Winther[134]). The argument goes as follows. Because of causality requirements, ε can only have poles in the lower half of the complex ω plane (e.g. Panofsky et al[166]). Deforming the path of integration in (3.30) to a large semicircle in the upper half plane, the poles of ε do not enter, and for sufficient large ω, ε approaches 1 like

$$\varepsilon \simeq 1 - \frac{\omega_0^2}{\omega^2} \qquad (3.31)$$

For this ε the integral can be carried out easily, and it gives the desired result (3.30).

One point to remember is that large oscillator strength corresponds to a large imaginary part of $1/\varepsilon$.

A useful formula for energy loss straggling (2.7) is found by the observation that $\hbar\omega$ in (3.27) corresponds to the energy loss (E_n-E_o) in (3.13). Therefore, we have

$$<(\Delta E-<\Delta E>)^2> = N\Delta x \; \frac{2\pi e_1^2 Z_2 e^2}{mv^2} \sum_n (E_n-E_o)^2 \int \frac{dQ}{Q^2} |F_n(q)|^2$$

$$= -\frac{e_1^2 \hbar}{\pi v^2} \Delta x \; Jm \int_o^\infty \frac{dk}{k} \int_o^{kv} 2\omega^2 d\omega \left(\frac{1}{\varepsilon(k,\omega)} -1\right)$$

(3.32)

This formula has been derived by Lindhard[129] by a rather more rigorous argument.

Up to this point, the dielectric theory mainly offers a compact notation for the Bethe theory, in that the evaluation of the stopping power and straggling is reduced to finding a material-dependent quantity $(\varepsilon(k,\omega))$ to the desired degree of sophistication. As a matter of fact, the dielectric constant of a medium, in particular a solid medium, is a key quantity that is of interest in a wide variety of applications including, of course, optical properties.

The unique properties of the dielectric treatment emerge when it is ·applied to a free electron gas, and combined with the Thomas-Fermi principle.

The free electron gas is a standard of many-body theory. Beginning with Drude's formula (2.29), you can find a wealth of more or less tractable expressions in the literature for the dielectric constant of such a system (see, e.g. Pines[171]). Recent treatments, in fact, go rather far in describing the dielectric properties of realistic solids. Calculations of stopping have mainly been based so far on expressions derived by Lindhard[129] for the longitudinal (and much less the transverse) dielectric constant of a Boltzmann gas or a degenerate Fermi gas, by use of either a classical Boltzmann transport equation with relaxation time, or a self-consistent quantal perturbation treatment. The latter treatment yields, in the general form,

$$\varepsilon(k,\omega) = 1 + \frac{2m^2\omega_0^2}{\hbar^2 k^2} \sum_n \frac{f(E_n)}{N} \qquad (3.33)$$

$$\left\{ \frac{1}{k^2 + 2\vec{k}\cdot\vec{k}_n - \frac{2m}{\hbar}(\omega + i\gamma/\hbar)} + \frac{1}{k^2 - 2\vec{k}\cdot\vec{k}_n + \frac{2m}{\hbar}(\omega + i\gamma/\hbar)} \right\}$$

where $f(E_n)$ is the Fermi (or Boltzmann) distribution, usually (but not necessarily) taken at T = OK, $\vec{k}_n$ are wave vectors characterizing the electron states, and γ is an infinitesimal damping constant. Eq.(3.33) follows from a Hartree treatment of the electron gas and, consequently, its accuracy is limited. It is not designed for a 1-percent type of theory.

The evaluation of the stopping power (3.27) by means of (3.33) in case of a degenerate Fermi gas has been studied in considerable detail by Lindhard & Winther[134], and subsequently by Erginsoy[76]. The first result is a useful series expansion of the stopping power of a Fermi gas,

$$\frac{dE}{dx} = \frac{4\pi e_1^2 Z_2 e^2}{mv^2} \left\{ \log \frac{2mv^2}{\hbar\omega_0} - \frac{3}{5}\frac{v_F^2}{v^2} - \frac{3}{14}\frac{v_F^4}{v^4} \ldots \right\} \quad (3.34)$$

where v_F is the Fermi velocity $v_F = (3\pi^2 n)^{1/3} \hbar/m$, and n the electron density. It is seen that deviations from Bethe's logarithmic term become significant at low velocities. These deviations, called inner-shell corrections, arise when v approaches v_F, i.e., when there are target electrons that do not move much more slowly than the projectile. In that case, the Bethe summation procedure -being based on the separability of two distinct regions $Q < Q_1$ and $Q > Q_1$ with clear specifications- becomes inaccurate.

Another significant result is that the double integral (3.32) in case of the degenerate Fermi gas splits up into two distinct parts, one being a true double integral over part of the (k,ω) plane, the other being a line integral along a curve $\omega = \omega(k)$ outside the region covered by the double integral. The latter part is interpreted as being due to energy losses into plasma resonance excitation- and hence with a definite dispersion- while the former corresponds to single-particle collisions. The somewhat loose distinction between distant and close collisions, necessary in Bohr's theo-

41

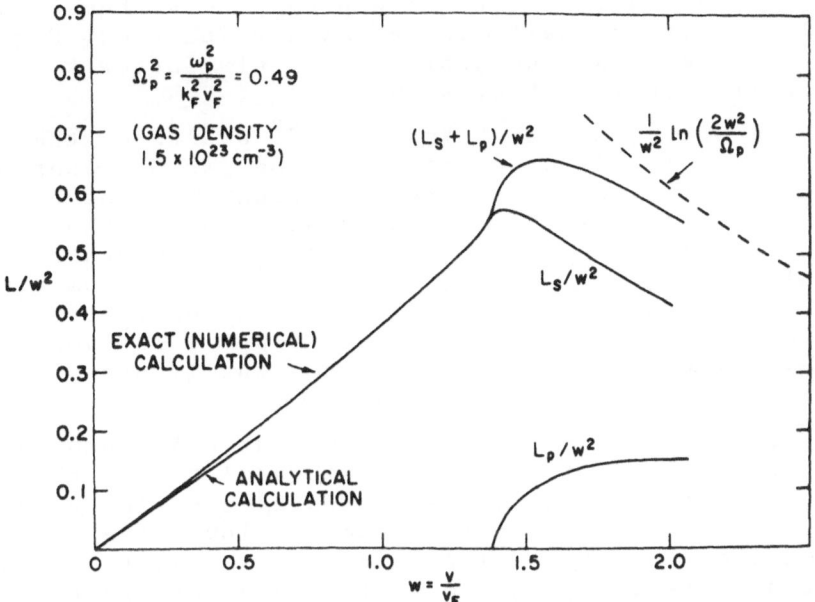

Fig.4 : Stopping power of a light ion in a homogenous Fermi gas,
based on Eqs.(3.27) and (3.33). Numerical evaluation by Erginsoy76.
The ordinate is $L=(v^2/e_1^2\omega_p^2)dE/dx$; $\omega_p^2=4\pi ne^2/m$; L_s refers to
close collisions ("single-particle excitation"), and L_p to reso-
nant ("plasma") excitation. The stipled line represents the first
term in Eq.(3.34) and corresponds to Bethe's result for a Fermi
gas. From Erginsoy[76].

ry and leading to rough equipartition in Bethe's theory, becomes a precise concept here. In fact, Lindhard & Winther[134] also showed that equipartition itself can be formulated as a precise concept, in that close and resonance collisions give equal contributions to the k-integral in (3.27) at fixed ω, at velocities that are high enough that resonance occurs. Since the integration region for ω increases with increasing v, it is asserted that also the stopping powers themselves will exhibit equipartition at high velocities. Erginsoy[76] showed that the approach to this latter type of equipartition is rather slow.

Recently, Pathak & Yussouff[167] evaluated the stopping power (3.27) of a free electron gas by means of a number of alternative approximations for the dielectric constant that contain exchange contributions and, to some extent, electron correlation. Although some results are quoted that differ drastically from those computed with the Lindhard dielectric constant, the agreement is excellent, within $\sim$ 5 pct., in those cases where the physical model has actually been improved (Hubbard[102], Langreth[122]).

Fig.4 shows a stopping power curve as computed along the scheme of Lindhard & Winther[134] by Erginsoy[76]. The upper curve represents the total stopping power at the particular gas density, and the lower one excludes resonance excitation. The stipled curve is the uncorrected Bethe stopping power. The resonance part drops to zero around the position of the maximum of dE/dx (this is another way of formulating the equipartition rule), and at low velocities the stopping power is proportional to velocity. We comment on this latter situation in sect.4.

The straggling formula (3.32) has been evaluated within the Lindhard-Winther scheme by Bonderup & Hvelplund[42], see sect.5. Lindhard[129] gave asymptotic expressions for high and low velocity, that cannot be applied at intermediate velocities, in the region where shell corrections are important.

3.4. Thomas-Fermi Arguments

In the Thomas-Fermi model of the atom (Gombas[92]), the electronic charge distribution is replaced by a Fermi gas of varying density, and the local energy density is taken to be equal to the sum of kinetic and potential energy density (sometimes including exchange and correlation energy) of an infinite Fermi gas with constant density. In the simple Thomas-Fermi model for

Table I   MEAN IONIZATION POTENTIALS

Substance		Fano (1963)		Recent Values	
	Z	I(eV)	I/Z	I(eV)	I/Z
H atomic	1	15.0(theor)	15.0		
		19 (theor)	19		
molecular		18.3±2.6			
in compounds		15 - 18			
He	2	42±3	21		
		41.8(theor)			
Li	3	40,38	13		12.5[c]
		45, 38.8(theor)			
Be	4	64			
		60,66(theor)	16		11.4[c]
C graphite	6	81	13.5		
in compounds		77 - 80			
N molecular	7	88	12.6		
in compounds		79 - 102			
O molecular	8	101			
in compounds		91-101	12.6		
Al	13	163	12.6		11.6[c]
Ar	18	190	10.6		11.5[c]
Fe	26	273	10.5	280.6±3.1[a]	10.8
Cu	29	315	10.9	319.9±3.2[b]	11.0
Kr	36	360	10.0		11.2[c]
Ag	47	471	10.0	469±8[b]	10.0
Au	79	761	9.6	771±20[b]	9.8
Pb	82	788	9.6	773 ±20[b]	9.4
U	92	872	9.5	839 ± 30[b]	9.1
Emulsion		323			
Air		85			
CH_4		45			

a – H.H.ANDERSEN, H.SØRENSEN & P.VAJDA, Phys.Rev.180, 373 (1969)

b – H.SØRENSEN & H.H.ANDERSEN, Phys.Rev. B (Sept.1973)

c – W.K.CHU & D.POWERS, Phys.Lett. 40A, 23, (1972).

neutral atoms (where exchange and correlation are not
included), the static properties of the atoms follow
some simple scaling laws, the scaling being determined
by the atomic number Z alone. It may be said that Z is
a dimensional quantity in this case. Bloch[31b] made an
attempt to establish a dynamical Thomas-Fermi model of
the atom, and formulated an equation to determine oscillator strengths. By applying "dimensional" arguments
he derived a scaling law for the mean ionization potential,

$$I = Z_2 \, I_o \qquad\qquad (3.35)$$

that can be rationalized in the following way. The unit
of distance in the Thomas-Fermi model is a $\propto Z^{-1/3}$ (eq.
2.42), and the energy of an atom $\propto Z^2 e^2/a \propto Z^{7/3}$, hence
the energy per electron $\propto Z^{4/3}$, and the velocity $\propto Z^{2/3}$,
thus the time unit becomes $\propto Z^{-1}$ and frequency is proportional to Z, as is stated by (3.35).

Table I shows values of I and I_o compiled by Fano[78]. It is seen that I_o is normally in the range of
10-15 eV, and that it tends to decrease slightly with
increasing atomic number. These values are based predominantly on experimental results. Ab initio calculations were only available in 1963 for very light atoms
(including Bethe's calculation[20] for atomic hydrogen),
but are becoming feasible now.

Lindhard & Scharff[132] made extensive use of dimensional arguments, in particular in order to establish semi-empirically a stopping-power plot in the
energy region where inner-shell corrections become important. Fig. 5 shows their compilation of stopping-
power data plotted as a function of v^2/Z. The remarkably good scaling not only established a considerable
credibility to the use of Thomas-Fermi arguments in
stopping theory, it moreover demonstrated very clearly
the significance of inner-shell corrections (the difference between the dashed and full-drawn line in fig.5)
to the Bethe formula even at rather high velocities.
Quantitative calculations of energy loss by means of
the Lindhard & Winther formula (3.34) have been performed by Bonderup[41], who took the average of the free-
electron stopping power over an atomic charge distribution following the Lenz-Jensen modification of the
Thomas-Fermi model (Gombas[92]). Although the averaging
process yields values of both the mean ionization potential and the effective inner-shell correction (the
v^{-4} term in (3.34) was ignored), various uncertainties
enter in such a way as to provide more accurate values

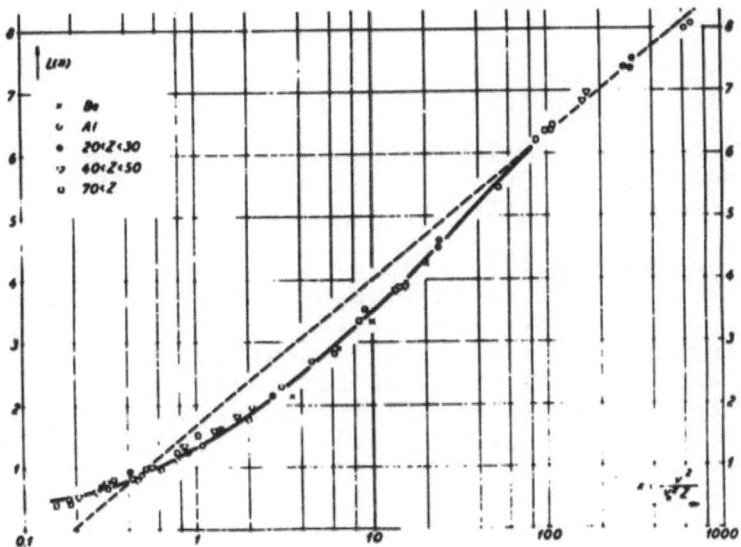

Fig.5 : Semi-empirical determination of the function $L(v^2/v_0^2 Z_2)=$ $\overline{(dE/dx)}(mv^2/(4\pi Ne_1^2 Z_2 e^2))$ by Lindhard and Scharff[132]. The dashed straight line represents Bethe's asymptotic result, Eq.(3.22) with Bloch's scaling (3.35) and $I_0 = 10eV$. For details on the experimental data, please check the original paper. From Lindhard & Scharff[132].

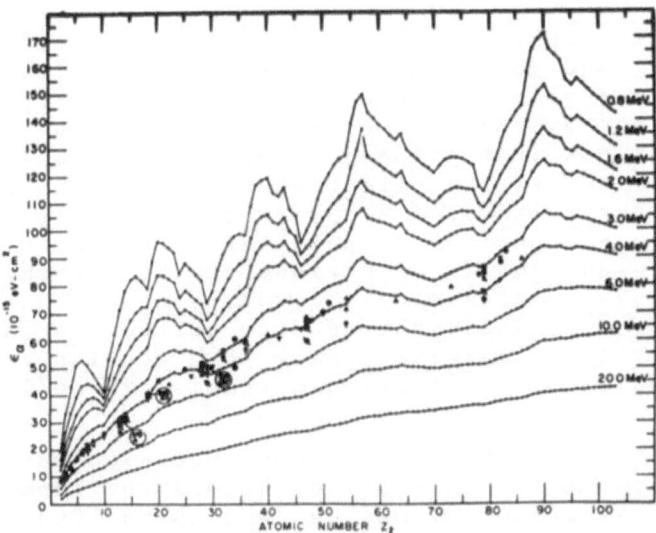

Fig.6 : Stopping cross sections for α-particles ($\varepsilon_\alpha \equiv S_e = (1/N)\frac{dE}{dx}$) as a function of target atomic number, for 0.8 MeV < E < 20.0 MeV. Calculated points, connected by curves, are due to Chu & Powers[61]. For details on the experimental data, please check the original paper. From Chu & Powers[61].

of the inner-shell correction than the ionization potential.

Bonderup's procedure has been used more recently in the calculation of stopping powers by means of Hartree-Fock-Slater charge distributions (Rousseau et al[179], Chu & Powers[61a], Ziegler & Chu[219]). The charge distributions were taken from Herman & Skillman's tables[98]. Fig. 6 shows calculated stopping cross sections as a function of target atomic number for α-particles in the energy range 0.8 - 20 MeV. Also included are experimental results of various authors, mostly for 4 MeV α-particles. Characteristic oscillations are found, that are most pronunced at the lowest velocities. Analysis of experimental stopping powers had indicated the possibility of structure in the dE/dx vs. Z_2 curves (Andersen et al[6], Chu & Powers[61b]).

3.5. Ionization Potentials

The complete stopping formula (Fano[78])

$$\frac{dE}{dx} = \frac{4\pi e_1^2 Z_2 e^2}{mv^2} \{\log \frac{2mv^2}{I} - \log(1 - \frac{v^2}{c^2}) - \frac{v^2}{c^2} - \frac{C}{Z} - \frac{\delta}{2}\}$$

(3.36)

contains the ionization potential I, the shell corrections C/Z, and the density correction δ as its input quantities ; roughly speaking, C/Z is a low-energy and δ a high-energy correction. However, both theoretical considerations (Fano[78]) and experimental results (see below) point into the direction that the density correction and the value of I are rather intimately correlated. From an experimental point of view, ionization potentials and shell corrections can be determined simultaneously by analysing the energy dependence of the stopping power, while the density correction in the low-energy limit mainly appears to give rise to a redefinition of the ionization potential (Fano[78], Brynjolfsson[58]).

Existing theoretical work on both ionization potentials and shell corrections beyond the Thomas-Fermi approach has been reviewed by Fano up to 1963. The rapidly expanding field of inner-shell excitations in ion-atom collisions (Garcia et al[88]) provides much evidence that may be useful in stopping theory. However apart from the case of very light target atoms that has been investigated in extreme detail both theoretically and experimentally, and reviewed recently by Inokuti[107], it appears still fair to claim that the numerical value of I, and hence the accurate value of the stopping power, is a quantity to be determined experimentally. On the other hand, one might admit that the requirements on accurate stopping-power data have been increasing continuously.

This is hardly the place to review experimental work on stopping powers in the Bethe-Bloch range : I might assure you that the literature is enormous. However, I might mention to you that a number of critical reviews and tabulations are contained in "Studies in Penetration of Charged Particles in Matter", Nat.Acad. Sci-Nat.Res.Council Publication 1133 ; further, there is the somewhat more recent report by Janni[110], and finally, there is Bichsel's contribution in the American Institute of Physics Handbook.

Most of the data contained in the above compilations have been found by suitable averaging procedures between different experimental results, and by use of semi-empirical fitting methods. Recent accurate measurements mainly on protons and deuterons at Risø (Andersen et al[2,3,4,5,6,7]) confirmed many of these interpolated results. Their experimental accuracy of $\sim$ 0.3 pct. was sufficient, despite their comparatively high ion energy (5-12 MeV) to detect experimentally a structure that may be of the type shown in fig.6. Apart from the more basic interest of determining ionization potentials, accurate stopping power data for light projectiles have received exceedingly much interest during the past few years because of their need in ion-beam surface-layer analysis. This applies in particular to 1-2 MeV α-particles. The currently achieved accuracy appears to be in the $\sim$ 5 pct. range. (Chu & Powers[61], Chu et al[62], Eisen et al[73], Ziegler & Brodsky[218], Lin et al[128]).

This, together with the fact that this energy region comes rather close to the maximum of the stopping power curve, makes the analysis in terms of atomic parameters exceedingly difficult. It has been suggested recently (Mitchell et al[147]) that an international organization ought to take the initiative of establising a committee for the purpose of standardizing stopping power data.

3.6. Bragg's Rule

Bragg's rule (Bragg & Kleeman[51]) states that the stopping power of a compound target is the weighted mean of the atomic stopping powers of the constituents i.e , following the argument leading to eq. (2.5),

$$\langle\Delta E\rangle = \sum_i N_i \Delta x \ S_i(E) \qquad (3.37)$$

where N_i is the number of <u>atoms</u> of type i per unit volume. Deviations from this rule may occur due to chemical effects, i.e. differences in the electronic structure between a free atom and an atom bound in a molecule or an alloy. Consequently, deviations from Bragg's
additivity rule should be expected mostly at comparatively low energies, where the relative contribution
from valence electrons to the stopping power becomes
relatively large, and, more important, for very light
elements where the valence electrons constitute a major
fraction of the total number of electrons. Fano[78] states that, in comparing atomic and molecular stopping
powers, it might be appropriate to use interpolated rather than measured values for the atomic stopping powers, since chemical aggregation tends to smear out
specific electronic configurations of the single atom.

The validity of Bragg's rule has recently been the
subject of a considerable number of experimental investigations (see, e.g., Mayer & Ziegler[143]), although the
variety of parameters varied (gas vs.solid, insulator
vs.conductor, ion type, ion energy) is still not large
enough to provide a consistent picture. The following
qualitative picture –which may soon be subject to correction– appears to arise from the experimental results.

i) Deviations from Bragg's rule in solid targets
seem to be small (few pct.) and hardly measurable in
metallic alloys.

ii) For 1-2 MeV α-particles, Bragg's rule is fulfilled for a large number of solid and gaseous compounds.

iii) Pronounced deviations (10-20 pct.) are found
for ∿ 1 MeV α-particles in gaseous compounds containing
hydrogen and carbon, or carbon and oxygen.

The last statement refers to work by Bourland et
al[49,50] and Powers et al[172,173]. These authors first
showed that stopping powers measured on several gaseous hydrocarbons did not fit with Bragg's rule when a-
tomic stopping powers were used as measured on solid
carbon and deduced from molecular hydrogen.

Most recently, by systematically varying the stoichiometry of the investigated compounds, they were able
to deduce atomic stopping powers for both carbon and

hydrogen, and it turned out that these values both dif-
fered from the anticipated ones, i.e. that the deviation from Bragg's rule was not due just to either hydrogen or carbon, as assumed earlier. At present, Powers et al[173] conclude that the stopping cross section
of a carbon atom in a triple-bond compound is greater
than that in a double-bond one, and even greater than
in a single-bond one. The reverse is found for hydrogen. Also, the magnitude of these variations varies
considerably from an α-particle energy of 0.5 MeV up
to 2 MeV, being rather small at the upper end.

Naturally, Bragg's rule must break down for any
compound at some high enough experimental accuracy. The
present experimental situation suggests more work to
be done at moderate energies and on compounds with
light constituents to establish more generally some
bounds on the validity of this rule.

3.7. Z_1^3 Effect

According to eq.(3.36), the stopping power in the
Bethe regime is proportional to e_1^2, i.e. the square of
the interaction force. This dependence is exact for
nonrelativistic Coulomb scattering between free collision partners, and follows from the first Born approximation in quantal stopping theory. In the unmodified
Bohr theory, the interaction also occurs under the logarithm. If the projectile is an ion, eq.(3.36) predicts
that for ions of the same velocity, only the ion charge enters the stopping power through e_1^2, but not the
mass. This behaviour has been checked in many measurements. For example, deuterons and protons of the same
velocity show the same stopping powers within very narrow error bounds ($\sim$ 0.3 pct.) (Andersen et al[5]). The
analogous statement can be made for ^{3}He and ^{4}He projectiles (Andersen et al[6]). However, when comparing stopping powers of He ions and deuterons, a systematic positive deviation from the expected factor of $2^2 = 4$
was observed. The relative deviation depended approximately like $\sim E^{-1}$ on energy. At a deuteron energy of
5 MeV, the relative deviation was almost 3 % for a tan-
talum target.

Similar effects had been observed earlier at both
lower and higher energies (reviewed by Jackson & Mc
Carthy[109]).

For heavy projectiles, like ions, the observed effects can be explained by going to one higher order in
perturbation theory. Ashley et al[13] performed a classical treatment using a uniformly moving point charge in-

teracting with an assembly of oscillators. The motion of the target electrons in the force field of the projectile is treated up to quadratic terms in e_1 (the corresponding terms in the motion of the ions would be smaller by a factor of $\sim m/M_1$). The frequency distribution of the oscillators is determined by the electron density distribution via the plasma frequency, similar to the Lindhard & Scharff[132] procedure. The close collisions were treated as in the Bohr theory, and the two approximations were joined together at a certain impact parameter that was used as an adjustable constant. Jackson & Mc Carthy[109] applied a very similar procedure to distant collisions, but investigated the scaling properties in more detail and fixed the limiting impact parameter in a more unique way. In addition, they included a relativistic higher-order effect arising from close collisions that was mentioned already in ch.3.2. Their low-velocity result can be written in the form

$$\left(\frac{dE}{dx}\right)_e = \left(\frac{dE}{dx}\right)_0 \left(1 + \frac{Z_1}{Z_2^{1/2}} F(V)\right) \tag{3.38}$$

where

$$V = \frac{v}{v_0} \frac{1}{\sqrt{Z_2(1-v^2/c^2)}} \quad , \tag{3.39}$$

F(V) a numerically given function, with a maximum value ~ 0.3 near $V \sim 1$, and $(dE/dx)_0$ the stopping power in first-order perturbation theory. In a more recent paper, Ashley et al[13] brought their result into essentially the same form as eq.(3.38), with a somewhat different function F(V). Both groups of authors achieved good agreement with the experimental results of Andersen et al[6] It was mentioned in ch.3.2 that a modified Bloch model calculation replacing the target electrons by quantal harmonic oscillators was also performed (Hill & Merzbacher[100]) and, according to Jackson & Mc Carthy, confirmed the classical results. A calculation based on the dielectric theory has not yet been published as far as I am aware.

3.8. Energy Loss and Channeling

Channeling is the common name for a number of phenomena that are observed when a charged-particle beam penetrates a crystalline target with the direction of motion coinciding within certain critical angles with

one of the major crystal axes or planes. Although a
pronounced reduction in average energy loss is a characteristic feature of ion channeling that served as a
clear experimental manifestation of the effect (Robinson & Oen[177], Piercy et al[170], Lutz & Sizmann[137], Dearnaley et al[68], Madden & Gibson[140]), the most dramatic
channeling effects are observed in the dependence of
close-encounter processes like Rutherford scattering
(Bøgh & Uggerhøj[35]) nuclear reactions (Bøgh et al[34]),
and inner-shell excitation of target atoms (Brandt et
al[52]) on the orientation of the crystal relative to the
ion beam. Within the scope of this lecture, however,
I shall only comment on energy loss.

Within the scheme of classification of Lindhard[129]
it is convenient to distinguish between A) axial channeling, B) planar channeling , and C) hyperchanneling.
In case of axial channeling (A), the particles are kept
away from such strings of atoms that are nearly parallel to the beam, but can otherwise move freely through
the crystal (Lindhard[129]). In case of planar channeling (B), particle trajectories are confined within
adjacent atomic planes and exhibit anharmonic oscillations at a wavelength that depends on the penetrated
depth and other parameters (Erginsoy[75], Robinson et
al[176]). The particle density peaks near the minimum
distance of approach to a plane (Lindhard[129]). Hyper-
channeled particles (C) are confined to the region near
the potential minimum between adjacent atomic strings
(Lehmann & Leibfried[125], Nelson & Thompson[156]), the oscillations being expected more harmonic than planar
ones. The critical angles ψ_c for these processes are
determined by the respective transverse potential barriers E_p through relations of the type

$$\psi_c \sim \sqrt{E_p/E} \quad ; \quad E_p << E \qquad (3.40)$$

E_p is largest for A) and smallest for C). Consequently,
at high energy, ψ_c is largest for axial channeling and
smallest for hyperchanneling. Channeling effects disappear gradually at low energies when the critical angles
become large and cause channels of different orientation to overlap. Because of the low barrier height ($\sim$
5 eV) hyperchanneling may be possible still at rather
low velocities (Robinson et al[177]). When a crystal is
hit by an external beam, part of the beam particles are
scattered into the "random beam", because their impact
parameters to the channel axis or plane are not compa-

tible with being channeled (Lindhard[129].)

The energy-loss spectra observed in transmission through single crystals differ more or less drastically from those found with random targets. In case of axial channeling (A), one normally observes two distinct groups or even well separated peaks (Dearnaley[68] Appleton et al[12], and many others), corresponding to random and channeled stopping, respectively. Practically all energy loss is electronic in this case, and the channeled energy loss is smaller because the trajectory samples a region of lower electron density than random. In case of planar channeling (B), similar effects are observed when the total transmitted beam is analysed. When only a small angular interval is analysed at a time, the spectrum splits up into a multiple peak structure (Lutz et al[136]). This has been attributed to the anharmonic nature of the oscillations in planar channels. Beam particles with different impact parameters to the atomic planes show different amplitudes, hence different wavelengths, and, therefore, have different phases when leaving the crystal. The phase determines the exit angle. Observation at a fixed exit angle samples the beam according to groups of distinct impact parameters, i.e. different spectral peaks in energy loss. Those peaks that have lower than random energy loss are clearly separable (Appleton et al[11]; and several others). Hyperchanneling (C) appears to produce a low-energy-loss group in the spectrum of axial energy loss at high energy (Appleton et al[10]). At low energy, the distinction between the different types of channeling becomes less clearcut, and nuclear stopping may need to be taken into account. As long as electronic stopping dominates, a clear separation of channeled and random particles may be possible (Eisen[72], Eriksson et al[77]), but has not always been observed.

After it became clear that nuclear energy loss decreases more rapidly with impact parameter than electronic energy loss (Lindhard[129]), theoretical efforts have concentrated on electronic energy loss under channeling conditions. In the present paragraph we deal with the velocity region $v \gg v_o$. A major effort has been made in developing suitable techniques of how to extract stopping powers and potentials from planar channeling measurements (Robinson[176], Gibson & Golovchenko[90]), but the basic theoretical problem is, of course, the determination from first principles of the energy loss as a function of impact parameter under channeling conditions.

The justification for an impact parameter treatment can be derived from Lindhard's demonstration[129] of the validity of a classical orbital picture for the channeled trajectory even at velocities $v \gg v_o$, together with the fact that such a treatment accurately describes electron excitation in the random case (Bloch[31]) Attempts to calculate stopping powers by modifying Bloch's theory have been made (Kitagawa & Ohtsuki[116], Dettmann & Robinson[70]). Brice[53] presented a stopping theory based on a plane-wave Born approximation. However, the range of validity of such a theory was not specified. It is not obvious whether the conditions for applicability of a plane-wave Born approximation to calculate ion trajectories on the one hand and electronic energy loss on the other hand are closely related.

Lindhard[129] based his estimates of stopping on the equipartition rule discussed previously. Since for $v \gg v_o$ resonance excitation takes place over several atomic distances d, it appears fair to assume that the resonant part of the interaction should be insensitive to the specific trajectory of an ion. Reversely, because of $(dE/dx)_e = NZ_2 S_e$, it is assumed that the stopping power for close collisions is proportional to the local electron density, so that,

$$\frac{dE}{dx}\bigg|_{\vec{r}} = S_e\{(1-\alpha)NZ_2 + \alpha n(\vec{r})\} \qquad (3.41)$$

where S_e is the stopping power <u>per electron</u> and n the electron density. α is a factor $(0<\alpha<1)$ that stems from equipartition and depends on energy. It may in principle be determined from graphs like fig.4. Obviously, eq. (3.31) is not rigorous. Upon averaging over all space one obtains the random stopping power, but for a channelled trajectory you have $<n(r)> < NZ_2$. In particular, there appears a lower limit

$$\frac{dE}{dx} \geq (1-\alpha)(\frac{dE}{dx})_{random} \qquad (3.42)$$

By means of a Thomas-Fermi-type charge distribution $n(\vec{r})$, Lindhard evaluated (3.41) for axial channeling and found

$$\frac{dE}{dx} = (\frac{dE}{dx})_{random}(1-\alpha e^{-\frac{2\psi^2}{\psi_1^2}}) \qquad (3.43)$$

where

$$\psi_1 = \sqrt{\frac{2Z_1Z_2e^2}{dE}} \qquad (3.44)$$

is the critical angle corresponding to a string direction with interatomic distance d.

In the velocity range where many stopping measurements have been made (0.1-1 MeV), it appears that α is still significantly greater than the asymptotic value 0.5 (fig.4). Consequently, for channels with low electron density, the ratio between channeled and random stopping power can become significantly smaller than 0.5. This has been documented in many measurements (Appleton et al[12], Clark et al[63], Eisen et al[73], and others).

3.9. Summary

i) The quantum theory of stopping in the form sketched in this paragraph is based on the lowest order of perturbation theory ; the basic stopping formula (Bethe's formula) can be derived by three equivalent procedures, straight Born approximation (Bethe), impact parameter treatment (Bloch) and dielectric treatment. (Lindhard).

ii) Bethe's treatment confirms several results that were found in the classical theory, in particular the connection between optical dispersion and stopping theory and the relativistic extension.

iii) Lindhard's dielectric description, when applied to a free electron gas and combined with the Thomas-Fermi principle, leads to reasonably accurate estimates of ionization potentials and, especially, inner-shell corrections for a major part of the periodic table. In addition, this description makes possible a precise distinction between distant and close collisions, the former leading to dipole resonant excitation, and a formulation of the equipartition of energy loss into the two types of excitation.

iv) The theory has been checked by numerous careful experiments employing light ions (especially hydrogen and helium), and accurate values of the critical parameters, in particular the ionization potential, are available for many elements. These values have mostly been obtained directly or indirectly from measurements rather than ab initio calculations.

v) In accurate experiments, higher-order perturbation effects (Z_1^3 effects) and deviations from Bragg's
additivity rule (for compound targets) need to be taken
into account. Reliable theoretical estimates are available for the former, and a few (by no means comprehensive) empirical rules for the latter.

vi) A cursory review has been given of the stopping power of single crystals under channeling conditions.

vii) Finally, it is mentioned that a number of
important aspects of the quantum theory of energy loss
have been left out from this presentation, such as the
detailed theory of inner-shell excitations by light-
ion (or electron) impact (see, e.g., Merzbacher & Lewis[145], Garcia et al[88]), the direct analysis of discrete energy losses in light gas targets (e.g. Inokuti[107])
and the extensive field of plasmon excitation by electron impact (e.g. Daniels et al[66]).

4. ENERGY LOSS OF HEAVY IONS AND LOW-VELOCITY STOPPING

General Reference : Northcliffe[159]

This section deals with two subjects that are somewhat related, the stopping of low-velocity ions and
the stopping of ions heavier than α-particles. In both
cases, pronounced deviations from the behaviour predicted by the Bethe formula may occur. Moreover, at constant ion energy, the velocity decreases with increasing
mass number. Hence, if you have an accelerator that
provides ions within a certain range of energy per ion,
light ions tend to have high velocity, and heavy ions
low velocity. For these and a number of other reasons,
it is convenient to treat the two topics jointly.

4.1. Charge States of Moving Heavy Ions

References : Betz[23]
Datz[67]

Protons penetrating through matter can be considered point charges to a good approximation, provided
that their velocity is large compared to the orbital
velocity v_0, i.e. E >> 25 keV. For heavier ions it may
be appropriate to carefully distinguish between the
nuclear charge of the ion, $e_1 = Z_1 e$, and the ionic charge $q = Z^* e$, where Z^* is the number of electrons stripped off. The ionic charge is a fluctuating quantity,
and the distribution of ionic charge is determined by
electron capture and loss processes that take place all
along the trajectory of the ion in a specific medium.

Often a dynamic equilibrium between capture and loss is established. This makes possible the introduction of a characteristic charge state (average, root mean square, most probable, etc.) that we simply call q in the following, without distinguishing between several possible definitions. In general, q depends mainly on Z_1 and Z_2, the velocity, and the density of the medium. Bohr[38b,c] and Knipp & Teller[117], estimating the ranges of fission fragments, suggested to use the usual stopping formulae and insert the ionic charge q for the projectile charge e_1 as a first approximation. This approach has since been used by many authors with considerable success, at not too low velocities. To show the qualitative effects on the stopping power, we can use a relatively simple model of charge states. The broader topic of charge states has been treated separately at this school in the lectures of Dr.Datz.

In general, the calculation of an equilibrium distribution of charge states requires detailed estimates of the cross sections for capture and loss of electrons (Bohr & Lindhard[40]). However, the average charge state can be estimated with reasonable accuracy by making the assumption that electrons with orbital velocities u less than the projectile velocity v are stripped after a small number of collisions, while electrons with u > v are kept since the collisions -seen, for example, from a reference system moving with the ion are adiabatic. Thus, the average charge state Z^* is equal to the number of electrons with orbital velocities u < v in this model. By use of a Thomas-Fermi estimate of u, sketched in appendix C, one obtains the result that Z^*/Z_1 should be a function of $(v/v_0)Z_1^{-2/3}$ only. More specifically, for $Z^* \lesssim Z_1/2$, Bohr[39]) obtained the result

$$Z^* \approx Z_1^{1/3} \, v/v_0 \qquad (4.1)$$

for the average charge state of heavy ions, in the velocity range where $1 << Z^* \lesssim Z_1/2$. This relation was found to fit reasonably well the experimentally measured charge states of fission fragments penetrating gas targets (Lassen[123]).

On the basis of a large amount of experimental data, Betz[23] favors the semiempirical expression

$$Z^* = Z_1(1-C \, e^{-\frac{v}{v_0 Z_1^{\gamma}}}) \qquad (4.2)$$

with adjustable parameters C and γ. These parameters turn out to be close to C = 1 and γ = 2/3. The expression (4.2) goes over into (4.1) for $Z^* \ll Z_1$ (if C $\approx$ 1 and $\gamma \approx$ 2/3) and shows an acceptable behaviour for $Z^* \to Z_1$. A slightly different expression for small Z^* is found in appendix C (eq.9).

As it stands, eq.(4.1) does not depend on the chemical or physical state of the medium. However, eq.(4.2) contains specific properties of the medium through the parameters C and γ. In particular, it has been established long ago that Z^* increases slightly with increasing gas pressure and that substantially higher average charge states are observed than predicted by eq.(4.1) for fast heavy ions emerging from solids (Lassen[123]). It is not quite clear at present whether the main source of the difference in observed charge states is electron loss at the target surface (Betz & Grodzins[24]), or whether the charge state of emerging ions essentially reflects the state of the ions inside the solid (Bohr & Lindhard[40]). The subject has been reviewed thoroughly (Betz[23]), and some of the later developments, in particular the use of X-ray high-resolution spectrometry to determine states of ions moving in solids, have been described in Dr Datz'lectures.

4.2. Charge State and Stopping Power

It follows from eq.(4.2) that a heavy ion can be expected to be stripped completely for

$$v \gg v_0 Z_1^{2/3} \qquad\qquad (4.3)$$

and, hence, the Bethe formula (3.36) should describe properly the stopping of such an ion with $e_1 = Z_1 e$, except for high-Z_1 projectiles where the Born approximation fails. In this latter case, the size of the necessary correction -as long as it is small- can be estimated from eq. (3.38).

When eq.(4.3) is not fulfilled, i.e. when $v \approx v_0 Z_1^{2/3}$, it appears tempting to insert $Z^* e$ for e_1 in the Bethe formula, and thus to obtain an expression for the stopping power in the region $v_0 \ll v \approx v_0 Z_1^{2/3}$.

Alternatively, one might employ an experimental (or more comprehensive theoretical) proton stopping power curve and use the scaling rule

$$\left(\frac{dE}{dx}\right)_{Z_1,v} = Z^{*2} \left(\frac{dE}{dx}\right)_{p,v} \qquad\qquad (4.4)$$

where the index p stands for "proton", and v indicates
that the stopping powers need to be taken at the same
particle velocity (or energy per mass unit). The advantage of eq.(4.4) is that it may contain inner-shell
corrections at low velocities.

In either case, the factor $(Z^*)^2$ will introduce a
factor $Z_1^{2/3} v^2/v_o^2$ at lower velocities. This indicates
that the energy dependence of the stopping power changes
over from an approximate $\ln E/E$ dependence to a slowly
varying $\ln E$ dependence, and that the variation with
projectile atomic number Z_1 will be less pronounced
than in the high-velocity limit. In other words, the
stopping power approaches a broad maximum around $v \sim v_o$
$Z_1^{2/3}$ and then decreases slowly with decreasing velocity. When the alternative low-velocity limit, eq.(9) of
appendix C is used for Z^*, the velocity dependence of
the stopping power becomes $\alpha v^{2/3}$ for $v_o \ll v \ll Z_1^{2/3} v_o$.

A scaling procedure of this type, together with
Lindhard & Scharff's scaling rules[132] for variation of
the stopping power with target atomic number Z_2, has
been used successfully for interpolation between experimental stopping powers (Northcliffe[159], Northcliffe
& Schilling[160]). The procedure appears feasible so long
as the required accuracy of heavy-ion stopping powers
is only in the 10 pct. range. Remind that a 1 pct. accuracy can be achieved in case of protons and α-particle stopping at not too low energies.

This scaling procedure has also been used extensively to analyse experimental stopping-power data in
terms of charge states (e.g.Pierce & Blann[169], Brown &
Moak[57]). By assuming (4.4) to be valid, comparison of
measured heavy-ion stopping powers with the corresponding proton stopping power at the same velocity yields
an effective charge state that turns out to show a rather similar behaviour as the average charge state measured directly. However, the scaling of stopping-power values in terms of charge states works equally well
for gasses and solids, and the extracted charge states
seem to agree with those measured directly on gasses.

Numerous semiempirical expressions have been proposed for the dependence of the effective charge on velocity. Most of them are of the type of eq.(4.2) ; however, the approach to complete stripping at high v is
somewhat uncertain, and not all proposed functions yield an exponential approach (e.g.Nikolaev & Dmitriev[158])
An exemple of excellent scaling of effective charge
states is shown in fig.7 which is taken from the work

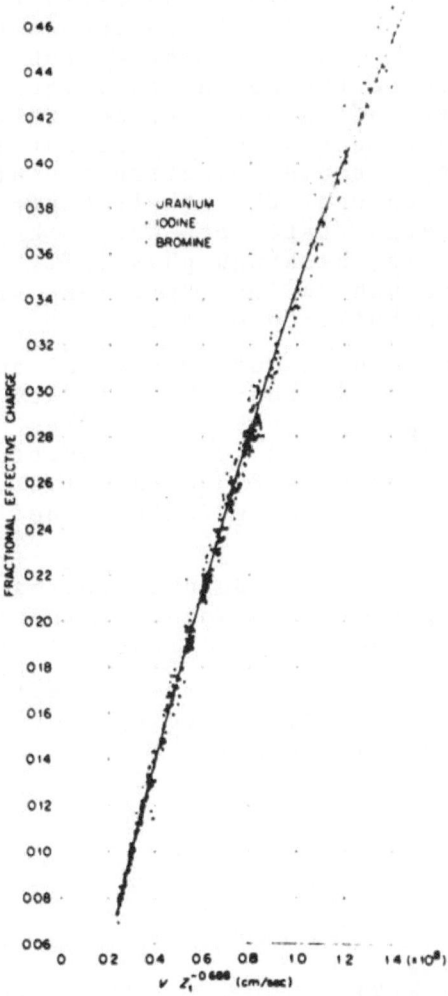

Fig.7 : Measured electronic stopping powers for uranium, iodine, and bromine ions in foils of carbon, aluminium, nickel, silver and gold (Brown & Moak[57]), plotted as Z^*/Z_1 vs. $vZ_1^{-0.688}$ where Z^* is defined by Eq.(4.4), with the proton stopping power taken from Northcliffe & Schilling[160]. The full-drawn line is a least-squares fit to Eq.(4.2) with C = 1.034 and p = 0.688. From Brown & Moak[57].

of Brown & Moak. With the success of the scaling rule
(4.4) in mind, Bloom and Sauter[32] and Sauter & Bloom[183]
made a new attempt to improve the early theoretical estimates of Z^* on the basis of the Thomas-Fermi model.
They decided to make use of a rather subtle Thomas-
Fermi description of the moving ion, including both
exchange and correlation effects, and they obtain a reasonably good agreement with "experimental" effective
charge states. A possible objection to this very usefull work seems, though, that their basic criterion to
determine the charge state does not go beyond Bohr's
original one ; this, although physically reasonable, is
hardly precise enough to justify an extensive Thomas-
Fermi calculation built up on it.

It has been pointed out by several authors (e.g.
Northcliffe[159], Betz[23]) that the validity of the scaling rule (4.4) can hardly be considered well justified
theoretically. Quite apart from the outstanding problem
of explaining the results found on solids, even with
gases a number of disturbing factors need to be taken
into account,

   i) deviation from the Born approximation
  ii) screening of the Coulomb interaction
 iii) electron promotion and related effects

A rough estimate of the first effect can be based
on eqs.(3.38) and (3.39), replacing Z_1 by Z^*. Then, the
relative error in (4.4) becomes :

$$\approx \frac{Z^*-1}{Z_2^{1/2}} F(V)$$

or, for $Z^* \gg 1$,

$$\approx Z_1^{1/3} V F(V) \qquad\qquad (4.5)$$

where $V = (v/v_0)Z_2^{-1/2}$. Since $F(V)$ goes approximately
like V^{-2}, eq.(4.5) imposes a lower limit on the velocity range where (4.4) should be applicable. In case of
$Z_1 \gg 1$, this limit will often be prohibitively high.

The two other effects disturbing the scaling law
are supposed to be less significant so long as $v \gg v_0$,
since, then, the stopping is determined mainly by collisions with impact parameters that are larger than the
ionic radius. The screening, when significant, is taken
into account effectively within the framework of dielectric theory. Screening as well as promotion effects be-

come important in case of strong mutual overlap of e-
lectron shells during the collisions.

4.3. Proton Stopping at Low Velocities

In the velocity range $v \lesssim v_0$ the Bethe theory of
stopping breaks down, as far as the important splitting
up into distant and close collisions is concerned.
"Inner"-shell corrections become important for all target electrons. Moreover, even a proton has a sizable
probability of becoming neutralized. For these reasons
a rather different approach to stopping theory has been
found appropriate.

In the dielectric theory, the difficulties mentioned can be overcome conveniently. In particular, proper consideration of the charge state is possible by
taking into account the screening of a point charge in
an electron gas.

Fermi & Teller[85] analysed the stopping and capture of muons in solid matter, in particular the stopping
in a Fermi gas at velocities $v \ll v_F$, the Fermi velocity. They came to the result that the stopping power
should be proportional to the projectile velocity.
Their qualitative argument holds as well for protons,
and goes as follows.

When a heavy, slow ion collides with an electron,
the collision is almost elastic in a reference frame
moving with the ion, i.e. the change of velocity of the
electron is of the order of v_F, and only those electrons
with velocities within $\sim v_F - v$ and v_F, i.e. $n' \propto \frac{v}{v_F}$ n electrons per volume, can absorb energy (n is the density
of the electron gas) ; the energy transfer is of the
order of $T \propto m v_F v$. For Coulomb scattering, the cross
section is of the order of $\sigma \sim b^2 \propto (e^2/m v_F^2)^2$, and
hence,

$$\frac{dE}{dt} \sim n' v_F T \sigma \sim \frac{e^2}{a_0^2} \frac{v^2}{v_0} \qquad (4.6)$$

or

$$\frac{dE}{dx} \sim \frac{e^2}{a_0^2} \frac{v}{v_0} \qquad (4.7)$$

Note that the density $n = (1/3\pi^2)(m v_F/\hbar)^3$ drops out.

Fermi & Teller also performed a simple transport
theoretical argument in order to obtain a more quantitative expression for the energy loss. This argument
was generalized by Lindhard (unpublished) and Trubnikov

& Yavlinskii[201]. The essential point is that the cross section σ in (4.6) becomes the transport cross section,

$$\sigma_{tr} = \int d\sigma (1 - \cos \theta) \qquad (4.8)$$

where θ is the scattering angle for an electron in the frame of reference where the ion is at rest. An approximate energy loss formula becomes, then,

$$\frac{dE}{dx} = nm \ v_F \ v \ \sigma_{tr}(v_F) \quad ; \quad v << v_F \qquad (4.9)$$

A derivation of eq.(4.9) has been sketched in appendix D. It is essential for (4.9) to be fulfilled that indeed $v << v_F$.

Fermi & Teller evaluated the transport cross section by assuming Rutherford scattering and a cut-off impact parameter

$$\lambda = (\hbar a_o / m v_F)^{1/2}$$

This is (almost) equivalent to applying the Born approximation to a screened Coulomb potential (Schiff[184]) : with the differential cross section

$$d\sigma = \frac{\pi b^2}{4} \frac{\cos \theta /2 \ \sin \ \theta/2 \ d\theta}{(\sin^2 \frac{\theta}{2} + (\frac{\hbar}{2\lambda p_F})^2)^2} \qquad (4.10)$$

$\sigma_{tr}(v_F)$ can be evaluated, and the stopping power reads

$$\frac{dE}{dx} = \frac{2}{3\pi} \frac{e^2}{a_o^2} \frac{v}{v_o} \{ \log (1 + (\frac{2\lambda p_F}{\hbar})^2) - \frac{1}{1 + (\frac{\hbar}{2\lambda p_F})^2} \} \qquad (4.11)$$

In (4.10) and (4.11), p_F is the Fermi momentum $m v_F$. Inserting the above expression for λ, and assuming that $v_F >> v_o$, one obtains

$$\frac{dE}{dx} = \frac{2}{3\pi} \frac{e^2}{a_o^2} \frac{v}{v_o} \log \frac{4 \ v_F}{v_o} \qquad (4.12)$$

which differs from Fermi & Teller's result only by the (unessential) factor 4 under the logarithm. The assumption $v_F \gg v_0$ is only fulfilled for solid targets with rather high atomic number Z_2.

It is seen from this derivation that the screening radius λ is an essential input parameter. In Lindhard's dielectric theory[129], where the properties of the medium enter via the dielectric constant, the screening is taken care of in a self-consistent way. In fact, evaluation of eq.(3.27) and (3.33) at low velocities, $v \ll v_F$, yields (Lindhard[129]).

$$\frac{dE}{dx} = \frac{4}{3\pi} \frac{e^2}{a_o^2} \frac{v}{v_o} C_1(\chi) \qquad (4.13)$$

where $C_1(\chi)$ is a well-defined function of the parameter

$$\chi = \sqrt{\frac{v_o}{\pi v_F}} \qquad (4.14)$$

The function $C_1(\chi)$ has been evaluated by Lindhard & Winther[134]. They obtain*

$$C_1(\chi) \approx \frac{1}{2(1-\frac{\chi^2}{3})^2} \{\log \frac{1+2\chi^2/3}{\chi^2} - \frac{1-\chi^2/3}{1+2\chi^2/3}\} \quad (4.15)$$

For $v_F \gg v_0$, or $\chi \ll 1$, one obtains eq.(4.12), again with a different numerical factor under the logarithm.

Note that the two procedures are almost equivalent through being based on first-order perturbation theory, but differ in the way how the screening enters.

Pathak & Yussouf[167] evaluated the stopping power of the Fermi gas numerically. They applied their calculation to the stopping of xenon ions ; this is hardly justified because of the breakdown of the Born approximation for high Z_1. However, since the projectile charge only enters through a factor, their calculations can certainly be applied to protons. As it stands, the numerical accuracy of their evaluation seems poor, in view of some seemingly unphysical wiggles around $v \approx 0.02 \, v_F$. Nevertheless, the overall result is an $E^{\sim 0.4}$

---

*A printing error pointed out by Ziegler & Chu[219] has been corrected in eq.(4.15)

dependence of the stopping power from $v/v_F = 0.01-0.024$. This indicates that the approach to the asymptotic $E^{1/2}$-dependence might be rather slow. As they did at higher velocities, the authors report only a very small difference between the results obtained with Lindhard's and Hubbard's[102] dielectric constant.

Trubnikov & Yavlinskii[201] also investigated the stopping of light ions by means of the Fermi-Teller and the Lindhard formalism (possibly not being aware of Lindhard's paper). They concentrated on the velocity-proportional regime, but investigated a few more physical situations. They report an instructive plot of the function $2C_1(\chi)$, eq.(4.13), and they made an attempt to compare with experimental results. The underlying assumption appears to be that only conduction electrons contribute to the stopping. They use standard values for the density of conduction electrons, and <u>effective</u> masses for the electron mass. They obtain good agreement with measured light-ion stopping powers, for Al and Ti, moderate agreement for Cu, and a drastic underestimate of the stopping power for Sn and Ag, the latter result is hardly unexpected.

Calculation of stopping powers within the statistical scheme implies averaging the function $C_1(\chi)$ in eq.(4.13) over the electron density distribution of single atoms, molecules, or solids. This is important since regions of high electron density (inner shells) have low χ, and hence large C_1. However, such averaging is not a trivial task, especially since at low enough velocities the ion <u>trajectory</u> needs to be taken into account, i.e. the decrease in probability to penetrate the inner parts of the atom. Rousseau et al[179] appear to have made some calculations using the charge distributions they used for higher-velocity calculations, but, unfortunately, they did not report any results.

A famous formula -the published derivation of which has been looked forward to for some time-predicts (Lindhard & Scharff[132])

$$\frac{dE}{dx} = \xi_e \, 8\pi e^2 a_0 \, \frac{Z_1 Z_2}{Z} \, \frac{v}{v_0} \qquad (4.16)$$

for the electronic stopping power at low velocities, where

$$Z^{2/3} = Z_1^{2/3} + Z_2^{2/3} \qquad (4.17)$$

and

$$\xi_e \approx z_1^{1/6} \qquad\qquad (4.18)$$

Eq.(4.16) is supposed to apply for arbitrary ions at $v < v_0 z_1^{2/3}$. It is the result of a fitting of a low-z_1 (Born approximation) and a high-z_1 form. It has proved to be a usefull relationship. However, it also appears likely that the proposed upper velocity limit may be too high.

Proton stopping at low velocities has been measured by Ormrod et al[164,165], Morita et al[152], White & Mueller[208] and others. (Many more experimental investigations deal with proton stopping around the maximum). Both Ormrod et al and Morita et al report an $E^{0.4}$ dependence of the stopping power up to $v \sim v_0$. The investigated velocity range, though, is rather small. The numerical values of the stopping power have normally been compared with eq.(4.16), and the agreement was often quite good, within better than $\sim$ 10 %. However, White and Mueller[208], comparing their own stopping power measurements with compilations of Whaling[207] concluded that the stopping power per atom vs. Z_2 had an oscillatory behaviour, with a maximum around $Z_2 = 18$ and a minimum around $Z_2 = 29$. An interpretation along similar lines as the one indicated in fig.6 appears plausible, but has not been accomplished yet, as far as I am a aware of. Note that Bernhard et al[19], who found oscillatory structure in stopping measurements with ions heavier than protons, did not find such an effect with protons although they were looking for it.

4.4. Electron Promotion

When you perform experiments with heavy ions, a number of spectacular effects occur. You ought to realize first, that for an Ar^+ projectile the adiabatic limit $v = v_0$ lies at an energy of 1 MeV. According to Bohr[39], electronic excitation effects ought to diminish rapidly below this velocity. The observations are that electrons from the Ar-L shell may be excited down at medium-keV energies, far below the adiabatic limit for L-shell electrons. The most direct evidence consists of detected Ar-L X-rays, or Auger electrons, but the mechanism had been identified earlier by direct measurements of the electronic energy loss in ion-gas collisions. Comprehensive reviews of experimental work in this field have become available recently (Garcia et al[88], Kessel & Fastrup[113]). Somewhat analogous to inner-shell excitation, also outer-shell excitation has been observed down in the medium eV region, even with rather heavy ions. This shows up, for example, in light emission (see, e.g. Tolk and White[200]). Neither inner-nor outer-shell excitation are small effects. Roughly, you deal with geometric cross sections, i.e. once the energy is high enough so the collision partners interpenetrate to the shell you want to excite, you have a fair chance (10-1-100) that excitation really occurs.

If you are less impressed by the fact that such remarkable processes occur with high probability below the adiabatic limit, you might appreciate the fact that the maximum energy transfer from an argon nucleus at 40 keV to an electron at rest is 2 eV - certainly not enough to excite the L-shell. Yet the effet has been observed above a threshold of $\sim$ 4 keV (Saris[181]). On the other hand, one of the first theorems all radiation damage physicists get punched into their heads is that an argon ion does not lose energy to electronic excitation below 40 keV. (Dienes et Vineyard [71], Seitz[186]). You should recognize that this latter theorem is incorrect both from a qualitative and a quantitative point of view.

A useful basis to explain electronic excitation in heavy-ion atom collision below the adiabatic limit appears to be the collision partners forming a quasi-molecule during the collision, the electron states becoming molecular orbitals (MO's) (Hund[103], Mulliken[155]) with the energy being a function of internuclear distance. There are well-defined rules how to correlate the energy levels of the separated atoms with those at close approach and, as it turns out, level crossings occur of many molecular orbitals with each other at various internuclear distances. As such crossings, there is finite probability for the system to switch diabatically from one MO to another ; as a result, the atoms will most often be in excited states after separation. The mechanism of such crossings has been investigated a long time ago (Landau[120] Zener[217], Stueckelberg[196]). Nevertheless, Firsov[86], in his paper on electron excitation in atomic collisions (to be discussed in the following paragraph), found that a quantitative treatment of electron promotion in particular for outer-shell excitation was an exceedingly complicated task, and, therefore, decided to develop his more qualitative and rather more intuitive, nevertheless highly succesfull approach. To a certain degree, Firsov's statement still appears valid. However, the theoretical description in particular of inner-shell excitations is improving rapidly (Barat and Lichten[16], Thulstrup and Johansen[199], Briggs and Macek[54]), in conjunction with experimental material accumulating. Interest is being extended to outer-shell effects (e.g. Bierman et al[27,28]).

The information that can be gained from typical energy-loss structure measurements in ion-gas scattering (e.g.Fastrup et al[81]) is more detailed than would be required for the purpose of stopping theory. This holds in particular for inner-shell effects.

On the other hand, a thorough study of recent experimental and theoretical work on excitation in heavy-ion atom collisions with a view at application in stopping theory appears to be a worth while task, Bierman's work quoted above being a promising start. I have refrained from going into details of the promotion model here, since quite a bit more work needs to be done before the needs of stopping theory may become satisfied. However, I wanted to mention this aspect because the field is developing so rapidly that it might possibly dominate low-energy stopping theory fairly soon.

4.5. Firsov Theory and Related Calculations

Reference : Firsov[86]

Let an ion 1 pass by an atom 2 at an impact parameter p, with velocity $v \ll v_0$ so the electrons can -to some extent- adjust their orbits to the instantaneous relative position of the nuclei. Consider the minimum equipotential surface between the collision partners; it is approximately a plane perpendicular to the "molecule" axis, the "Firsov plane". At least classically, you can say that the Firsov plane divides the electrons into groups belonging to atom 1 and atom 2, respectively. There is, however, a flux of electrons each way through the Firsov plane.

Now, an electron passing from 2 to 1 needs to be accelerated to the velocity v, in order to belong to particle 1 ; it picks up a linear momentum mv from the nucleus, and thus contributes to stopping. An electron passing from 1 to 2 transfers its momentum to the nucleus 2 in order to get captured, but does not slow down the nucleus 1. Thus, the force on the ion 1 is given by

$$F = mv \int_S \frac{1}{4} nv' \, dS \qquad (4.19)$$

where the integral represents the number of electrons per unit time that pass the Firsov plane S from 2 to 1.

n is the electron density around the areal element dS, and v' is the Thomas-Fermi velocity connected to n via

$$v' = \frac{3}{4} (3\pi^2 n)^{1/3} \frac{\hbar}{m} \qquad (4.20)$$

The factor $\frac{1}{4}$ arises as follows. Assume an isotropic velocity distribution ; only one half of all particles with velocity v' move from 2 to 1, the other half going the opposite way. Since only the velocity component perpendicular to the Firsov plane counts, you get another factor 1/2 from $<\cos \alpha>_{hemisphere} = 1/2$. If v is assumed constant (soft collision), the energy loss per collision is

$$Q = mv \int_{-\infty}^{\infty} dx \int_S \frac{1}{4} nv' dS \qquad (4.21)$$

As it stands, eq.(4.21) depends on the projectile velocity only through the factor v, while the surface integral contains the spatial coordinates. Thus, the energy loss and hence the stopping power becomes strictly proportional to velocity.

Firsov's evaluation of eq.(4.21) makes extensive use of the Thomas-Fermi method. For v' he uses the average speed of an electron in a Fermi gas, eq.(20) and the connection between n and the potential ϕ is given by the standard Thomas-Fermi expression

$$n = \frac{(2\phi/ea_0)^{3/2}}{3\pi^2} \qquad (4.22)$$

for neutral systems. For the potential at the Firsov plane, an expression

$$\phi(r) = \frac{Z_0}{r} \chi(r/a') \qquad (4.23)$$

is used, with

$$Z_0 = Z_1 + Z_2 \qquad (4.23a)$$

and

$$a' = 0.8853 \, a_0 Z_0^{-1/3} \qquad (4.23b)$$

The somewhat unusual screening radius (4.23b) has probably been inserted to account qualitatively for po-larisation of the overlapping electron clouds in the quasi-molecule. It appears that Kishinevskii's expression[115] for ϕ, which is the Thomas-Fermi potential corresponding to the superimposed, undisturbed atomic charge distributions, is superior to (4.23). Kishinevskii also selects the plane S by minimizing the potential, while Firsov, in his evaluation, places S midways in between 1 and 2.

This approximation limits the range of applicability to systems with similar Z. Thus, Kishinevskii's procedure allows evaluation of the Firsov energy loss even for collision partners very different in mass.

In the evaluation of (4.21), Firsov assumes a straight-line trajectory, while Kishinevskii includes the deflection of the projectile approximately. The evaluation is done numerically by both authors. Firsov obtains

$$Q = \frac{0.35 \, Z_0^{5/3}}{(1+0.16 \, Z_0^{1/3} \frac{R_0}{a_0})^5} \frac{e^2}{a_0} \frac{v}{v_0} \qquad (4.24)$$

where R_0 is taken to be the distance of closest approach, which is identical with the impact parameter in this approximation. Kishinevskii obtains

$$Q = \frac{0.24 \, Z_1 Z^{1/2} (Z_1^{1/6} + Z_2^{1/6})}{[1+0.67 \, Z_1^{1/2} (Z_1^{1/6} + Z_2^{1/6}) \frac{R_0}{a}]^3}$$

$$\times (1-0.68 \frac{V(R_0)}{E}) \frac{e^2}{a_0} \frac{v}{v_0} \qquad (4.25)$$

where $Z = (Z_1^{1/2} + Z_2^{1/2})^{2/3}$ and $a = 0.8853 a_0 Z^{-1/3}$.

Although designed originally for analysing single-collision energy losses, the Firsov theory has mostly been used to interpret stopping data. The reason for the limited use in single-collision phenomena is the fact that a calculation of an average energy loss does not account for the observed multiple-peak structure (see e.g. Kessel et al[113]). (Actually, when plotting Q vs. R_0 according to eq.(4.24) for the Ar^+-Ar system, Finnemann (private communication) obtained a rather close correlation with the data of Morgan & Everhart[151])

The extensive use of eq.(4.24) can be justified by its apparent success and its simplicity. The theory has been extended by various authors. However, as in the case of Bohr's theory of charge states (ch.4.1), any substantial improvement warrants an answer to the question whether the simplified basic assumptions justify the labor of expanding the details of the evaluation. In view of the mentioned shortcomings of Firsov's derivation, I find Kishinevskii's approach acceptable in principle, but would not advise you to go much further without starting over again at the basic principles.

Fig.8 shows measured electronic stopping powers on thin carbon foils at several fixed velocities as a function of atomic number of the ion (Hvelplund et al[105]) The curve labelled "Firsov" has been found from eq.(4.24) by integration, (Teplova et al[198])

$$\left(\frac{dE}{dx}\right)_e = N \int_0^\infty 2\pi R_0 dR_0 Q \qquad (4.26)$$

At low velocities, (4.26) must overestimate the average energy loss since interpenetration to $R_0 = 0$ has been allowed. It appears that the overall agreement between the experimental results and eq.(4.26) is rather good, about as good as with the Lindhard & Scharff formula (4.16). The oscillatory behaviour of the experimental data needs special consideration (see below).

Brice[53] has examined the Firsov theory and expanded it on two essential points. First, instead of (4.19) he presented a quantum mechanical calculation of the particle flux and, second, a correction for deviations from adiabatic behaviour was made.

The first correction includes the motion of the Firsov plane, and was performed on the basis of a Heitler-London model of the quasi-molecule, using hydrogenic wave functions and an (adjustable) effectivecharge. The second correction was done by introducing a trial function containing two adjustable parameters. As it turns out, the resulting stopping formula reproduces measured stopping powers quite accurately in the low-, intermediate-, and high-energy region, and Brice states that this is the main merit of his formula. From a more basic point of view, this approach appears less satisfactory. First, I find that there is a lack of balance in that the one of two corrections of comparable magni-

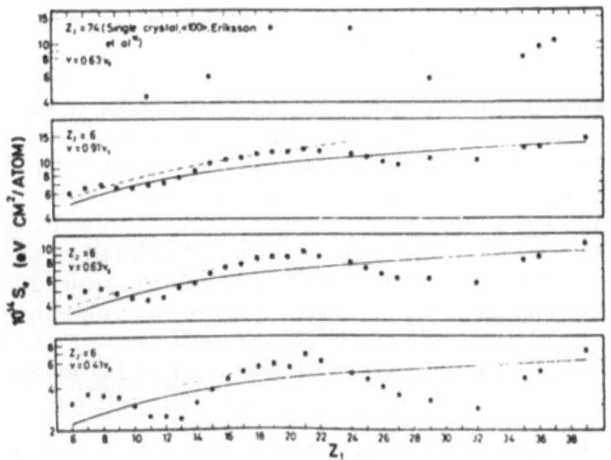

Fig.8 : Measured and calculated electronic stopping cross sections
$\overline{S_e}$ = (1/N)dE/dx, as a function of the atomic number of the ion.
The upper graph (Erikson & al[77]) refers to range measurements on
single crystals of tungsten. The lower three graphs (Hvelplund &
Fastrup[105]) refer to energy-loss measurements on thin carbon foils
Data of Ormrod & al[164,165]) were included. The full-drawn line
corresponds to Eq.(4.16), and the dashed one to Eqs.(4.46) and
(4.26). From Hvelplund & al[105].

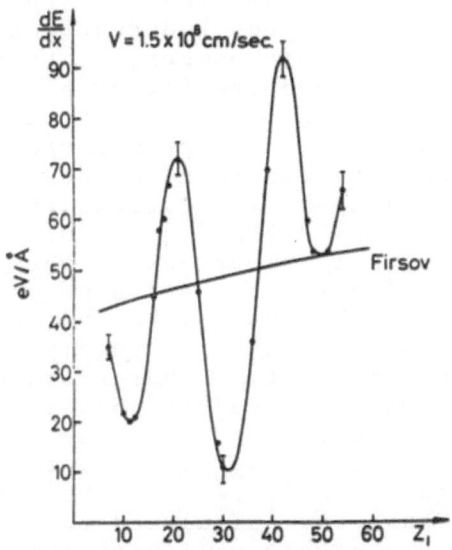

Fig.9 : Similar plot as fig.8 for gold single crystal foils
(Bøttinger & Bason[46]). The curve labelled "Firsov" is based on
Eq.(4.24) and an impact parameter corresponding to a trajectory
confined to the center of a <110> axial channel. From Bøttinger
& Bason[46].

tude is done by a fudge factor, and the other by an extensive quantal calculation. Second, the Firsov plane is consistently placed midways in between the collision partners, even for rather different atomic numbers. Since Brice applies his formula mostly to hydrogen and helium projectiles, this assumption is expected to lead to severe errors. Third, once adiabatic states and hydrogenic wave functions are used, it would seem appropriate to avoid the Firsov plane altogether, and define the charge transfer quantum mechanically. This would simplify the treatment and make it more consistent.

A number of investigations deal with modified Firsov models with the specific aim of explaining the oscillatory structure of the stopping power vs.Z_1, as indicated in fig.8 and reported in a number of experimental investigations (Teplova et al[198], Ormrod et al[164,165], Fastrup et al[80,82], Hvelplund & Fastrup[105], Macdonald et al[139], Ormrod[163]). As it turns out, the oscillations are most pronounced under channeling conditions (Eriksson et al[77], Eisen[72], Bøttiger & Bason[46]) as is shown in fig.9.

The various modifications of the Firsov theory are all based upon the fact that the electronic charge distribution of an atom expands and contracts periodically as a function of atomic number, so that the density of a quasimolecule around the Firsov plane exhibits similar oscillations, once such a density is calculated beyond the Thomas-Fermi approximation. With the density, also the electron flux and hence the stopping power shows oscillations. The various treatments differ in details. Winterbon[213] kept eq.(4.20) for the velocity, but used charge distributions of various charge states, thus allowing for establishing an average charge. Cheshire et al[59,60], calculate the velocity as a root-mean-square value quantum mechanically, but take only into account the initial charge state. Bhalla et al[26] applied a similar procedure, but separated the effects of different shells and made use of several sets of tabulated charge distributions. Winterbon and Cheshire et al compared their results with single-crystal data, and found that the calculated phases in the oscillations agreed reasonably well with the measured ones, while the calculated amplitudes dreastically underestimated the measured effects. In the comparison, it might be worth while noting that the experimental data refer to "the leading edge", i.e. those ions that, after penetrating the crystal, have suffered least ener-

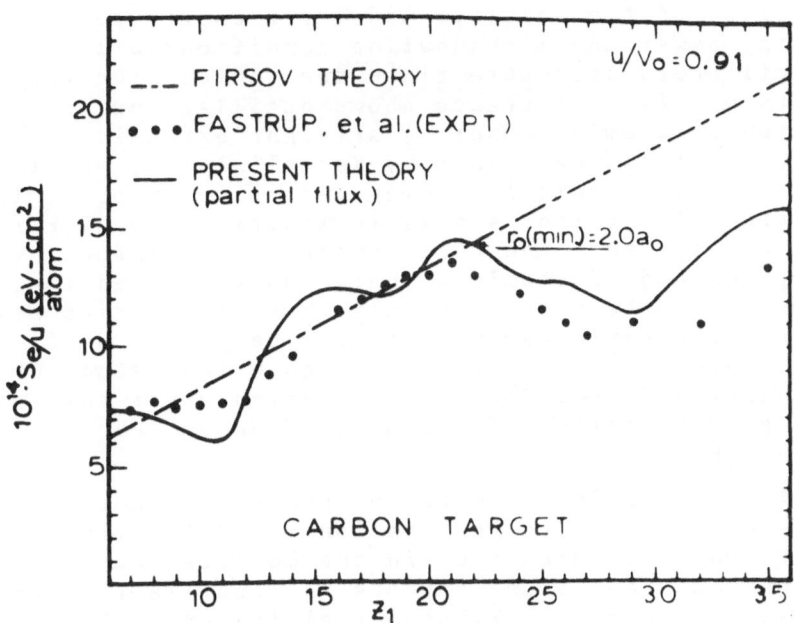

Fig.10 : Similar plot as fig.8 for amorphous carbon. Experimental
points from Fastrup et al[82]. u=velocity. Full-drawn curve from
Bhalla & al[25] dot-dashed curve from Eqs. (4.46) and (4.24). Hartree-Fock-Slater electron distributions were used in the evaluation of Eq.(4.19). The term "partial flux" refers to the exclusion
of certain subshells from contributing to the electron flux (4.19)
because of binding. The Firsov plane was placed at the potential
minimum.

gy loss. Similarly, in the calculated energy loss the assumption was made that the ion moves on a straight line, at maximum (and constant) impact parameter, so that eq.(4.24) after modification can be used without integration. Winterbon and Bhalla et al evaluated the case of random stopping, i.e. eq.(4.26) plus modifications, and obtained fairly good agreement with experimental results (fig.10). Harrison[97] pointed out that, regardless of the specific model for energy loss, the stopping power under channeling conditions will show an oscillatory structure since the radial electron density at a fixed distance shows oscillations as a function of atomic number. A straight extension of the Firsov model has been made by Kessel'man[114], who introduced a nonmonotonic screening radius in an otherwise analytic calculation. A drastic modification of Firsov's theory, based on the concepts of an effective charge instead of atomic number, and a reduced impact parameter in the Firsov formula -both quantities being used as adjustable parameters- has been done by El-Hoshy & Gibbons[74]. An extensive analytical study of the modified Firsov model has appeared after this manuscript was finished (Komarov & Kumakhov, Phys.Stat. Sol. b, 58, 389 (1973).

In view of the symmetry of the Firsov formula (4.24) with regard to permutation of projectile and target, which is preserved in the modified Firsov models quoted above*, one expects Z_2-oscillations in electronic stopping. (Wilson et al[212]). Such effects have been known for some time at intermediate and high energies, and were discussed briefly in sect.3. We also mentioned the proton data of White & Mueller[208] in sect.4.3. Z_2-oscillations with heavier bombarding ions were found experimentally by Bernhard et al[19] and Apel et al[9], who bombarded a number of polycrystalline targets with 40 keV lithium ions. They obtain a pronounced oscillatory effect with an amplitude of almost a factor of two, a minimum near copper and a maximum near vanadium. These oscillations are not identical with the usual Z_1-oscillations, but are not completely out of phase either. Pronounced variations of the stopping power per atom have been reported by Whitton[209], who determined maximum penetration depths of potassium ions

---

* A (slight) asymmetry occurs in Winterbon's scheme[213] in view of the different treatment of charge states.

in several single crystals under channeling conditions. As in case of Z_1-oscillations, the channeling effect strongly enhances the oscillations in stopping power per atom, (up to a factor of 5-6), and the phases closely coincide with those observed in Z_1-variations of the stopping in gold single crystals (Bøttiger et al., 1970). Again it is most likely that a modified Firsov model of the type described in this chapter will not be sufficient to describe these pronounced oscillatory effects.

4.6. Fermi-Gas Treatment of Low-Velocity Heavy-Ion Stopping

For low-energy electrons scattered on heavy ions, the Born approximation is not valid in general (Schiff[184]). Therefore, the basic equations of the Lindhard[129] dielectric treatment of energy loss -that are based on the ion being a perturbation- become questionable, and conclusions drawn, for example, from eq.(4.13) may be invalid. This almost certainly applies to the stopping powers for xenon atoms computed by Pathak & Yussouff[167], and, to a lesser extent, to Bhalla & Bradford's[25] attempt to scale eq.(4.13) in such a way as to predict the Z_1-oscillations of heavy-ion stopping powers in amorphous media.

The Fermi-gas treatment of low-velocity heavy-ion stopping in random targets has been outlined by Lindhard & Scharff about 15 years ago, before any direct stopping -power measurements- and very few range measurements -were available. In view of the obvious complexity of the problem, these estimates were fairly crude, and were bypassed later by rather accurate experiments in the early 1960's. Nevertheless, the Lindhard & Scharff stopping formula (4.16) had an enormous impact on the field of low-velocity penetration phenomena, and it is regrettable, therefore, that the complete derivation was never published. To give you an indication of how you can arrive at a result like eq.(4.16), start first at the definition, $(dE/dx)_e = N \int T d\sigma$. As may be concluded from Schiøtt's[185] summary, the Thomas-Fermi scaling rules (2.40) and (2.42a) are also relevant in case of "quasi-elastic" collisions. Hence, $(dE/dx)_e$ is approximately proportional to the energy unit $Z_1 Z_2 e^2 /a$ and the cross section unit πa^2. According to eq.(4.13), $(dE/dx)_e$ should also be proportional to velocity. This can be arranged by a dimensionless factor $v/(Z^{2/3} v_0)$, where $Z^{2/3} v_0$ is a scaled Thomas-Fermi velocity. From this, one obtains

$$\left(\frac{dE}{dx}\right)_e \propto N\frac{Z_1Z_2e^2}{a} \cdot \pi a^2 \frac{v}{Z^{2/3}v_o} \propto N\pi e^2 a_o \frac{Z_1Z_2}{Z} \frac{v}{v_o} \qquad (4.47)$$

An alternative expression, valid for a light ion, can be found by use of Lindhard & Scharff's scaling law[132] (cf.fig.5).

$$\frac{dE}{dx} = \frac{4\pi Z_1^2 e^4}{mv^2} NZ_2L\left(\frac{v^2}{Z_2 v_o^2}\right) \quad v \gg v_o \qquad (4.48)$$

that is essentially the Bethe formula, including Bloch scaling and shell corrections. If you assume the stopping power to approach proportionality with v at low velocities, you have to assume $L(\xi) = const \; \xi^{3/2}$, and hence

$$\frac{dE}{dx} = const \; 8\pi e^2 a_o \frac{NZ_1^2 Z_2}{Z_2^{3/2}} \frac{v}{v_o} \qquad (4.49)$$

Eq.(4.49) takes into account part of the screening of the inner shells of the target atoms.
For a heavy ion, at $v \gg v_o$, you might try eq.(4.48) including the charge state (4.1), and obtain

$$\frac{dE}{dx} = \frac{4\pi Z_1^{2/3} e^4}{mv_o^2} NZ_2L\left(\frac{v^2}{Z_2 v_o^2}\right) \qquad (4.50)$$

If this is going to be proportional with velocity, you have to assume $L(\xi) = const \; \xi^{1/2}$, and, hence,

$$\frac{dE}{dx} = const \; 8\pi e^2 a_o \frac{NZ_1^{2/3} Z_2}{Z_2^{1/2}} \frac{v}{v_o} ; \begin{matrix} v \gg v_o \\ Z_1 \gg 1 \end{matrix} \qquad (4.51)$$

Neither (4.49) nor (4.51) take full account of projectile screening. Because of symmetry, it is reasonable to replace Z_2 in the denominator of (4.49) and (4.51) by :

$$Z = (Z^{2/3} + Z_2^{2/3})^{3/2} \qquad (4.52)$$

the screening parameter for electric interaction. You can do this replacement in different ways, but you can argue that you are correcting for a volume effect, i.e.

you want to get a factor $a^3 \propto Z^{-1}$ into (4.52). In order
to obtain the Lindhard-Scharff formula, you have to
multiply (4.51) by $(Z_1 Z_2)^{1/2}/Z$. Since this is within
the uncertainty of the present derivation, I won't argue for this factor.

I should stress that the original considerations
leading to eq.(4.16) were more sophisticated than the
present ones. Nevertheless, you should note that one
key point in the derivation is the proportionality of
stopping power with v at low velocities.

One might ask how well this proportionality has
been verified experimentally and theoretically. To
start with the latter first, we note that the existing
derivations, i.e. Fermi & Teller's[85], Lindhard's[129],
and Firsov's[86], all apply to the velocity regions
$v \ll v_0$ or $v \ll v_F$, and that none of them goes beyond the
first term in v/v_0 or v/v_F. Some indication of the size of deviations can be found in Pathak & Yussouff's
work[167] but, as was pointed out earlier, there may be
inaccuracies in the numerical evaluation. Brice's corrections to the Firsov theory, despite the reservations
made earlier, appear necessary and substantial. Finally, as will be mentioned below, the transport cross
section (4.8) may depend sensitively on the velocity
(Ramsauer effect), and hence even for small values of
v/v_F, substantial velocity-dependent corrections may
occur in eq.(4.9). Thus, despite the seemingly convincing argument that the three rather different published
low-velocity stopping theories all predict velocity-
proportional stopping, there is little theoretical evidence on the magnitude of the leading correction terms
and even less on the accuracy of the extrapolation to
real systems. In experimental work it has been emerging
from the early stopping measurements (e.g. van Wijngaar-
den & Duckworth[211], Ormrod et al[164,165], Fastrup et
al[82], and especially Hvelplund & Fastrup[105]) that the
stopping power can be represented over a reasonably large energy interval by a power dependence $\propto E^p$, with p
close to $1/2$, but showing oscillatory behavior as a
function of Z_1 and, as shown more recently (Apel et al[9],
Bernhard et al[19]), of Z_2. In particular in measurements
under channeling conditions, p can vary between 0.2 and
0.9 (Bøttiger & Bason[46]). This fact seems to limit
strongly the validity of the modified Firsov theories
to explain the oscillatory behaviour of electronic stopping cross sections both as a function of Z_1 and Z_2.

Finally, some indications should be made about the
Lindhard & Finnemann theory of low-velocity stopping.

This theory, which is still unpublished, is based upon eq.(4.9), and the transport cross section is evaluated along the conventional lines of the theory of electron-atom scattering (Mott & Massey[154]). From experiments with gases there is ample material available on angular and energy dependence of the cross sections (e.g. Kollath[118]). It is well known that diffraction effects cause the total cross section to approach zero when the de Broglie wavelength of the scattered particle increases above atomic dimensions. (Ramsauer effect). The total cross section may show a pronounced peak in the low-eV region. The transport cross section behaves qualitively similar, since the angular variation of the differential cross section is much less pronounced than at high energies. Indeed, for isotropic scattering, you have $\int d\sigma \cos\theta = 0$ and $\sigma_{tr} = \int d\sigma = \sigma_{tot}$. When a Thomas-Fermi type potential is used to determine a transport cross section via phase-shift analysis, the transport cross section exhibits oscillations as a function of the atomic number of the "target", which is the projectile in eq.(4.9). These oscillations can be quite pronounced (Finnemann[85a], Briggs & Pathak[55]). It is essential to notice that the calculated oscillations depend to some degree on whether you apply statistical or Hartree-Fock-type electron distributions, but their very occurrence does not hinge on a built-in shell-structure in the charge distribution, contrary to the modified Firsov theories. You may also note that Z_2 -oscillations- or, more precisely -Z_2- structure is also built into the Lindhard -Finnemann model through the sensitivity of the transport cross section to changes in electron velocity, i.e. Fermi velocity. This very sensitivity, however, appears to be one of the main complications in the quantitative theory, since correction terms to eq.(4.9) may become appreciable, and the Thomas-Fermi averaging process may involve greater uncertainties than in the computation of ionization potentials and shell corrections. No calculated stopping powers for ion-gas or ion-solid systems have become available yet.

4.7. Summary

1. For heavy ions at velocities $v \gg v_0$, the Bethe formula can be modified by including an effective charge state to yield expressions for the stopping power. The scheme appears moderately accurate, but needs some more theoretical justification.

2. The theory of proton stopping at low velocities appears to be well described by the dielectric theory.

Accurate evaluations of the stopping power in the transition region from low to intermediate velocities appear to be still missing, as well as average stopping
powers computed for atomic charge distributions.

3. Electron promotion may be relevant to stopping
theory, and exploratory studies appear promising.

4. The Firsov and Lindhard & Scharff stopping formulae describe well a great number of experimental data on ranges and stopping powers within a 20-30 % accuracy or better. In going beyond the Firsov approximation, one reasonable requirement is to keep within the
limits of complexity and detail that are set by the basic assumptions.

5. The stopping power at constant velocity shows
oscillations as a function of projectile atomic number
and structure as a function of target atomic number.
Within the modified Firsov models, the two types of variation are closely related effects. By comparison of
calculated with measured variations, it appears that
the modified Firsov models do not account for the measured effects, at least in case of channeling. The Lin-
dhard-Finnemann model might be more powerful in explaining the observed effects.

6. For recent compilations of range and stopping
data for heavy ions see Northcliffe & Schilling[160] and
Mayer et al[142]. For calculation of ranges see Lindhard
et al[133] and Sigmund[191].

5. <u>MULTIPLE SCATTERING AND ENERGY-LOSS DISTRIBUTIONS</u>
<u>STOPPING MEASUREMENTS</u>

5.1. <u>Introduction</u>

This section deals with some of the theoretical
problems that arise in stopping measurements. Take, for
example, the standard technique of stopping-power measurement, i.e. analysis of the energy spectrum of an
initially monoenergetic, well-collimated beam after
passage through a foil of thickness Δx. You will measure a shift in the energy spectrum towards lower energies and a broadening. You might be tempted to determine the peak of the spectrum, and identify the peak
shift with the average energy loss, $N\Delta x S(E)$, and sometimes you may even be right in doing so. Prominent reasons why you are not always right are the following :
1) The transmitted energy spectrum need not be symmetric around the peak. Therefore, average and most probable energy loss do not necessarily coincide. 2) Beam

particles are scattered more or less away from their
initial direction of motion. This makes the travelled
path length longer than Δx and may, dependent on the
geometry, cause a larger or smaller fraction of beam
particles to miss the detector. If the latter is true,
the measured energy spectrum need not be representative for the stopping of all beam particles. 3) Your foil
may be so thick that the linear relation between foil
thickness and energy loss breaks down. You can usually
find out by use of simple estimates whether or not these effects or others need to be considered. The problem
is rather how to do the corrections when they are expected to be substantial, and how to find an alternative approach when the corrections are large enough to
dominate the picture.

The simple estimates you need are those of the a-
verage energy loss, straggling, and multiple-scattering
Since we have not considered the latter effect so far,
I shall give you a brief outline of that topic first.
The detailed theory of energy-loss and multiple-scattering distributions is based on transport theory. It was
developed predominantly for electron beams -since the
effects turn out to be more pronounced there- but with
the increasing interest for accurate data on ion stopping, there is a need for detailed estimates also for
ion scattering and stopping. As in the previous sections
I shall mainly talk about ions, although the techniques
and part of the results also apply to electrons.

A short survey of experimental techniques concludes this lecture.

5.2. <u>Multiple Scattering</u> - <u>Simples Estimates</u>

Reference : Bohr[39]

In the remainder of these lectures, the term "scattering angle" will normally mean a laboratory rather
than a center-of-mass scattering angle. Denoting the
former by the symbol ϕ, the connection with the latter
is given by :

$$ tg \; \phi = \frac{m_2 \; \sin \theta}{m_1 + m_2 \; \cos \theta} \qquad (5.1) $$

or, for small angles

$$ \phi \approx \frac{m_2}{m_1 + m_2} \; \theta \qquad (5.2) $$

Let a beam of, say, α-particles hit a thin foil of thickness Δx. Then, the fraction of beam particles scattered over a certain minimum scattering angle ϕ^* is given by (2.2),

$$P(\phi^*) = N' \, \Delta x \int_{\phi > \phi^*} d\sigma \qquad (5.3)$$

provided that $P(\phi^*) \ll 1$. We use the notation of sect. 2. If the projectile is an ion, the scattering is dominated by the target <u>nuclei</u>, since the maximum angle for scattering on electrons is extremely small. In case of electron projectiles, the target electrons contribute a fraction $1/Z_2$ to the scattering by the nuclei, as you may see by inserting Rutherford's cross section (2.11) into (5.3).

By applying the simple estimates of stopping (2.14), you may also verify that the probability $P(\phi^*)$ for deflection over an appreciable angle, say $\phi^* > \pi/2$, is very small in case of α-particles even when you insert the full penetration depth for Δx. You are all familiar with the beautiful straight-line trajectories of α-particle tracks in a cloud chamber. On the other hand, for Rutherford scattering, eq.(2.11), $P(\phi^*)$ behaves like $(\phi^*)^{-2}$ at small ϕ^*, so $P(\phi^*)$ becomes large at small ϕ^*, even if we choose Δx and N' very small. The limit $P(\phi^*) = 1$ specifies an approximate boundary between the domains of single and multiple collisions.

In the domain of multiple scattering, you often deal with small angles. This is usually a simplification. However, each scattering event is characterized by a polar and an azimuthal angle. In the small-angle approximation, you may assume the longitudinal velocity component constant, and represent the deflections as 2-dimensional vectors $\vec{\phi}$ in the plane perpendicular to the beam direction. With n_i deflections characterized by a scattering angle $\vec{\phi}_i$ the total deviation becomes

$$\vec{\psi} = \Sigma \, n_i \, \vec{\phi}_i \qquad (5.4)$$

and the average deviation of the beam is

$$\vec{\psi} = \Sigma < n_i \vec{\phi}_i > = 0 \qquad (5.5)$$

because of azimuthal symmetry. By use of the argument

leading to (2.7) and azimuthal symmetry again, you obtain

$$< \psi^2 > = \Sigma \; P_i \; <\phi_i^2> \; = N'\Delta x \int \phi^2 d\sigma \qquad (5.6)$$

where the integration is conveniently taken over the interval $0 < \phi < \phi^*$. As a first approximation, you can represent the angular distribution of your beam by a gaussian, i.e.

$$F(\psi) \approx \frac{1}{\sqrt{2\,\pi<\psi^2>}} e^{-\frac{\psi^2}{2<\psi^2>}} \quad ; \; \psi \lesssim \phi^* \qquad (5.7a)$$

and

$$F(\psi) \approx P(\psi) = N'\Delta x \int d\sigma \; \delta(\phi - \psi) \; ; \; \psi \gtrsim \phi^* \qquad (5.7b)$$

Again you can convince yourself that, for projectile <u>ions</u>, it is mainly the target <u>nuclei</u> that are responsible for multiple scattering. Moreover, since $\phi^2 \propto \theta^2 \propto T$, the nuclear recoil energy, the expression (5.6) shows the same low-energy behavior as the nuclear stopping power. This makes it important to properly incorporate the screened Coulomb interaction.

Eqs.(5.7) and (5.6) constitute the Bohr-Williams theory of multiple scattering. It is appropriate for rough estimates. In the following, we shall go into considerably more detail.

5.3. <u>Transport Equations</u>

Multiple collision processes in a random medium are conveniently treated by means of the integral equations of transport theory. The multiple scattering of ions of not too low energy is ideally suited for such a treatment, since the underlying assumptions seem fulfilled at least as well as in kinetic gas theory. However, under channeling conditions, the collisions with lattice atoms are not at random. A different treatment is required there.

Transport equations have been used in many branches of physics, and in particular in the theory of multiple scattering and slowing down of electrons and other charged particles. So many prominent physicists have made contributions (in view of the importance in elementary-particle physics) that I do not dare to give you a chronology : I might miss one of them.

You should be aware of the forward and the backward form of the transport equation. The forward form
is the ordinary Boltzmann equation for a medium with
stationary scattering centers,

$$- \frac{\partial F}{\partial t} - \vec{v}.\vec{\nabla}F = N' \int d^3v' \{vFK(\vec{v}, \vec{v}') - K(\vec{v}', \vec{v})v'F'\} \qquad (5.8)$$

where $F(\vec{r}, \vec{v}, t)d^3r d^3v$ is the probability to find the
projectile in the volume $(\vec{r}, d^3r)$ moving with velocity
$(\vec{v}, d^3v)$ at time t, and $d\sigma(\vec{v}, \vec{v}') = K(\vec{v}, \vec{v}')d^3v'$ is the
differential cross section for scattering from $\vec{v}$ to
$(\vec{v}', d^3v')$. We use the abreviation $F' = F(\vec{r}, \vec{v}', t)$.
Eq. (5.8) is derived by considering the change in time
of $F(\vec{r}, \vec{v}, t)$ at fixed $\vec{r}$ and $\vec{v}$. If the ion starts with
a velocity $\vec{v}_0$ in $\vec{r} = 0$ at $t = 0$, you have the initial
condition

$$F(\vec{r}, \vec{v}, 0) = \delta(\vec{r})\delta(\vec{v} - \vec{v}_0) \qquad (5.8a)$$

The backward form, or propagator, relates spatial
and velocity distributions over a finite amount of ti-
me. Let $F(\vec{r}, \vec{v}, \vec{v}_0, t) d^3r d^3v_0$ be the probability to
find the ion in $(\vec{r}, d^3r)$ moving with velocity $(\vec{v}_0, d^3v_0)$
(sic !) at time t, if it starts in $\vec{r} = 0$ with velocity
$\vec{v}$ at time $t = 0$, then

$$- \frac{\partial F}{\partial t} - \vec{v}.\vec{\nabla}F = vN' \int d^3v' K(\vec{v}, \vec{v}')\{F - F'\} \qquad (5.9)$$

where $F' = F(\vec{r}, \vec{v}', \vec{v}_0, t)$. The initial condition is
the same as in (5.8a). You can find a derivation of a
similar equation in my paper on sputtering (Sigmund[191]).
The connection between the two types of equations, (58)
and (5.9), has been discussed in very general terms
by Lindhard & Nielsen[130].

The two equations are equivalent, but their respective domains for convenient applicability are different.
Eq. (5.9) contains the velocity variable v_0 as a parameter rather than a variable. If you wish to calculate
the total number of particles transmitted through a
foil, or the transmitted energy, then (5.9) is what you
conveniently use.

The simplest problems that can be treated on the
basis of eqs. (5.9) and (5.8) are the angular spread
of a beam as a function of time (the standard multiple-
scattering problem), the energy spectrum vs. time, and

the lateral spread. You can also study the correlated distributions of energy and angle, or lateral spread and angle, or lateral spread and energy, and you can study the full 6-dimensional distribution. You also have the option of going over to planar geometry, thus reducing the number of independent variables. For details you may consult Scott[185].

We now introduce a coordinate system with the x-axis parallel to the original beam direction. If you are not interested in the lateral spread, you integrate over the y-z coordinates, and obtain from (5.8)

$$- \frac{\partial F}{\partial t} - v_x \frac{\partial F}{\partial x} = N' \int d^3v' \{vFK(\vec{v},\vec{v}') - K(\vec{v}',\vec{v})v'F'\} \quad (5.10)$$

where now $F = F(x,\vec{v},t)$. If the foil is "thin", i.e., when the multiple-scattering angle is small, the depth scale x is equivalent to the time scale vt, so you only need one of them. In the same approximation, $v_x = v$.

If you ignore all directional dependence, you obtain the equation for the energy spectrum,

$$- \frac{\partial G}{\partial x} = N' \int dE' \{GK(E,E') - K(E',E)G'\} \quad (5.11)$$

where

$$G(x,E)dE = v^3 dv \int F(x,\vec{v})d^2e \quad (5.11a)$$

and $\vec{e} = \vec{v}/v$. Eq.(5.8a) reads

$$G(0,E) = v\delta(E-E_0) \quad (5.11b)$$

If you, instead, ignore all energy variation, and keep the directional variables, you obtain the multiple-scattering distribution;

$$- \frac{\partial F}{\partial x} = N' \int d^2e' \{FK(\vec{e},\vec{e}') - K(\vec{e}',\vec{e})F'\} \quad (5.12)$$

where

$$F(x,\vec{e})d^2e = d^2e \int v^2 dv \, F(x,\vec{v}) \quad (5.12a)$$

and

$$F(0,\vec{e}) = \delta(\vec{e}-\vec{e}_0) \quad (5.12b)$$

with $\vec{e}_0 = \vec{v}_0 / v_0$.

In the following two chapters, we shall study a number of simple solutions of eqs.(5.11) and (5.12). The spatial problem has been discussed in detail by Scott[185] and in a forthcoming paper by Marwick and myself. A few comments on thick-target effects will be made subsequently.

5.4. Multiple-Scattering Distributions

Eq.(5.12) can be integrated easily within the small-angle approximation. If you assume that F falls off rapidly towards zero within a small angular interval $|\vec{e} - \vec{e}_0|$ -which implies that the cross section shows a pronounced peak near $|\vec{e} - \vec{e}'| = 0$, you can extend the $\vec{e}'$- integration in (5.12) over an infinite plane instead of the unit sphere. Noting that

$$K(\vec{e},\vec{e}') \equiv K(|\vec{e}-\vec{e}'|) \equiv K(\vec{e}',\vec{e})$$

(since all energy loss was ignored), and taking Fourier transform of (5.12) and (5.12b), you obtain a differential equation that is solved easily. Taking the back transform, you obtain

$$F(x,\vec{e}) = \frac{1}{(2\pi)^2} \int d^2k e^{i\vec{k}\cdot(\vec{e}-\vec{e}_0)} \cdot e^{-N'x\sigma(k)} \qquad (5.13')$$

where

$$\sigma(k) = \int d^2e' K(\vec{e},\vec{e}')(1-e^{i\vec{k}\cdot(\vec{e}'-\vec{e})}) \qquad (5.13a)$$

or, carrying out the azimuthal integrals,

$$F(x,\psi) = \frac{1}{2\pi} \int_0^\infty k dk \, J_0 \, (k\psi) e^{-N'x\sigma(k)} \qquad (5.13)$$

and

$$\sigma(k) = \int_0^\infty d\sigma(\phi) \, (1-J_0(k\phi)) \qquad (5.13a)$$

J_0 is a Bessel function

$$J_0(z) = \frac{1}{2\pi} \int_0^{2\pi} d\chi e^{iz \cos \chi} \qquad (5.14)$$

Eqs.(5.13) and (5.13a) are due to Bothe[45], while the derivation from the transport equation was found by

Bethe[20]. As was noted in ch. (5.2), it is mainly the nuclear collisions that determine the multiple scattering. Hence, eq.(5.13) need to be evaluated first of all for screened Coulomb interaction. Such an evaluation has been performed by Molière[148], and by a number of authors later on (for references cf. Sigmund & Winterbon[193]. Meyer[146] evaluated (5.13) applying the cross section eq.(2.44). In dimensionless units,

$$\tau = N\pi a^2 x \quad ; \quad \tilde{\psi} = \frac{Ea}{2Z_1 Z_2 e^2} \psi \qquad (5.15)$$

eqs.(5.13) and (5.13a) read

$$F(x,\psi)d^2\psi = \frac{d^2\tilde{\psi}}{2\pi} \int_0^\infty z\,dz \; J_0(z\tilde{\psi})^{-\tau\tilde{\sigma}(z)} \qquad (5.16)$$

and

$$\tilde{\sigma}(z) = \int_0^\infty d\eta \; \frac{f(\eta)}{\eta^2}\cdot(1-J_0(z\eta)) \qquad (5.16a)$$

Here, $f(\eta)$ is a function determining the differential scattering cross section. (eq.2.44).

It is instructive to give an example of an analytic evaluation first. If you set $f(\eta) = C = $ const.in eq.(5.16a), you have an approximate scattering law for a r^{-2} potential function (Lindhard et al[131]). The integral then becomes

$$F(x,\psi)d^2\psi = \frac{d^2\tilde{\psi}}{2\pi} \frac{\tau C}{((\tau C)^2 + \tilde{\psi}^2)3/2} \qquad (5.17)$$

This result is exact for the given cross section within the small-angle approximation. It was first derived by Bothe[45]. Note that eq.(5.17) goes over into the single-scattering distribution $\propto \tilde{\psi}^{-3}$ at small thicknesses ($\tau C << \tilde{\psi}$). The reduced angle τC corresponds to the limiting angle ϕ^* in ch.(5.2), below which the distribution is approximately gaussian.

The analytical form of eq.(5.17) is rather specific for the underlying scattering function $f(\eta)=C$. Meyer[146] has tabulated the more general distribution as a function of τ and $\tilde{\psi}$, together with another function that is supposed to account for the finite range of the interaction potential, but actually appears to be an artefact of the evaluation (Sigmund & Winterbon[193]).

This function should be ignored. Meyer mentions some
inaccuracies of his tabulations, and it turns out that
these may be substantial, up to 30 pct. at small thicknesses. We have, therefore, redone the numerical computation and extended the tables such as to include the
scattering on very thin layers over a very wide range
of ion-target combinations and energies, and for two
different screened -Coulomb potentials.

When the multiple-scattering angle is not small
on an absolute scale, you have to apply a different
procedure. It is customary and convenient, then, to
expand the angular distribution in terms of Legendre
polynomials (Lewis[127], Goudsmit & Saunderson[93]). This
is more important for electrons than for ions, (although caution is required for light ions in the medium
and low keV region). An exact result has been derived
recently. For the simple scattering function $f(\eta) = C$
introduced above, Lindhard & Nielsen[130] and Gott[94]
found

$$F(x,\psi)d^2\psi = \frac{d^2\tilde{\psi}}{4\pi} \frac{1-e^{-2\tau C}}{[(1-e^{-\tau C})^2 + 2e^{-\tau C}(1-\cos\tilde{\psi})]^{3/2}} \qquad (5.18)$$

which goes over into (5.17) for small $\tilde{\psi}$ and τC.

I have reviewed briefly some recent multiple-
scattering measurements (Sigmund[191]). It turns
out that the scaling rules predicted by the Thomas-
Fermi expression (5.16) are not always fulfilled. In
the one year that has passed, no agreement has yet been
reached as to whether the apparent descrepancies are
due to deviations from Thomas-Fermi scaling in the basic scattering law or some different effect.

5.5. Transmitted Energy Spectra : Thin Absorber

There is a close analogy between the small-angle
approximation in multiple scattering and the thin-ab-
sorber approximation in energy loss. If the total energy loss ΔE is small compared to the initial energy E,
the cross section can be assumed constant as a function
of ion energy,

$$K(E,E') \tilde{\approx} K(E-E') \equiv K(T) \qquad (5.19)$$

where T is the energy loss in a single collision. It
is convenient to introduce the total energy loss ΔE
as the variable in eq. (5.11) instead of E. The solution is then found by Laplace transform in complete a-

nalogy to the Fourier transform applied in the previous chapter. It reads

$$G(x, \Delta E) = \frac{1}{2\pi i} \int_{c-i\infty}^{c+i\infty} dp e^{p\Delta E} \cdot e^{-N'x\sigma(p)} \qquad (5.20)$$

where

$$\sigma(p) = \int d\sigma (1 - e^{-pT}) \qquad (5.20a)$$

Although eq.(5.20) is very similar to Bothe's formula (5.13'), it was apparently first derived about 20 years later by Landau[120].

It is again instructive first to investigate a simple analytic example. We consider nuclear energy loss according to essentially the same power potential as in the previous chapter. In terms of energy loss, this cross section function reads

$$d\sigma = C \frac{dT}{T^{3/2}} \qquad (5.21)$$

and, since we deal with small energy loss, we boldly extend the range of validity of (5.21) for $T = 0 \dots \infty$. Then, eq.(5.20) has the solution

$$G(x, \Delta E) = \frac{NxC}{(\Delta E)^{3/2}} \cdot e^{-\frac{N^2 \pi x^2 C^2}{\Delta E}} \qquad (5.22)$$

a result derived by Lindhard & Nielsen[130]. The main point to be made is the highly non-symmetric (and non-gaussian) nature of the energy-loss distribution. (fig. 11). The most probable energy loss is given by

$$(\Delta E)_{peak} = \frac{2\pi}{3} (NxC)^2 \qquad (5.23)$$

while the average energy loss diverges due to the allowance of infinite energy transfer, as can be seen already from (5.21). If you write (5.23) in the form

$$\frac{(\Delta E)_{peak}}{Nx} = \frac{2\pi}{3} NxC^2 \qquad (5.23')$$

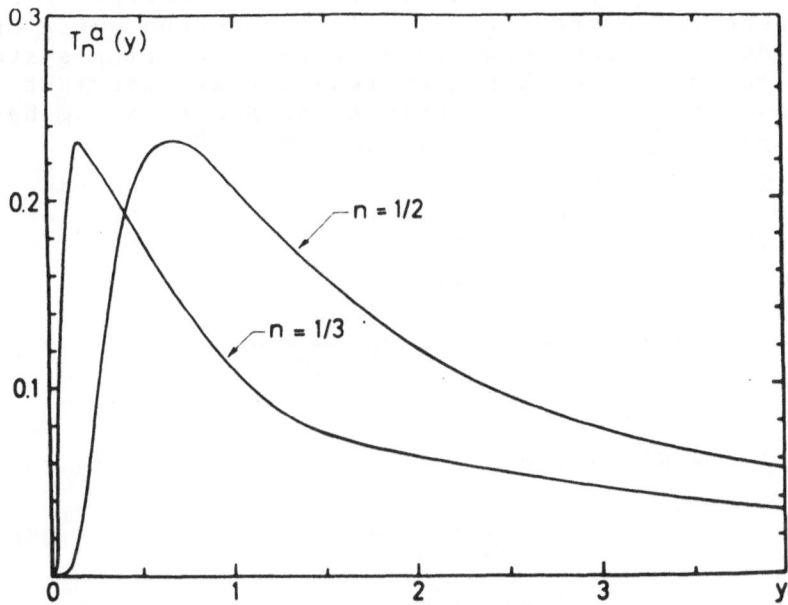

Fig.11 : Calculated energy distributions after transimission
through thin foils. n = 1/2 refers to Eqs. (5.22) and (5.21).
n=1/3 refers to a cross section $d\sigma = CdT/T^{4/3}$. The spectra are normalized to unity. For n=1/2, $y=\Delta E/(\pi(NCx)^2)$. For n=1/3, y=const
$x/NCx)^3$. From Lindhard & Nielsen[130].

you arrive at the remarkable result that the (wrongly defined !) stopping cross section is proportional to the foil thickness.

Effects like (5.23') are observed in experiments done at sufficiently low ion energies where nuclear stopping dominates (Högberg[101]). The quantitative theory is more involved, since in case of nuclear stopping, the particles are also deflected ; the energy spectrum and especially the high-ΔE tail may depend quite heavily on the acceptance angle of the detecting system. Calculations of combined energy-angular distributions in case of pure nuclear scattering and stopping have been performed recently by Winterbon[216].

We now go back to the more general equation (5.20) It is useful to note that taking averages $\langle \Delta E^n \rangle$ over the distribution function $g(x, \Delta E)$ yields

$$\langle \Delta E^n \rangle = (-)^n \frac{n!}{2\pi i} \int_{c-i\infty}^{c+i\infty} dp \frac{e^{-N'x\sigma(p)}}{p^{n+1}} \qquad (5.24)$$

or

$$\langle \Delta E \rangle = N'x \int T d\sigma \qquad (5.25a)$$

$$\langle (\Delta E - \langle \Delta E \rangle)^2 \rangle = N'x \int T^2 d\sigma \qquad (5.25b)$$

$$\langle (\Delta E - \langle \Delta E \rangle)^3 \rangle = N'x \int T^3 d\sigma \qquad (5.25c)$$

Note in particular that Bohr's relation (5.25) for the straggling also results from the present derivation, and that (5.25c) in general introduces a skewness in the spectra. Note, however, the approximate character of eq.(5.20) upon which eqs.(5.25a-c) are based.

You may realize that the higher moments of the ΔE-spectrum are determined more by the large energy transfers T, i.e. close collisions become important. In case of electronic energy loss, you may assume classical Rutherford scattering at nonrelativistic velocities and obtain, for $v \gg v_0$,

$$\int_0^{T_{max}} T^3 d\sigma = \frac{1}{2} T_{max} \int_0^{T_{max}} T^2 d\sigma$$

Since $T_{max} = 2mv^2$, the skewness of the distribution is characterized by the ratio,

$$\frac{<(\Delta E - <\Delta E>)^3>}{<(\Delta E - <\Delta E>)^2>^{3/2}} = \frac{mv^2}{\sqrt{N'x} \int T^2 d\sigma} \qquad (5.26)$$

Eq.(5.26) says that the profile is very skew for so small target thickness where the width of the spectrum is small compared to the maximum energy loss in a single collision ; but with increasing thickness the profile becomes more and more symmetric. Only at very large x, when ΔE becomes comparable to the initial energy, another type of skewness, caused by the energy dependence of the cross section, may be induced.

A number of procedures have been applied·in the literature to evaluate eq.(5.20) more explicitly (Landau[120], Symon[197], Vavilov[204]).

Vavilov's procedure yields the first correction term to the gaussian energy-loss profile of the Bohr theory. He expands $\sigma(p)$, eq.(5.20a),in powers of p, so eq.(5.20) reads

$$G(x,\Delta E) = \frac{1}{2\pi i} \int_{c-i\infty}^{c+i\infty} dp \qquad (5.27)$$

$$\exp\{p(\Delta E - N'xS_1) + p^2 \frac{N'x}{2} S_2 - p^3 \frac{N'x}{6} S_3\}$$

where

$$S_i(E) = \int T^i d\sigma \qquad (5.27a)$$

When the S_3-term is neglected, (5.27) reduces to a gaussian with the width (5.25b) around the mean energy loss (5.25a). When S_3 is taken into account, (5.27) becomes an Airy integral,

$$G(x,\Delta E) = \frac{1}{\eta\sqrt{\pi}} e^{-at-a^3/3} V(t) \qquad (5.27b)$$

where

$$t = \frac{\Delta E - N'xS_1}{\eta} + a^2 \; ; \; a = \eta \frac{S_2}{S_3}$$

$$\frac{1}{\eta} = (\frac{2}{N'xS_3})^{1/3}$$

and

$$V(t) = \frac{1}{\sqrt{\pi}} \int_0^\infty dy \cos(yt+y^3/3)$$

Numerical evaluations of the Vavilov distribution (e.g. Seltzer & Berger[187]) have dealt mainly with the case of high-speed light ions. Although the classical expressions for the second and third moments of the cross section could be used, proper description of resonance collisions is usually aimed at (Blunck & Leisegang[33], Shulek et al[190], Clarkson & Jarmie[64]). Measurements with light projectiles, mostly protons in the medium and upper MeV region (Maccabee[138], Hilbert & al[199], Baily et al[14]) show excellent agreement between measured energy losses and those calculated from the Vavilov theory for thin solid layers and gas targets (typically, $\Delta E/E \sim 10^{-3}$).

An extensive theoretical and experimental study of transmitted energy spectra at lower energies, based on the Vavilov approximation as well as the (more general) Landau equation, eq.(5.20), has been performed by Hvelplund[104] . This includes a Thomas-Fermi calculation of the nuclear-energy-loss spectrum and a discussion of the contribution of electronic and nuclear energy losses to keV heavy-ion energy-loss profiles. The experimental part of this work includes determinations of straggling for both heavy (Hvelplund[104]) and light (Bonderup & Hvelplund[42]) ions. Fig.12 shows the results of straggling measurements in gas targets as a function of the atomic number of the ion. As in case of the stopping power, there is a tendency towards oscillatory behavior around an average curve that was calculated from Firsov's differential expression for electronic energy loss.

Bonderup & Hvelplund performed calculations of straggling for swift protons along similar lines as in Lindhard & Scharff's[132] and Bonderup's work[41] on stopping power. Fig.13 shows their result for the straggling in a free electron gas. Note that, except for $v < v_F$, Bohr's classical result eq. (2.14d) is accurate to better than 10 pct. Their experimental results (for 100-500 keV protons in seeveral gases) show significant deviations (up to $\sim$ 40 pct. higher) from the calculated curves. An explanation for this behavior -which

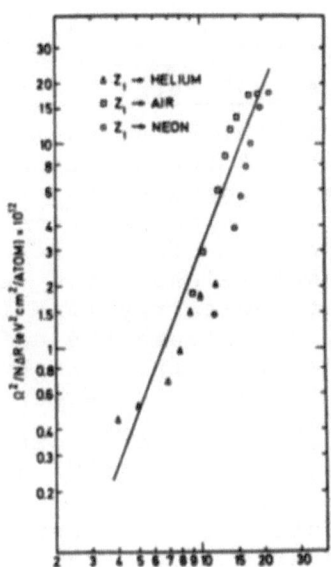

Fig.12 : Relative straggling $<(\Delta E - <\Delta E>)^2/(N\Delta x) = \int(T_e + T_n)^2$ dσ for keV ions after passing gas targets of helium, neon and atmospheric air. The calculated expression is based on Eq.(4.24) and a correction for multiple scattering. From Hvelplund[104].

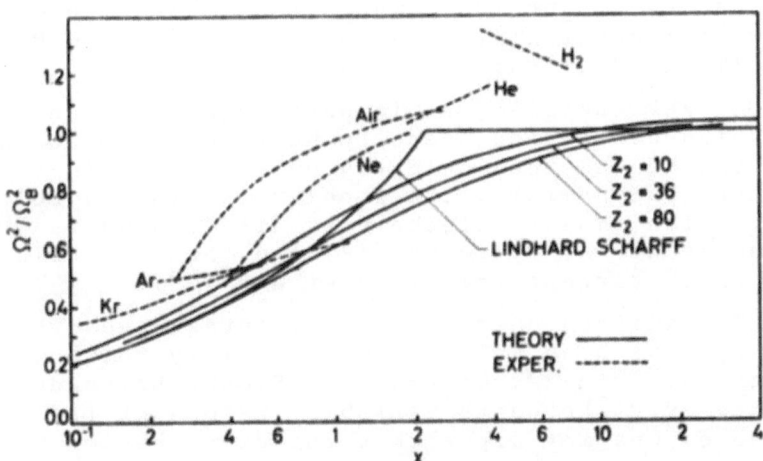

Fig.13 : Calculated values of Ω^2/Ω_B^2 for a point charge penetrating a Fermi gas. Ω^2 is the straggling evaluated from Eqs.(3.32) and (3.33). Points are taken from numerical calculations, curves represent analytic asymptotic expressions. $\chi = (v_0/\pi v_F)^{1/2}$. From Bonderup & Hvelplund[42].

has been observed by other authors, too (e.g. Ramirez & al[175])- was not given.

5.6. Transmitted Energy Spectra : Thick Absorber

The notion of a "thick absorber" refers to a situation where the width of the spectrum of transmitted particles is not small compared to the initial energy. The Landau and Vavilov approximations are then no longer applicable, since they do not take into account the dependence of the energy-loss cross section on energy. Moreover, in case of low-energy ions (and electrons), it may be important to include the deflection of the projectiles, in order to obtain the proper relation between transmitted energy and penetrated depth.

The problem has been treated extensively by Symon[197], and recently by Tschalär[202] and Payne[168]. All these authors started from the basic equation (5.11) and determined moments $<\Delta E^n>$ without applying the assumption of small energy loss. From the moments, the energy distributions can be reconstructed by using certain standard techniques or trial functions (Shohat & Tamarkin[189], Cramer[65], Johnson[112]). A distinction was made between thick absorbers and extremely thick absorbers. In the latter case, the attenuation of the beam because of complete stopping of ions is appreciable. We do not go into details, but mention that the three authors quoted above arrived at rather similar results, and that their results compare favorably with measurements on light-ion slowing down in the 10-100 MeV region (Tschalär & Maccabee[203]).

At keV energies, and in particular in case of light ions, ion deflection needs to be taken into account. This can be done conveniently by expansion in terms of Legendre polynomials. The problem is closely related to that of ion ranges (Lindhard et al[133], Sanders[180], Winterbon et al[214], Winterbon[215]) and that of reflection from thick targets (Bøttiger et al[47,48]).

Transmitted energy spectra, integrated over all angles, have been calculated explicitly for low-velocity ions by Brice[53]. Contrary to Symon, Brice determined moments over the depth variable instead of the energy variable. One may argue that the distribution $G(x,\Delta E)$ is close to gaussian shape as a function of x, even though it may deviate drastically from gaussian behavior as a function of ΔE. This assertion is supported, for example, by the model distribution eq.(5.22). By assuming the profile of ions to be strictly gaussian at a fixed slowing-down energy as a function of travelled path

length, and applying approximate corrections for the
difference between path length and depth, Brice arrived at a usefull set of transmitted-energy profiles
that have been utilized mainly in the theory of energy deposition by ion beams (Brice[53]).

5.7. Stopping Measurements

You ought to have at least a slight impression of
the various techniques for measuring stopping powers
and related quantities. These notes will be rather
fragmentary, especially for the experimentalists among
you.

There is not much more to be said about the me-
rits of direct measurement of transmitted energy spectra through thin foils. It is an excellent technique
in many situations. It is appropriate for measurements
with electrons over a wide energy range, and it has
been used extensively for ions of velocities ranging
over several orders of magnitude, both on solid and
gaseous targets. Examples are the straggling measurements discussed in the previous chapter with protons
in the upper MeV range, and, in the other end, elec-
tronic-and nuclear-stopping measurements with heavy
ions in the lower and medium keV range. In the electronic-stopping region, for large enough thickness so
that the energy profile is symmetric, the average energy loss can be determined accurately from the peak value (to $\sim$ 1 pct. for $\sim$ 1 MeV α-particles ; Bourland
& al[49]), and the limiting factor is often the determination of the target thickness, and thickness inhomo-
geneity. At low velocities, e.g. for keV heavy ions,
thin films in the thickness range of few hundred Angs-
troms or less may be needed for precise measurements.
This puts a natural limit on measurements with solids.

Very accurate stopping-power measurements on foils
(Andersen et al[3]) were based on a different technique.
The energy loss ΔE is measured on the target instead
of the beam by means of the heat release that it gives
rise to ; by comparison with the residual energy E-ΔE
that is absorbed in a thick collector, the relative energy
loss <ΔE>/E is found. Target thickness inhomogeneities
are not critical since an average is taken over a comparatively large foil area. By measuring the heat release at liquid helium temperature, the experimental
accuracy has been pushed to 0.3 %.

In the following I shall briéfly mention a number of techniques that can be used when transmission
through thin, self-supporting films does not appear
feasible.

Range measurements have through more than half of this century been the main tool for determining stopping powers (Whaling[207]). When the range has been measured as a function of energy,

$$R(E) = \int_0^R dx = \int_0^E \frac{dE'}{dE'/dx} \qquad (5.28)$$

the stopping power can be found by differentiation. In recent years, range measurements have provided much information on stopping of keV ions in gases (e.g. Lassen et al[124]), solids (e.g. Mayer et al[142]), and especially in single crystals (e.g. Eriksson et al[77]). A large number of nondestructive and destructive techniques are available for range measurements in solids (reviewed by Mayer et al[142]). Because of a number of complications (competition of nuclear and electronic stopping ; path length correction ; appreciable width and skewness of the profile ; channeling), the results of low-velocity range measurements are usually discussed in terms of the degree of agreement with theoretical predictions ; direct inversion has been made mostly for relatively narrow profiles.

A whole family of techniques makes use of thin stopping layers evaporated on heavy substrates. One method was developed by Warters[205], and applied recently by Chu & Powers[69]. The energy spectra of, e.g. α-particles backscattered from a thick target of a heavy element into a narrow angular interval show a well defined edge at the energy corresponding to scattering at the surface. A thin layer of a light element on top does not cause much backscattering itself, but shifts the edge towards smaller energies. The shift corresponds to the energy loss for twice passing the stopping layer. The geometric relations have been worked out by Warters.

According to Chu & Powers, the main source of error stems from the nonuniformity of the evaporated films (2 pct.) and the energy measurement (2 pct.). White & Mueller[208] and Ziegler & Brodsky[218] deposited the stopping layer on a thin heavy backing on top of a light substrate. In this way, α-particles (or photons) reflected from the layer of heavy atoms form a peak in the energy spectrum instead of a continuum with an edge. This appears to make possible a more accurate determination of the shift in energy. Johansen & al[111] employed the shift in the yield of proton-induced characteristic X-rays from a gold backing to measure the energy loss of protons in a stopping layer of light atoms.

Despite the accuracy of better than 5 pct. in stopping-power measurements on evaporated layers, discrepancies of 15 % or more have been reported between the results found by different groups, cf. the discussion of Ziegler & Brodsky[218]. Target preparation is only one of several important issues in this connection.

Stopping power measurements on heavy target materials can hardly be done by the techniques just mentioned. However, backscattering spectra taken on thick, homogenous samples also contain stopping power information. (Wenzel & Whaling[206]). For protons and α-particles in the upper keV region and above, backscattering is essentially a single-scattering phenomenon, hence the intensity backscattered into a fixed solid-angle interval $d\Omega$ is proportional to

$$N\Delta x \; d\sigma \;\; = \frac{\Delta E'}{S_e(E')} \; \frac{d\sigma}{d\Omega} \; d\Omega \qquad\qquad (5.29)$$

where Δx is a path-length segment and E' the energy at which a scattering takes place. Thus, at a constant counting geometry, the observed spectrum is inversely proportional to the stopping power and proportional to the Rutherford cross section. Eq. (5.29) can be applied as it stands (apart from geometry factors) for collisions near the surface, i.e. near the edge of the observed spectrum. Otherwise corrections have to be made for the energy loss ; this means that the stopping power enters in a more complicated manner and that direct inversion becomes slightly more involved (Behrisch & Scherzer[17], Chu et al[62], Foster et al[87], Leminen[126], Sirotinin et al[194]). Since also multiple-scattering effects may influence those parts of the spectrum that lie far below the edge, it appears wise not to go too far below the energy of the edge in the simple inversion procedure unless sufficient accuracy of the analysis has been ascertained. If the stopping power is needed over a greater energy range, it appears safer to vary the ion energy.

The technique just described requires good accuracy of the beam current measurement and geometry factors. To avoid the uncertainties connected with absolute measurements, Feng et al[83] used a double layer consisting of two materials of sufficiently different masses and low thicknesses to separate the reflected ions into two nonoverlapping distributions. From the relative heights of the spectra one obtains the ratio between the two stopping-powers.

I should like to conclude this chapter by making the remark that the number of techniques you might use to measure stopping powers (and, to a lesser extent, straggling) is almost unlimited. Just as there is a great need for stopping data in many fields of Physics and Chemistry (e.g. nuclear reactions, chemical analysis by particle beams), you can often consider the reverse situation. Nuclear physicists have used the Doppler-shift attenuation of γ decays following nuclear reactions to measure lifetimes of excited nuclei, and used heavy-ion stopping powers as the main input parameter in the data treatment (Blaugrund[30], Broude[56], Winterbon[215]). However, with the lifetime known, you can vary the stopping medium and invert the analysis, and thus obtain the Z_2-variation of the stopping power (Neuwirth et al[157]), check Bragg's rule, etc. Similarly, Rutherford scattering has been used for many years to determine heavy impurities in light substrates (Powers & Whaling[174], Mayer et al[142], Mayer & Ziegler[143]) and again the stopping power is the key input parameter. However, if you control your impurities, e.g. by having implanted a known dose with an isotope separator, you should have a fair chance, by Rutherford scattering to determine reasonably accurate stopping powers (Björkqvist & Domeij[29])

5.8. Summary

i) The significance of multiple-scattering corrections in stopping measurements can be estimated from the Bohr-Williams theory.

ii) Extensive tabulations are available of calculated multiple-scattering distributions for the purpose of more detailed estimates. Some uncertainty remains with regard to the range of validity of Thomas-Fermi scaling.

iii) Energy spectra of transmitted particles are skew for very thin layers (Vavilov regime), approximately gaussian at intermediate thickness (Bohr regime), and skew again when the energy loss is comparable with the initial energy. The theory is well developed in the high-velocity regime, while little published information is available for low-speed ions.

iv) At low velocities, in particular when nuclear

stopping is significant, the transmitted energy spectra depend on direction, and, for moderately thick targets, a path-length correction need to be made.

v) A variety of techniques for stopping-power measurements has been developed. The standard method is analysis of transmitted-energy spectra from thin-target experiments. Alternatively, one has measured the heat release in the target. When the penetration depth is low, measurements on thick targets are to be preferred. In addition to range measurements, one has a number of techniques based on energy analysis of reflected particles. Secondary effects that are sensitive to stopping power can often be utilized for stopping-power measurements.

APPENDIX A

Fermi's Golden Rule

Consider an atom, atomic number Z, governed by a hamiltonian H, with stationary states $|n\rangle$ so that

$$H|n\rangle = E_n|n\rangle \qquad (1)$$

and a projectile, mass m_1, velocity v, interacting with the atom via a potential V that is assumed to be "small". Consider a constant flux of incident particles so that, for $V = 0$, the total wave function of the system would be

$$\psi^{(o)} = e^{i\vec{k}\cdot\vec{r}} |0\rangle \qquad (2)$$

where $\hbar\vec{k} = m_1\vec{v}$, and $\vec{r}$ is the position vector of the projectile. $|0\rangle$ refers to the ground state of the atom. The stationary Schrödinger equation reads

$$(\frac{p^2}{2m_1} + H+V-E-E_o) \Psi = 0 \qquad (3)$$

where $\vec{p} = -i\hbar\vec{\nabla}$ and $E = m_1v^2/2$. Expanding Ψ in terms of V,

$$\Psi = \psi^{(o)} + \psi^{(1)} + \ldots \qquad (4)$$

and collecting only terms up to first order, we obtain from (3), using (1) and (2),

$$(\frac{p^2}{2m_1} + H-E-E_o) \psi^{(1)} = -V\psi^{(o)} \qquad (5)$$

Expanding $\psi^{(1)}$ according to

$$\psi^{(1)} = \Sigma u_n (\vec{r})|n\rangle \qquad (6)$$

multiplying (5) by $\langle m|$, and using orthonormality of the state vectors $|n\rangle$, we obtain,

$$(\frac{p^2}{2m_1} + E_n-E_o-E) u_n = -e^{i\vec{k}\cdot\vec{r}}\langle n|V|0\rangle \qquad (7)$$

This is a wave equation with a source term. Denoting

$$E-(E_n-E_0) = \frac{\hbar^2 k'^2}{2m_1} \quad (>0) \qquad (8)$$

the solution reads (Jackson[108])

$$u_n(\vec{r}) = -\frac{1}{4\pi} \int d^3r' \frac{e^{\pm ik'|\vec{r}-\vec{r}'|}}{|\vec{r}-\vec{r}'|} \frac{2m_1}{\hbar^2} <n|V(\vec{r}')|0>e^{i\vec{k}.\vec{r}'} \qquad (9)$$

At large distances from the atom, i.e. $|\vec{r}| >> |\vec{r}'|$, you expand $|\vec{r}-\vec{r}'| = r-\vec{e}'.\vec{r}'$, where $\vec{e}' = \vec{r}/r$. Inserting the result into (6), the first-order solution reads

$$\psi^{(1)} = -\frac{m_1}{2\pi\hbar^2} \sum_n \frac{e^{ik'r}}{r} |n><\vec{k}',n|V|\vec{k},0> \qquad (10)$$

where $\vec{k}' = k'\vec{e}'$ is an outgoing wave vector, and the $\vec{k}$ in the state vector refers to $_e ik.\vec{r}$.

For those events where the atom has been excited into the state $|n>$, the number of projectiles passing through an element $d\Omega'$ of solid angle per unit time is given by

$$\frac{\hbar k'}{m_1} d\Omega' (\frac{m_1}{2\pi\hbar^2})^2 |<\vec{k}',n|V|k,0>|^2 \qquad (11)$$

dividing by the incoming particle flux $\hbar k/m_1$, and using (2.3), we obtain the cross section

$$d\sigma = \frac{k'}{k} d\Omega' (\frac{m_1}{2\pi\hbar^2})^2 |<\vec{k}',n|V|\vec{k},0>|^2 \qquad (12)$$

for excitation of the target into the state n and scattering of the projectile into the solid angle $(\vec{e}',d\Omega')$. Observing (8), we can add a factor $\delta(E'-E+E_n-E_0)dE'$ to (12), where $E' = \hbar^2 k'^2/2m_1$ is the outgoing projectile energy. Replacing $dE' = (\hbar^2/m_1)k'dk'$, and going over to momentum instead of wave number variables, (12) reads

$$d\sigma = \frac{2\pi}{\hbar v}|<\vec{p}',n|V|\vec{p},0>|^2 \delta(E'-E+E_n-E_0) \frac{d^3p'}{h^3} \qquad (13)$$

which is usually called Fermi's golden rule. It applies to both $n \neq 0$ and $n = 0$. Both the result and its derivation are contained in Born's[44] original treatment.

APPENDIX B

Bethe's Sum Rule   Reference : Bethe[20d]

Take the notation of appendix A. From the definition (3.14)

$$F_n(\vec{q}) = Z^{-1/2} \sum_{i=1}^{Z} <n|e^{i\vec{q}\cdot\vec{r}_i/\hbar}|0> \qquad (1)$$

we obtain, by use of (A1)

$$(E_n - E_0)F_n(\vec{q}) = Z^{-1/2} <n|[H, \sum_i e^{i\vec{q}\cdot\vec{r}_i/\hbar}]|0> \qquad (2)$$

where the [] brackets indicate the commutator. Since the spatial coordinates commute, only the kinetic-energy part of H needs to be taken into account. Hence,

$$[H, \sum_i e^{i\vec{q}\cdot\vec{r}_i/\hbar}] = \sum_i \frac{1}{2m} [p_i^2, e^{i\vec{q}\cdot\vec{r}_i/\hbar}]$$

$$= \sum_i \frac{1}{2m} e^{i\vec{q}\cdot\vec{r}_i/\hbar} (q^2 + 2\vec{q}\cdot\vec{p}_i) \qquad (3)$$

and

$$(E_n - E_0)F_n(\vec{q}) = Z^{-1/2} \frac{1}{2m} \{q^2 (\sum_i e^{i\vec{q}\cdot\vec{r}_i/\hbar})_{no}$$

$$+ 2\vec{q}\cdot(\sum_i e^{i\vec{q}\cdot\vec{r}_i/\hbar}\vec{p}_i)_{no}\} \qquad (4)$$

Multiplying with $F_n^*(\vec{q})$, and summing over all n yields

$$\sum_n (E_n - E_0)|F_n(\vec{q})|^2 = \qquad (5)$$

$$\frac{1}{Z} \frac{1}{2m} \sum_{ij} \{q^2 <0|e^{i\vec{q}\cdot(\vec{r}_i-\vec{r}_j)/\hbar}|0> + 2\vec{q}\cdot<0|e^{i\vec{q}\cdot(\vec{r}_i-\vec{r}_j)/\hbar}\vec{p}_i|0>\}$$

The terms with i=j yield

$$\frac{1}{Z} \frac{1}{2m}(Zq^2 + 2\vec{q} \cdot \langle 0|\sum_i \vec{p}_i|0\rangle)$$

where the second term vanishes since the atom is assumed at rest in the state $|0\rangle$. The terms with $i \neq j$ can be rearranged by interchanging indices and partial integration that the total expression becomes equal to its negative, hence zero.
Thus, (5) reads

$$\sum(E_n - E_0)|F_n(\vec{q})|^2 = \frac{q^2}{2m} = Q \qquad (6)$$

which is Bethe's sum rule.

Furthermore, for small q, eq.(1) can be expanded

$$F_n(\vec{q}) \cong \delta_{no} Z^{1/2} + Z^{-1/2} \frac{iq}{\hbar}\langle n|\sum x_i|0\rangle \qquad (7)$$

Inserting this into (6) yields, to lowest order in q,

$$\frac{1}{Z}\sum_n (E_n - E_0) \frac{q^2}{\hbar^2}|\langle n|\sum x_i|0\rangle|^2 = \frac{q^2}{2m} \qquad (8')$$

or

$$\sum_n \frac{1}{Z} \frac{2m}{\hbar^2}(E_n - E_0)|\langle n|\sum x_i|0\rangle|^2 = 1 \qquad (8)$$

which is the usual oscillator-strength sum rule (3.21)

APPENDIX C

Charge State of a Moving Ion

We wish to estimate the number of electrons Z^* in a Thomas-Fermi atom (atomic number Z) that have orbital velocities u less than a certain velocity v. Let those electrons be located at distances r greater than some distance r_0. Then,

$$\int_{r_0}^{\infty} d^3r \, \rho(r) = -Z^* e \qquad (1)$$

where $\rho(r)$ is the charge density, and

$$\frac{z^* e^2}{2 r_0} \sim \frac{1}{2} m v^2 \qquad (2)$$

Eqs.(1) and (2) can be solved for z^* and r_0 if $\rho(r)$ is known. From Laplace's equation and the Thomas-Fermi potential

$$V(r) = \frac{Ze}{r} \phi_{z^*}(r/a) \qquad (3)$$

where $\phi_{z^*}(x)$ is the screening function and a the screening radius, eq.(1) reads,

$$\int_{r_0/a}^{\infty} dx \, x \frac{d^2}{dx^2} \phi_z^*(x) = \frac{z^*}{Z} \qquad (4)$$

From (4) we obtain a connection $r_0 = af(z^*/Z)$, with some function f that needs to be specified. By use of $a = a_0 Z^{-1/3}$ and eq.(2) we obtain

$$z^* = Z \cdot g(\frac{v}{v_0 Z^{2/3}}) \qquad (5)$$

For not too large values of Z, we can use the power approximation for the screening function of a <u>neutral</u> atom (Lindhard et al[131]).

$$\phi_z^*(x) \sim \phi_0(x) \sim \frac{k_s}{s} x^{1-s} \qquad s=1,2,3\ldots \qquad (6)$$

Then, eq.(4) reads

$$\frac{r_0}{a} = (\frac{z^*}{k_s Z})^{\frac{1}{1-s}} \qquad (7)$$

Inserting this into (2), we obtain

$$z^* = k_s \, Z^{\frac{4-s}{3s}} (\frac{v}{v_0})^{2-2/s} \qquad (8)$$

For $1 << z^* \lesssim Z/2$, we have $s \sim 2$, so

$$z^* \sim Z^{1/3} \frac{v}{v_0} \qquad (9)$$

where we have set $k_2 \sim 1$. This is Bohr's result.

For smaller values of Z^*, the estimate based on $s = 3$ might be more appropriate. Then,

$$Z^* \sim k_3 Z^{1/9} (\frac{v}{v_o})^{4/3} \tag{10}$$

APPENDIX D

Energy Loss of Slow Particles in Electron Gas

Let a heavy particle move with velocity v through a free, degenerate electron gas. The velocity distribution of the electron is given by $f(\vec{v}')d^3v'$, such that

$$\int f(\vec{v}')d^3v' = n \tag{1}$$

In a system of reference moving with the ion, an electron with lab-velocity $\vec{v}'$ has gotten the velocity

$$\vec{w} = \vec{v}' - \vec{v} \tag{2}$$

If such an electron is scattered at the projectile at an angle θ, it transfers, in the average, a momentum

$$m\vec{w} (1 - \cos \theta) \tag{3}$$

to the projectile. Take all electrons in the velocity interval $(\vec{v}', d^3v')$. The total momentum transferred in the time interval dt is, according to (3),

$$d\vec{P} = f(\vec{v}')d^3v'|\vec{w}| dt \; m\vec{w} \int (1-\cos\theta)d\sigma(w,\theta) \tag{4}$$

Integrating this, inserting (2), and noticing that $dP/dt = dE/dx$, we obtain from (4)

$$\frac{dE}{dx} = \tag{5}$$
$$m \left| \int f(\vec{v}')d^3v' |\vec{v}'-\vec{v}| (\vec{v}'-\vec{v}) \int (1-\cos\theta)d\sigma(|\vec{v}'-\vec{v}|,\theta) \right|$$

This formula has first been published by Trubnikov & Yavlinskii[201]. The argument is the same as Fermi & Teller's[85], and has been used in Lindhard's lectures. Lindhard (unpubl.) points out that eq.(5) automatically satisfies the Pauli principle : if the collisions of the electrons are elastic in the moving system, a scattering event that ends up in an occupied state will have a counterpart with equal probability, and opposite momentum transfer. Hence, the θ-integral can be carried out without further restrictions, and yields the transport cross section

$$\int (1-\cos\theta)\,d\sigma(|\vec{v}'-\vec{v}|,\theta) = \sigma_{tr}(|\vec{v}'-\vec{v}|) \qquad (6)$$

For $v \ll v_F$ the scattering is due to electrons near v_F. Hence, $v \ll v'$ in eq.(5) and, up to first order in v,

$$\frac{dE}{dx} = nmv\ v_F\ \sigma_{tr}(v_F) \qquad (7)$$

since $\int f(\vec{v}')\vec{v}'d^3v' = 0$ for symmetry reasons. Eq. (7) is due to Lindhard (unpublished). Deviations for finite ratio v/v_F have been calculated by Finnemann[85a].

Acknowledgements

During preparing these lectures I was drawing heavily on the experience I got over a number of years from conversations with H.H.Andersen, N.Andersen, R. Behrisch, J.A.Davies, J.Lindhard, J.W.Mayer & K.B.Winterbon. In particular, a series of lectures by J.Lindhard during the academic year 1964/65 proved to be a major source of inspiration.

I should like to thank W.Brandt, W.K.Chu, D.Powers, R.H.Ritchie, and M.T.Robinson for providing information in advance of publication. I am indebted to M.Sc.J.Jepsen for his helpfull and efficient typewriting in an emergency situation, and to A.Johansen and L.Sarholt-Kristensen for their support. Finally, my thanks are due to royal court singer, Mrs L.Lamprecht, for the opportunity to write these notes at Røvballehus, Salthammeren, Bornholm, and to Pia, Nina and Ole for their constant encouragement.

REFERENCES

1. I.Amdur , H.Inouye, A.J.H.Boerboom, A.N.v.d.Steege, J.Los, & J.Kistemaker, 1969. Physica 41, 566

2. H.H.Andersen, J.Bøttiger, & H.Knudsen, 1972. Rad. Effects 13, 203. 1973. Phys.Rev.A7, 154

3. H.H.Andersen, A.F.Garfinkel, C.C.Hanke, & H.Sørensen 1966. Mat.Fys.Medd.Dan.Vid.Selsk. 35, n°4

4. H.H.Andersen, C.C.Hanke, H.Simonsen, H.Sørensen, & P.Vajda, 1968. Phys.Rev. 175, 389

5. H.H.Andersen, C.C.Hanke, H.Sørensen, & P.Vajda, 1967. Phys.Rev. 153, 338

6. H.H.Andersen, H.Simonsen, & H.Sørensen, 1969. Nucl. Phys.A.125, 171

7. H.H.Andersen, H.Simonsen, H.Sørensen, & P.Vajda, 1969. Phys.Rev. 186, 372

8. H.H.Andersen, H.Sørensen, & P.Vajda, 1969.Phys.Rev. 180, 373

9. P.Apel, U.Müller-Jahreis, G.Rockstroh, & S.Schwabe, 1970. Phys.Stat.Sol. (a), 3, K 173

10. B.R.Appleton, J.H.Barrett, T.S..Noggle, & C.D.Moak, 1972, Rad.Effects

11. B.R.Appleton, S.Datz, C.D.Moak, & M.T.Robinson, 1971, Phys.Rev. B4, 1452

12. B.R.Appleton, C.Erginsoy, & W.M.Gibson, 1967, Phys. Rev. 161, 330

13. J.C.Ashley, R.H.Ritchie, & W.Brandt, 1972, Phys.Rev. B.5, 2393. 1973. Phys.Rev. (in press).

14. N.A.Baily, J.E.Steigerwalt, & J.W.Hilbert, 1970. Phys.Rev. B2, 577

15. J.Bang, & J.M.Hansteen, 1959. Mat.Fys.Medd.Dan.Vid. Selsk. 31, n°13

16. M.Barat, & W.Lichten, 1972. Phys.Rev. A6, 211

17. R.Behrisch & B.M.U.Scherzer, 1973. Thin Solid Films (in press)

18. R.J.Bell & A.Dalgarno, 1965.Proc.Phys.Soc.86, 375
                            1966.Ibid .89, 55

19. F.Bernhard, U.Müller-Jahreis, G.Rockstroh, & S.Schwabe, 1969.Phys.Stat.Sol. 35, 285

20. H.A.Bethe, 1930. Ann.Phys. (5), $\underline{5}$, 325
             1932. Z.f.Physik $\underline{76}$, $\overline{2}93$
             1953. Phys.Rev. $\underline{89}$, 1256
             1964. Intermediate Quantum Mechanics
                   Benjamin, New-York, Amsterdam.Ch.15

21. H.A.Bethe, & J.Ashkin, 1953. $\underline{in}$ Experimental Nucle-
    ar Physics (E.Segré, Ed.).Wiley New York, London
    p.166

22. H.A.Bethe & W.Heitler, 1934.Proc.Roy.Soc. $\underline{A.146}$,
    83

23. H.D.Betz, 1972. Rev. Mod.Phys. $\underline{44}$, 465

24. H.D.Betz & L.Grodzins, 1970. Phys.Rev.Lett.$\underline{25}$, 211

25. C.P.Bhalla & J.N.Bradford, 1968. Phys.Lett.$\underline{27A}$, 318

26. C.P.Bhalla, J.N.Bradford, & G.Reese, 1970. $\underline{in}$ Ato-
    mic Collision Phenomena in Solids (D.Palmer et al.,
    Ed.). North Holland, Amsterdam.p.361

27. D.J.Bierman & W.C.Turkenburg, 1970. Phys.Rev.Lett.
    $\underline{25}$, 633

28. D.J.Bierman & D.Van Vliet, 1972. Physica $\underline{57}$, 221

29. K.Björkqvist & B.Domeij, 1972. Rad.Effects

30. A.E.Blaugrund, 1966. Nucl.Phys. $\underline{88}$, 501

31. F.Bloch, 1933 a.Ann.Phys. (5), $\underline{16}$, 287
             1933 b.Z.f.Physik $\underline{81}$, $\overline{3}63$

32. S.D.Bloom & G.D.Sauter, 1971, Phys.Rev.Lett.$\underline{26}$, 607

33. O.Blunck & S.Leisegang, 1950. Z.Physik $\underline{128}$, 500

34. E.Bøgh, J.A.Davies, & K.O.Nielsen, 1964.Phys.Lett.
    $\underline{12}$, 129

35. E.Bøgh & E.Uggerhøj, 1965. Phys.Lett. $\underline{17}$, 116

36. Aa.Bohr, 1948. Mat.Fys.Medd.Dan.Vid.Selsk. $\underline{24}$, n°19

37. N.Bohr, 1913. Phil.Mag. (6), $\underline{25}$, 10

38.aN.Bohr, 1915. Phil.Mag. (6), $\underline{30}$, 581
      b,1940. Phys.Rev. $\underline{58}$, $\overline{6}54$
      c,1941. ibid. $\underline{59}$, $\overline{2}70$

39. N.Bohr, 1948. Mat.Fys.Medd.Dan.Vid.Selsk. $\underline{18}$, n°8

40. N.Bohr & J.Lindhard, 1954. ibid. $\underline{28}$, n°7

41. E.Bonderup, 1967. ibid. $\underline{35}$, n°17

42. E.Bonderup & P.Hvelplund, 1971.Phys.Rev.A$\underline{4}$, 562

43. F.Bonsignori & A.Desalvo, 1970. J.Phys.Chem.Solids

$\underline{31}$, 2191

44. M. Born, 1926. Z.f.Physik $\underline{38}$, 803

45. W.Bothe, 1921. Z.f.Physik $\underline{5}$, 63

46. J.Bøttiger & F.Bason, Rad.Effects $\underline{2}$, 105

47. J.Bøttiger, J.A.Davies, P.Sigmund, & K.B.Winterbon, 1971. Rad.Effects $\underline{11}$, 69

48. J.Bøttiger, H.W.Jørgensen, & K.B.Winterbon, 1972. Rad.Effects $\underline{11}$, 133

49. P.D.Bourland, W.K.Chu, & D.Powers, 1971a.Phys.Rev. $\underline{B3}$, 3625

50. P.D.Bourland & D.Powers, 1971b,Phys.Rev.$\underline{B3}$, 3635

51. W.H.Bragg & R.Kleeman, 1905.Phil.Mag.$\underline{10}$, S 318

52. W.Brandt, J.M.Khan, D.L.Potter, R.D.Worley, & H.P. Smith, 1965. Phys.Rev.Lett. $\underline{14}$, 42

53. D.K.Brice, 1968.Phys.Rev. $\underline{165}$, 475
    1971.Rad.Effects $\underline{6}$, 77
    1972.Rad.Effects
    1972.Phys.Rev.$\underline{A6}$, 1791
    1973.to be published

54. J.S.Briggs & J.H.Macek, 1972.J.Phys. $\underline{B5}$, 579
    1973.ibid. $\underline{B6}$, 982

55. J.S.Briggs & A.P.Pathak, 1973. J.Phys.$\underline{C6}$, L 153

56. C.Broude, 1967.Can.J.Phys. $\underline{45}$,3415

57. M.D.Brown, & C.D.Moak, 1972.Phys.Rev.$\underline{B6}$, 90

58. A.Brynjolfsson, 1973.Doctoral Thesis, Univ.Copenhagen

59. I.M.Cheshire, G.Dearnaley, & J.M.Poate, 1968.Phys. Lett. $\underline{27A}$, 304

60. I.M.Cheshire & J.M.Poate, 1970. in Atomic Collision Phenomena in Solids (D.Palmer et al.,Ed.). North Holland, Amsterdam. p.351

61. W.K.Chu & D.Powers, 1969.Phys.Rev. $\underline{187}$, 478
    1971.ibid.$\underline{B4}$, 10
    1972.Phys.Lett.$\underline{38A}$, 267

62. W.K.Chu, J.F.Ziegler, I.V.Mitchell, & W.D.Mackintosh, 1973, Appl.Phys.Lett.$\underline{22}$, 437

63. G.J.Clark, D.V.Morgan, & J.M.Poate, 1970 in Atomic Collision Phenomena in Solids (D.Palmer et al. Ed.).North Holland, Amsterdam.p.388

110

64. R.G.Clarkson & N.Jarmie, 1971.Comp.Phys.Comm. $\underline{2}$, 433

65. H.Cramér, 1945. Mathematical Methods of Statistics, Princeton University Press.

66. J.Daniels, C.V.Festenberg, H.Raether, & K.Zeppenfeld, 1970. Springer Tracts of Modern Physics $\underline{54}$, 77

67. S.Datz, 1973. This school

68. G.Dearnaley, 1964. IEEE Trans.Nucl.Sci.$\underline{11}$, 243

69. G.Della Mea, A.V.Drigo, S.Lo Russo, P.Mazzoldi, & G.G.Bentini, 1972. Rad.Effects $\underline{13}$, 115

70. K.Dettmann & M.T.Robinson, 1973. To be published

71. G.J.Dienes & G.H.Vineyard, 1957. Radiation Effects in Solids.Academic Press, New York

72. F.H.Eisen, 1968. Can.J.Phys. $\underline{46}$, 561

73. F.H.Eisen, G.J.Clark, J.Bøttiger, & J.M.Poate ,1972. Rad.Effects $\underline{13}$, 93

74. A.H.El-Hoshy & J.F.Gibbons, 1968. Phys.Rev. $\underline{173}$, 454

75. C.Erginsoy, 1965. Phys.Rev.Lett. $\underline{15}$, 360

76. C.Erginsoy, 1967. in Proc.Internat.Conf.on Solid State Physics Research with Accelerators (A.N.Goland, Ed). Brookhaven Report, BNL-50083(C-52),p.30

77. L.Eriksson, J.A.Davies, & P.Jespersgaard, 1967. Phys.Rev. $\underline{161}$, 219

78. U.Fano, 1960. in Penetration of Charged Particles in Matter (E.A. Uehling, Ed). NAS-NRC Publ.$\underline{752}$, p. 158.
        1963. Ann.Rev.Nucl.Sci. $\underline{13}$, 1

79. U.Fano & W.Lichten, 1965. Phys.Rev.Lett. $\underline{14}$, 627

80. B.Fastrup, A.Borup, & P.Hvelplund, 1968.Can.J.Phys. $\underline{46}$, 489

81. B.Fastrup, G.Hermann, & K.J.Smith, 1971. Phys.Rev. $\underline{A3}$, 1591

82. B.Fastrup, P.Hvelplund, & C.A.Sautter, 1966. Mat. Fys.Medd.Dan.Vid.Selsk. $\underline{35}$, n°10

83. J.S.Feng, W.K.Chu, M.A.Nicolet, & J.W.Mayer, 1973. Thin Solid Films(in press)

84. E.Fermi, 1924. Z.f.Physik 29, 315
           1940. Phys.Rev.57, 485
           1949. Nuclear Physics.The University of
           Chicago Press.Ch. TTA

85. E.Fermi and E.Teller, 1947.Phys.Rev. 72, 399

85a. J.Finnemann, 1968, M.Sc.Thesis, Univ.Aarhus unpu-
     blished.

86. O.B.Firsov, 1957a. Zh.ETF.32, 1464. Engl.Transl.
    Sov. Phys. JETP 5, 1192 (1957)
               1957b. ibid. 33, 696. Engl.Transl. ibid.
    6, 534 (1958)
               1959. Zh.ETF 36, 1517. Engl.Transl.Sov.
    Phys. JETP 9, 1076 (1959)

87. C.Foster, W.K.Kool, W.F.v.d.Weg, & H.E.Roosendaal,
    1972. Rad.Effects 16, 139

88. J.D.Garcia, R.J.Fortner, & T.M.Kavanagh, 1973. Rev.
    Mod.Phys. 45, 111

89. V.I.Gaydaenko & V.K.Nikulin, 1970. Chem.Phys.Lett.
    7, 360

90. W.M.Gibson & J.Golovchenko, 1971.Phys.Rev.Lett. 28,
    1301

91. T.L.Gilbert & A.C.Wahl, 1967. J.Chem.Phys. 47, 3425

92. P.Gombas, 1949. Die Statistische Theorie des Atoms.
    Springer, Wien. 1956. Encyclopedia of Physics (S.
    Fluegge, Ed.). Vol.36, 109. Springer, Berlin, Göt-
    tingen, Heidelberg.

93. S.Goudsmit & J.L.Saunderson, 1940.Phys.Rev.57, 24

94. Yu.V.Gott, 1971.Zh. ETF 60, 1291. Engl.Transl.Sov.
    Phys. JETP 33, 698 (1971)

95. R.J.Gould, 1972. Physica 62, 555

96. O.Halpern & H.Hall, 1948. Phys.Rev. 73, 477

97. D.E.Harrison, 1968. Appl.Phys.Lett. 13, 279

98. F.Herman, & S.Skillman, 1963. Atomic Structure Cal-
    culations. Prentice Hall, Englewood Cliffs.

99. J.W.Hilbert, N.A.Baily, & R.G.Lane, 1968. Phys.Rev.
    168, 290.

100. K.W.Hill & E.Merzbacher, 1971. Bull.Am.Phys.Soc.
     16, 1349

101. G.Högberg, 1971. Phys.Stat.Sol.(b), 48, 829

102. J.Hubbard, 1957. Proc.Roy.Soc. A243, 336

103. F.Hund, 1927. Z.f.Physik 40, 742

104. P.Hvelplund, 1968. Prize Essay, Univ.Aarhus
     1971. Mat.Fys.Medd.Dan.Vid.Selsk.
     $\underline{38}$, n°4

105. P.Hvelplund & B.Fastrup, 1968. Phys.Rev. $\underline{165}$, 408

106. P.Hvelplund & G.Sørensen, 1971. Amsterdam Confe-
     rence on the Physics of Electronic and Atomic
     Collisions. Book of Abstracts p.1003

107. M.Inokuti, 1971. Rev.Mod.Phys. $\underline{43}$, 297

108. J.D.Jackson, 1962. Classical Electrodynamics.Wi-
     ley, New-York. Ch. 13

109. J.D.Jackson, & R.L.Mc.Carthy, 1972. Phys.Rev. $\underline{B6}$,
     4131

110. J.E.Janni, 1966. Air Force Weapons Laboratory Re-
     port, AFWL-TR-65-150

111. A.Johansen, S.Steenstrup, & T.Wohlenberg, 1971.
     Rad.Effects $\underline{8}$, 31

112. N.L.Johnson, 1949. Biometr. $\underline{36}$, 149

113. Q.C.Kessel & B.Fastrup, 1973. Case Studies in A-
     tomic Physics $\underline{3}$, 137

114. V.S.Kessel'man, 1971.Zh.TF. $\underline{41}$, 1708. Engl.Transl.
     Sov. Phys.-Techn.Phys.$\underline{16}$, 1346 (1972)

115. L.M.Kishinevskii, 1962. Izv.Akad.Nauk SSSR- Ser.
     Fiz. $\underline{26}$, 1410. Engl.Transl.Bull.Acad.Sci. USSR,
     Physical Ser. $\underline{26}$, 1433 (1963)

116. M.Kitagawa & Y.H.Ohtsuki, 1972. Phys.Rev. $\underline{B5}$,
     3418

117. J.Knipp & E.Teller, 1941. Phys.Rev. $\underline{59}$, 659

118. R.Kollath, 1958. Encyclopedia of Physics (S.Flueg-
     ge, Ed.), Vol. $\underline{34}$, 1. Springer, Berlin, Göttingen
     Heidelberg.

119. H.A.Kramers, 1947. Physica $\underline{13}$, 401

120. L.D.Landau, 1932. Phys.Z.Sowjetunion $\underline{2}$, 46
     1944. ibid. $\underline{8}$, 204

121. L.D.Landau & E.M.Lifshitz, 1960. Electrodynamics
     of Continuous Media. Ch.12

122. D.C.Langreth, 1969. Phys.Rev. $\underline{181}$, 753

123. N.O.Lassen, 1951. Mat.Fys.Medd.Dan.Vid.Selsk. $\underline{26}$,
     n° 5

124. N.O.Lassen, N.O.Roy Poulsen, G.Sidenius, & L.Vis-

tisen, 1064. ibid. $\underline{34}$, n°5

125. C.Lehmann & G.Leibfried, 1963. J.Appl.Phys. $\underline{34}$, 2821

126. E.Leminen, 1972. Ann.Acad.Sci.Fenn. $\underline{A6}$, 386, 1

127. H.Lewis, 1950. Phys.Rev. $\underline{78}$, 526

128. W.K.Lin, H.G.Olson, & D.Powers, 1973a.J.Appl.Phys. (Aug. 1973) 1973b. Phys.Rev.B (Sept.1973)

129. J.Lindhard, 1954. Mat.Fys.Medd.Dan.Vid.Selsk. $\underline{28}$, n° 8. 1965. ibid. $\underline{34}$, n°14

130. J.Lindhard & V.Nielsen, 1971. ibid. $\underline{38}$, n°9

131. J.Lindhard, V.Nielsen, & M.Scharff, 1968. ibid. $\underline{36}$, n°10

132. J.Lindhard & M.Scharff, 1953. ibid. $\underline{27}$, n°15 1961. Phys.Rev. $\underline{124}$, 128

133. J.Lindhard, M.Scharff, & H.E.Schiøtt, 1963. Mat. Fys. Medd.Dan.Vid.Selsk. $\underline{33}$, n°14

134. J.Lindhard & Aa.Winther, 1964. ibid. $\underline{34}$, n°4

135. P.Loftager & G.Claussen, 1969. M.I.T. Conference on the Physics of Electronic and Atomic Collisions. Book of Abstracts, p.518

136. H.O.Lutz, S.Datz, C.D.Moak, & T.S.Noggle, 1966. Phys.Rev.Lett. $\underline{17}$, 285

137. H.Lutz & R.Sizmann, 1963, Phys.Lett. $\underline{5}$, 113

138. H.D.Maccabee, M.R.Raju, & C.A.Thomas, 1968. Phys. Rev. $\underline{165}$, 469

139. J.R.Macdonald, J.H.Ormrod, & H.E.Duckworth, 1966. Z. Nat.forschung $\underline{21a}$, 130

140. T.C.Madden & W.M.Gibson, 1964, IEEE Trans.Nucl. Sci. $\underline{11}$, 254

141. Yu.V.Martynenko, 1967. F.T.T. $\underline{9}$, 3646. Engl.Transl. Sov.Phys.-Solid State $\underline{9}$, 2879 (1968)

142. J.W.Mayer, L.Eriksson, & J.A.Davies, 1970. Ion Implantation in Semiconductors. Academic Press, New York.

143. J.W.Mayer & J.F.Ziegler (Ed.) 1973. International Conference on Ion-Beam Surface Analysis. Thin Solid Films (in press).

144. W.A.Mc Kinley & H.Feshbach, 1948. Phys.Rev.$\underline{74}$,

114

1759

145. E.Merzbacher & H.W.Lewis, 1958. Encyclopedia of Physics (S.Fluegge Ed.), Vol.34, 166. Springer, Berlin, Göttingen, Heidelberg.

146. L.Meyer, 1971. Phys.Stat.Sol.44, 253

147. I.V.Mitchell, K.B.Winterbon, & F.Brown, 1973. To be published.

148. G.Molière, 1948. Z.Nat.forschung 3a, 78

149. C.Møller, 1932.Ann.Phys. (5), 14, 531

150. S.H.Morgan, & P.B.Eby, 1973. Nucl.Inst.Meth. 106, 429

151. G.H.Morgan & E.Everhart, 1962. Phys.Rev. 128, 667

152. K.Morita, H.Akimune, & T.Suita, 1967. J.Phys.Soc. Jap. 22, 1503. 1968. ibid. 25, 1525

153. N.F.Mott, 1929. Proc.Roy.Soc. A 124, 425

154. N.F.Mott & H.S.W.Massey, 1971. The Theory of Atomic Collisions. Clarendon Press. Oxford

155. R.S.Mulliken, 1928. Phys.Rev. 32, 186

156. R.S.Nelson, & M.W.Thompson, 1963. Phil.Mag. 8, 1677

157. W.Neuwirth, U.Hauser, & E.Kühn, 1969. Z.f.Physik 220, 241

158. V.S.Nikolaev & I.S.Dmitriev, 1968. Phys.Lett. 28a, 277

159. L.C.Northcliffe, Ann.Rev.Nucl.Sci. 13, 67

160. L.C.Northcliffe & R.F.Schilling, 1970. Nuclear Data Tables 7A, 233

161. P.Nozières & D.Pines, 1958. Nuov.Cim. 9, 470

162. Y.H.Ohtsuki, M.Mizuno, & M.Kitagawa, 1971. J.Phys. Soc.Japan 31, 1109

163. J.H.Ormrod, 1968. Can.J.Phys. 46, 497

164. J.H.Ormrod, & H.E.Duckworth, 1963, Can.J.Phys. 41, 1424

165. J.H.Ormrod, J.R.Macdonald, & H.E.Duckworth, 1965. Can.J.Phys. 43, 275

166. W.K.H.Panofsky & M.Phillips, 1955. Classical Electricity and Magnetism. Addison Wesley, Reading, Mass

167. A.P.Pathak, & M.Yussouff, 1972. Phys.Stat.Sol.(b) 49, 431

168. M.G.Payne, 1969. Phys.Rev. 185, 611

169. T.E.Pierce & M.Blann, 1968, Phys.Rev.173, 390

170. G.R.Piercy, F.Brown, J.A.Davies, & M.Mc.Cargo, 1968. Phys.Rev.Lett.10, 399

171. D.Pines, 1963. Elementary Excitations in Solids. Benjamin, New York

172. D.Powers, W.K.Chu, R.J.Robinson, & A.S.Lodhi, 1972. Phys.Rev. A6, 1425

173. D.Powers, A.S.Lodhi, W.K.Lin, & H.L.Cox, 1973. Thin Solid Films (in press)

174. D.Powers & W.Whaling, 1962.Phys.Rev. 126, 61

175. J.J.Ramirez, R.M.Prior, J.B.Swint, A.R.Quinton, & R.A.Blue, 1969. Phys.Rev. 179, 310

176. M.T.Robinson, 1969. Phys.Rev. 179, 327
                  1970. Tables of Classical Scatte-
                        ring Integrals. Oak Ridge Re-
                        port, ORNL-4556
                  1971. Phys.Rev. B4, 1461

177. M.T.Robinson, & O.S.Oen, 1963. Appl.Phys.Lett. 2, 30. 1963.Phys.Rev. 132, 2385

178. B.Rossi, 1952. High Energy Particles. Prentice Hall, Englewood Cliffs, N.J.

179. C.C.Rousseau, W.K.Chu, & D.Powers, 1971. Phys. Rev. A4, 1066

180. J.B.Sanders, 1968. Can.J.Phys.46, 455

181. F.W.Saris, 1971. Physica 52, 290

182. F.Sauter, 1934. Ann.Phys. (5), 20, 404

183. G.D.Sauter, & S.D.Bloom, 1972. Phys.Rev.B6, 699

184. L.I.Schiff, 1955. Quantum Mechanics, ch.26. Mc Graw-Hill, New York

185. H.E.Schiøtt, 1973. in Interaction of Energetic Charged Particles with Solids (AN Goland, Ed.), Brookhaven National Laboratory Report, BNL 50336

185! W.T.Scott, 1963. Rev.Mod.Phys. 35, 231

186. F.Seitz, 1949. Disc.Far.Soc. 5, 271

187. S.Seltzer & M.J.Berger, 1964. NAS-NRC Publ.1133, 187

188. I.Shimamura & T.Watanabe, 1973. J.Phys.Soc. Japan
     $\underline{34}$, 483

189. J.A.Shohat, & J.D.Tamarkin, 1943. The Problem of
     Moments. Amer.Math.Soc. New York

190. P.Shulek, B.M.Golovin, L.A.Kulyukina, S.V.Medved'
     & P.Pavlovich, 1967. Sov.J.Nucl.Phys. $\underline{4}$, 400

191. P.Sigmund, 1969. Phys.Rev.$\underline{184}$, 383
                 1972a. Rev.Roum.Phys. $\underline{17}$, 823
                 1972b. ibid. $\underline{17}$, 969
                 1972c. ibid. $\underline{17}$, 1079
                 1972d. in Physics of Ionized Gases (M.
                 Kurepa,Ed.) p.137. Institute of
                 Physics, Belgrade.

192. P.Sigmund & U.Lillemark, 1973. To be published

193. P.Sigmund & K.B.Winterbon, 1973. To be published

194. E.I.Sirotinin, A.F.Tulinov, A.Fiderkerich, & K.S.
     Shishkin, 1972. Rad.Effects $\underline{15}$, 149

195. R.M.Sternheimer, 1952. Phys.Rev. $\underline{88}$, 851.
                       1960. Ibid. $\underline{117}$, 485

196. E.C.G.Stückelberg, 1932. Helv.Phys.Acta $\underline{5}$, 369

197. K.R.Symon, 1948. Thesis, Harvard University (un-
     published). Summaries in Rossi (1952) and Payne
     (1969).

198. A.Teplova, V.S.Nikolaev, I.S.Dmitriev, & L.N.Fa-
     teeva, 1961. Zh.E.T.F. $\underline{42}$, 44. Engl.Transl.Sov.
     Phys.JETP $\underline{15}$, 31 (1962)

199. E.W.Thulstrup & H.Johansen, 1972. Phys.Rev.A$\underline{6}$,
     206

200. N.H.Tolk & C.W.White, 1973. Rad.Effects.(in press)

201. B.A.Trubnikov & Yu.N.Yavlinskii, 1965. Zh.E.T.F.
     $\underline{48}$, 253. Engl.Transl.Sov.Phys. JETP $\underline{21}$, 167 (1965)

202. C.Tschalär, 1968a. Nucl.Inst.Meth. $\underline{61}$, 141
                  1968b. ibid. $\underline{64}$, 237

203. C.Tschalär, & H.D.Maccabee, 1970. Phys.Rev. B$\underline{1}$,
     2863

204. P.V.Vavilov, 1957. Zh.E.T.F.$\underline{32}$, 920. Engl.Transl.
     Sov.Phys. JETP $\underline{5}$, 749 (1957)

205. W.D.Warters, 1953. Ph.D.Thesis, Caltech, Pasadena
     (unpublished)

206. W.A.Wenzel & W.Whaling, 1952. Phys.Rev.$\underline{87}$, 499

207. W.Whaling, 1958. Encyclopedia of Physics (S.Fluegge, Ed.). Vol.$\underline{34}$, 193. Springer, Berlin, Göttingen, Heidelberg

208. W.White & R.M.Mueller, 1967. J.Appl.Phys. $\underline{38}$, 3660
    1968. Phys.Rev. $\underline{187}$, 499

209. J.L.Whitton, 1973. Can.J.Phys. (in press)

210. A.v.Wijngaarden & W.E.Baylis, 1973. Phys.Rev. $\underline{A7}$, 937

211. A.v.Wijngaarden & H.E.Duckworth, 1962. Can.J.Phys. $\underline{40}$, 1749

212. W.D.Wilson, R.D.Hatcher, & C.L.Bisson, 1971. Bull. Am.Phys. Soc. $\underline{16}$, 333

213. K.B.Winterbon, 1968.Can.J.Phys. $\underline{46}$, 2429

214. K.B.Winterbon, P.Sigmund, & J.B.Sanders, 1970. Mat.Fys.Medd.Dan.Vid.Selsk. $\underline{37}$, n°14

215. K.B.Winterbon, 1972a. Rad.Effects $\underline{13}$, 215
    1972b. Can.J.Phys. $\underline{45}$, 3415

216. K.B.Winterbon, 1973. To be published

217. C.Zener, 1932. Proc.Roy.Soc. $\underline{A137}$, 696

218. J.F.Ziegler & M.H.Brodsky, 1973. J.Appl.Phys.$\underline{44}$, 188

219. J.F.Ziegler & W.K.Chu, 1973. Thin Solid Films (in press)

ON THE STATES OF IONS PENETRATING SOLIDS[*]

S.DATZ

Oak Ridge National Laboratory
Oak Ridge, Tennessee 37830
USA

INTRODUCTION

Energetic ions penetrating matter undergo a series
of collisions in which electrons are captured or lost
by the ion and in which excitation and ionization of
both the ion and the atoms of the medium occur. These
events are responsible for the slowing of the ion (e-
lectronic stopping) and for the generation of free e-
lectrons and radiation by the penetrating particle.

In this series we shall discuss what is known a-
bout the mechanisms for these processes from studies of
isolated atomic collisions and combine this knowledge
with experimental data obtained from collisions in solids to estimate ionic states in condensed media (both
in random solids and in crystal channels). For example,
for most light ions at moderate energies "charge state
equilibrium" is attained in a dilute gas through a series of electron capture and loss collisions involving
the outer electrons of ground state ions interacting
with outer electrons ground state target atoms. As the
medium becomes more dense the time between collisions
is too short to allow relaxation of excited states so
that one must think in terms of "excitation equilibrium"
as well as "charge state equilibrium". Ions channeled

---

[*] Research sponsored by the U.S. Atomic Energy Commis-
sion under contract with the Union Carbide Corpora-
tion.

in single crystals have a restricted set of impact parameters with target atoms so that both their "charge state" and "excitation" equilibria will differ from either a dilute gas or a random solid.

As the energy of the light ion increases, its cross section for ionizing outer shell electrons of the target decreases and the cross section for inner shell ionization increases. Inner shell ionization via Coulomb excitation can then become the dominant energy loss process. On an atomic scale this is signalled by x-ray and Auger electron emission and can lead to significant changes in charge state equilibria as well as in the energy spectrum of electrons released along the path of the penetrating ion.

As the z of the penetrating ion increases, two new effects come into play. First, multiple ionization by Coulomb excitation in a single collision becomes dominant, and second, a new process, quasi-molecular excitation, becomes possible. This latter phenomenon (also called "Pauli excitation") occurs because the atomic orbitals of two colliding particles are merged through molecular orbitals to the atomic orbitals of the united atom. When level crossings occur, inner shell electrons can, in some cases, be promoted to higher unoccupied states and, when the atoms disengage, inner shell vacancies can occur. The probability of these events occuring is governed by the condition of the outer shell at the time the collision takes place and hence studies of x-rays and electrons coming from solids gives information relating to steady state outer shell conditions in the penetrating ion. Studies of these phenomena when taken with other recent experimental data on radiative electron capture, ejected electron energy spectra, charge changing collisions in channels, and dynamical screening effects on x-ray production cross sections yield much new information on the states of energetic ions in condensed media.

CHARGE CHANGING COLLISIONS

To begin, let us consider starting with an energetic neutral atom and passing it into a medium. In the first collision that it makes it will in all probability lose an outer electron, in the second collision it may lose a second one, and so on. The cross section for electron loss σ_ℓ will, in general, decrease with increasing charge state q. In the meantime the probability of electron capture σ_c by the penetrating ion increases with increasing charge state and ultimately a

steady state between capture and loss is attained where the most probable "equilibrium" charge is attained,
and it is unaffected by additional collisions, i.e.,
after traversing a given target thickness ρ the fractional population ϕ_q of charge state q depends on relative cross sections for production and loss. Considering only single electron capture and loss

$$d\phi_q/d\rho = \phi_{q-1}\phi_\ell(q-1)+\phi_{q+1}\sigma_c(q+1)-\phi_q[\sigma_\ell(q)+\sigma_c(q)] \quad (1)$$

At equilibrium the derivative is zero, and if σ_c increases with q at about the same rate that σ_ℓ decreases the most probably charge state q is that for which
$\sigma_c \sim \sigma_\ell$. It now behooves us to consider in some detail
individual charge changing collisions.

The study of the phenomenon of charge exchange has
a long and distinguished history dating back to Henderson's observation of electron capture by α particles
in 1923. Since that time many major scientists dealing
with atomic physics have contributed to this area of
physics. Among the many results coming from these studies are a number of very elegant approximations for
treating the transfer of an electron from a ground state hydrogen atom to a proton. An excellent summary of
theoretical methods is contained in the recent book by
R.A. Mapleton[1]. The most readily applicable methods are
those of the plane wave Born Approximation (PWBA) (at
high velocities) and its classical analogue, the binary encounter approximation (BEA). These will prove especially useful for innershell ionization in later
talks.

For dealing with the charge changing events occuring with a great variety of ions in many materials we
will use as reference frame the simple theories developed by Bohr and Lindhard[2]. This and similar theories
are discussed in a recent review by Betz[3]. For the purpose of these talks we will start with the approach of
Bohr and Lindhard. They treated the problem of electron
capture and loss in highly simplistic manner. However,
their treatment gives a good physical representation
of the problem. The fundamental units here are the Bohr
radius and velocity : $a_o = h^2/me^2$ and $v_o = e^2/h$. An e-
lectron in an ion or atom of nuclear charge Z has an
orbital velocity v_e defined in terms of the electron
binding energy $I = (1/2)mv^2$ and in reduced units has
a radius and velocity

$$a = a_o(v^2/n), \quad v_e = v_o(n/v)$$

where (Z−n) is the number of electrons with orbital radius smaller than $\underline{z}$ (i.e., the screened nuclear charge). The quantity ν is an "effective quantum number" based on a Thomas Fermi statistical model. In the case of heavy atoms the most firmly bound electrons move in a Coulomb field with approximately the total nuclear charge and $\nu \simeq 1$ and 2 for the K and L shells respectively. For outer electrons where the effective nuclear charge is ~ 1, ν is again $\simeq 1$. Over a large intermediate region ν will have a flat maximum at a value close to $Z^{1/3}$. From this model an approximate expression for the volocity distribution for most of the electrons in the atom is obtained

$$dn = Z^{1/3} \, dv_e/v_o \qquad (2)$$

Anticipating at this point the final result we will find that on an ion having a velocity v the most loosely bound electron remaining attached to the ion in its most probable equilibrium charge state q will have a velocity $v^* \simeq V$ from which

$$\bar{q} = Z^{1/3} \, v/v_o \qquad (3)$$

The process of electron loss is treated as simple ionization in which a bound electron is removed to the continuum. Starting with the cross section for transfer of energy greater than T by collision between a free electron at rest and a heavy particle with charge z^*e and velocity v

$$\sigma = 2\pi a_o^2 \, z^{*2} \left(\frac{v_o}{v}\right)^2 \left(\frac{mv_o^2}{T} - \frac{mv_o^2}{T_{max}}\right) \qquad (4)$$

where the maximum transferred energy is $T_{max} = 2mv^2$. Introducing for each ion electron $T = mv^2/2$ and summing over all ion electrons through Eq.2 they obtain

$$\sigma_e = \pi a_o^2 z^{*2} Z^{1/3} (v_o/v^*)^3 \qquad (5)$$

where v^*, the binding energy of the most loosely bound electron, is in the order of V and z^* is the nuclear charge for light gases or the core charge for heavy ones. In order to estimate the dependance of the loss cross section on ion charge q, v^* is approximated by $v_o q Z^{-1/3}$ and

$$\sigma_e = \pi a_o^2 z^{2/3} Z^{4/3} q^{-3} (v/v_o)^2 \qquad (6)$$

Note here that the quadratic dependence on velocity is only assumed valid near $(v/v^{x}) \simeq 1$; i.e., for q in the region of $\bar{q}$ and does not apply to more strongly bound electrons.

The process of electron capture is much more delicate since in this case we are moving into a well-defined bound state and energy conservation restrictions on the motion become more stringent. Bohr and Lindhard considered capture in the following manner. During the approach of the highly charged ion, the electron will be exposed to a strong field of force, giving rise to an increasing polarization of the binding, which may subsequently lead to its rupture. In order to estimate when electron release takes place, we note that, at a distance R between the two systems given by

$$qe^2/R^2 = mv_e^2/a \qquad\qquad (7)$$

the force from the ion and the atomic binding force are approximately equal. Still, it has to be taken into account that the possibility for electron release is not only determined by a comparison between the forces, but that the completion of the process will require a time of the order a/v, and that therefore, especially in the case of the more loosely bound atomic electrons, the ion may have travelled a distance comparable with R before the electron is liberated from the atomic field.

After the release from the atom, the electron will be captured if its total energy relative to the ion has a negative value. In his estimate of capture cross sections, on similar lines as followed here, Bell assumed that an atomic electron is released at a distance R from the ion with velocities corresponding to the momentum distribution in its original binding state. It must, however, be taken into consideration that, under the combined action of the atom and ion fields, the electron velocity distribution will have changed considerably from that in the isolated atom, and that we must expect the velocity of the electron to be largely reduced during the gradual loosening of the atomic binding. At the completion of the release process, we may thus in first approximation assume that the velocity of the electron relative to the ion will not differ essentially from the ion velocity. On such assumptions, the condition for capture is that the process of electron release is effectively completed at a distance from the ion smaller than R', determined by

$$qe^2/R' = 1/2 \ mv^2 \qquad\qquad (8)$$

Assuming, in first approximation, that the release takes place at the distance R, we find that, if R < R', capture occurs with a cross section πR^2, while for R > R' there will be no capture. According to (7) and (8), it is seen that on this assumption only strongly bound atomic electrons can contribute to capture. Actually, in a heavy atom, the contribution will arise mainly from a comparatively narrow region or orbital velocities around v/2. Summing over the electrons in the atom, we obtain

$$\sigma_c = \pi a_o^2 q^2 z^{1/3} (v_o/v)^3 \qquad\qquad (9)$$

for the total capture cross section for atoms in which a considerable part of the electrons have velocities comparable vith v. Note here that capture of electrons from light target atoms which have no electrons of sufficiently high velocity to be captured would have essentially zero probability for capture in this formulation. In order to explain capture from low Z targets they point out that electron release is a gradual process. Thus there is a small probability that a loosely bound electron will remain with the target atom until a highly charged ion approaches closely enough so that capture can occur.

Assuming that the probability of electron release from the target atom within a distance smaller than R' is of the order $(R'/v)(v_e/a)$ they obtain

$$\sigma_c = \pi a_o^2 q^3 (v_o/v)^7 (n_e'^2/\nu'^3) \qquad\qquad (10)$$

for very loosely bound electrons with binding characterized by a screened nuclear charge n_e' and an effective quantum number ν'.

EXCITED STATES

In the discussion above we have been considering only ground state ion configurations. In practice we must also consider the effect of excited states on both capture and loss processes, especially in condensed media. For example, the excitation cross section for a given (outer shell) electron on an ion is of the same order as the ionization cross section and the ionization cross section for the excited state is, in general, higher than that for the ground state. In a dilute gas radiative decay will occur between collisions but

as the target becomes more dense the effective loss
cross section will be increased by the presence of excited states.

Capture into excited states for light ions (e.g.,
protons and alpha particles) has been extensively studied both theoretically and experimentally. Oppenheimer
showed that alpha particles capture electrons from hydrogen atoms mainly into s states with cross sections
which, for sufficiently high velocity, are simply given
by $\sigma_\nu = \sigma_1/\nu^3$ $(v >> v_o)$ where ν is the principal quantum
number and σ_1 is the cross section for capture into 1s.
Hence in these cases only $\sim$ 20 % will be captured into
higher ν states. For heavy ions at velocities $v \simeq v_o$
it is anticipated that capture into excited states will
predominate since these involve lower and more closely spaced bound states for the captured electron. Here
again in a dilute gas target excitation will be relaxed
but the presence of an excited electron in a subsequent
collision will reduce the effective capture cross section for two reasons : first, for the same reason given in the case of excitation from the ground state,
the ionization cross section for the excited electron
is higher and second, the presence of one excited electron at the time of capture of a second electron can
lead to an auto-ionizing state of the ion in which the
total excitation energy of the ion goes into removal
of one of the electrons.

RADIATIVE ELECTRON CAPTURE

In the above discussions of electron capture we
have considered only capture from bound states in which
energy and momentum are conserved by the motion of the
particles involved in the exchange. Clearly in the case of free (or almost free) electrons these descriptions
will not suffice. An alternative capture mechanism is
that in which an electron is captured from a free to a
bound state with the emission of a photon whose energy
E_γ, is the sum of the binding energy of the electron,
E_B, in the final state plus the relative kinetic energy of the ion electron system $E - 1/2mV^2$. Bethe and Saltpeter[5] considered this problem and derived the following total cross section for capture into a 1s orbit
of a bare ion.

$$\sigma_{Rc} = 9.1 \times 10^{-21} (E_B E'/E_\gamma)^2 [\exp(-4E' \tan^{-1} E'^{-1})]$$
$$[1 - \exp(-2\pi E')] (cm^2) \tag{11}$$

For capture into higher states the cross section decreases as ν^{-3}. The magnitude of this cross section is small ($\sim 10^{-22}$ cm^2) in heavy targets for charge states near equilibrium. Nonetheless in light gases and in crystal channels this process may contribute appreciably.

EXPERIMENTAL MEASUREMENTS ON CHARGE CHANGING COLLISIONS OF HEAVY IONS IN GASES

Before a consideration of the data a brief word about how measurements are made. The experimental arrangement for charge state studies at Oak Ridge is fairly typical and is shown in fig.1. In this arrangement a beam of particles of known energy and ionic charge from a Tandem accelerator enters on the left. Either foil targets or gas targets may be used. When charge distributions from crystals channels are measured, a goniometer holding the thin crystal is substituted for the foil holder. The gas target in this case has four 1 mm diameter apertures which form the differential pumping system and collimate the emerging particles. After passage through the foil or gas target the emerging ions pass through a low-resolution electrostatic analyzer which gives a spatial separation of the particles according to charge state and they are counted on a position-sensitive surface barrier detector.

The data shown in fig.2 were taken with 110 MeV I ions. The spectra show the charge states well separated but they are close enough so that all the charge states may be recorded simultaneously ; in this way the need to normalize peaks to each other in separate runs is completely avoided.

To obtain charge changing cross sections the gas density is kept sufficiently low to insure single collisions or the charge distributions are studied as a function of cell pressure and the cross sections are obtained from an analysis of the growth curves.

Considering the crudeness of the model discussed in the preceeding section the match to experimental results is qualitatively good. But there are differences which are observed which significantly affect the ion state and in many cases are not completely understood. As an example consider the cross sections obtained for electron capture by 13.9 and 25 MeV Br ions in H_2, He and Ar shown in figs.3 and 4[6]. As expected, single-capture probabilities $\sigma_c(q)$ increases with the charge state q, but in all cases characteristic anomalies were found which were clearly outside the limits of error.

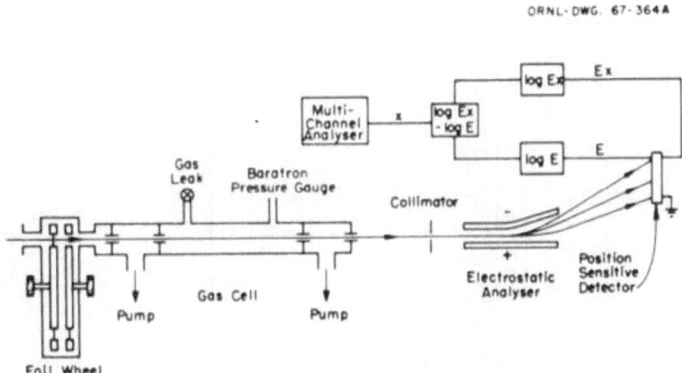

ORNL-DWG. 67-364A

Fig.1 : Experimental apparatus for measuring charge state distribution.

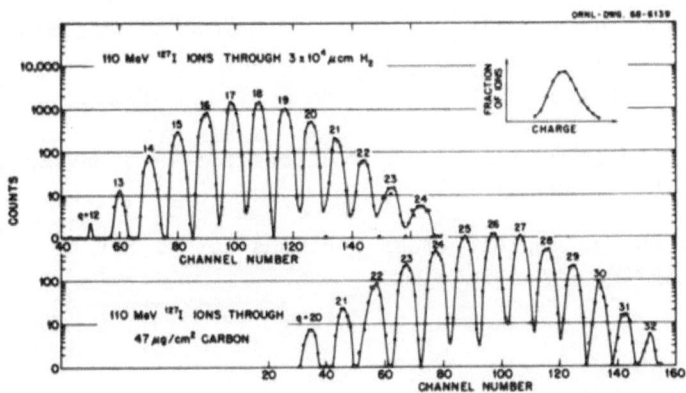

Fig.2 : Charge state population data obtained on a position sensitive detector for 110 MeV I in H_2 and in C.

128

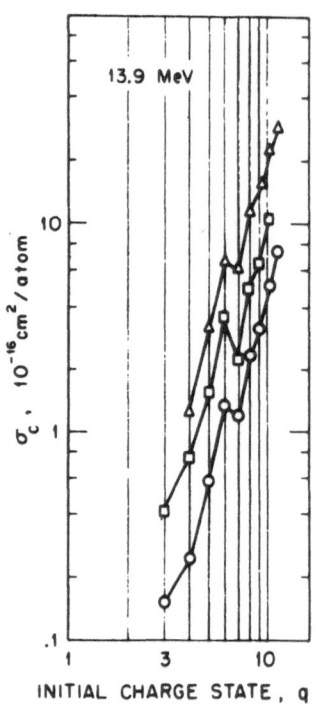

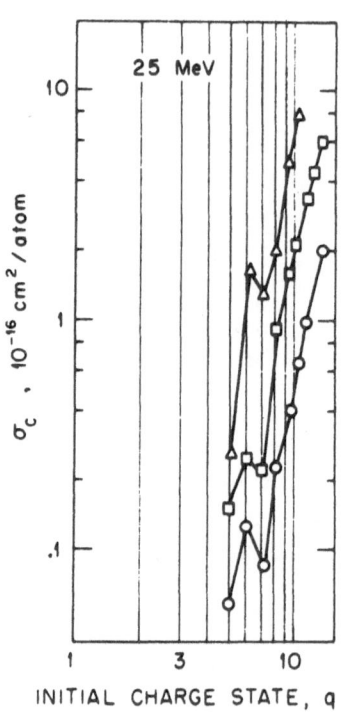

Fig.3   Fig.4

Fig.3 : Charge capture cross sections for 13,9 MeV Br in H_2(o),
         He($\square$) and Ar (Δ)

Fig.4 : Charge capture cross sections for 25 MeV Br in H_2(o),
         He($\square$) and Ar (Δ)

Compared with a smooth trend in $\sigma_c(q)$, the cross sections $\sigma_c(6)$ and to some extent $\sigma_c(8)$ are above the trend whereas $\sigma_c(7)$ is clearly below. In fact, in all cases, $\sigma_c(7)$ is smaller than $\sigma_c(6)$.

In bromine ions with charge $q = 5$, all five 4p electrons are removed. The slight inflection from $q = 5$ to $q = 7$ corresponds to removal of the two 4s electrons and the jump at $q = 8$ corresponds to removal of the first electron from the 3d shell. One might then use quasichemical reasoning and attribute the high value of $\sigma_c(6)$ to closing the 4s subshell, the value of $\sigma_c(7)$ to beginning a new subshell, and the high value $\sigma_c(8)$ to closing the 3d subshell. With this reasoning, however, one would certainly expect the largest effect to have occured at $\sigma_c(8)$. On the basis of energetics and of density states, one might expect capture to occur principally to highly excited levels of the lower-charge-state ion. This effect would tend to blur any clear-cut shell boundaries in electron capture. An alternative explanation can be made if one proposes the necessity for small excitation of a bound electron to stabilize electron capture. In this case the availability of a 4s electron enhances the $\sigma_c(6)$, while the necessity for exciting an electron out of a closed shell depresses $\sigma_c(7)$. In any case, a more detailed knowledge of the actual capture process is required in order to fully understand the shell effects observed in the cross sections.

Another question of interest is the over-all dependence of the single-capture cross sections σ_c on the charge q of the ions and the nuclear charge Z of the target gas. The smoothed increase of σ_c may be approximated by a power function

$$\sigma_c(q) \sim q^{\bar{\alpha}} \qquad (12)$$

where the average exponent $\bar{\alpha}$ is close to 3 and 4 at 13.9 and 25 MeV, respectively. Our data indicate no systematic dependence of $\bar{\alpha}$ on q. Also, the target gas does not seem to have a strong influence on $\bar{\alpha}$. Measurements with light ions resulted in values of $\bar{\alpha}$ ranging from 1.5 to 3, but stronger powers can not be ruled out ; an increase of $\bar{\alpha}$ with velocity v of the ions was also observed. Theoretical calculations lead to the same proportionality Eq.(9), where the prediction is $\bar{\alpha} = 2$.

An important effect which occurs in heavy ion collisions with heavy targets is multiple ionization. In fig.5 we have plotted the charge changing cross sections for 110 MeV I^{12+} ions in H_2, He and Ar^7. The point at 13+ is the single electron loss cross section ; the point at 14+ the double loss cross section, etc. Here it can be seen that in the Ar target the cross section for loss of as many as 12 electrons in a single collision is still measurable and that the sum of the multiple loss cross sections is actually greater than the single loss cross section. The cross section for ionization to states higher than 25+ drops sharply since it requires removal of an electron by the cross sections given for 162 MeV I^{17+} in O_2. Note also that multiple ionization in light target gases is much less significant. The implication here is that multiple ionization involves multiple electron interaction. We will discuss this more fully later.

As we increase target thickness by adding more gas to the cell the charge state distribution approaches the equilibrium value as shown in fig.6. Here we start with a beam of 13.9 MeV Br at charge 10+ and plot the fraction in a given charge state as a function of thickness. Equilibrium is attained for this case at about 5×10^{16} atoms/cm^2 in H_2. Note that in a solid target this corresponds to only about ten atomic layers.

The effect of multiple ionization is manifested in the high charge state tail observed in equilibrium charge state distributions shown in fig.7. The curves for 110 MeV I in H_2 and He are symetric whereas those for the heavier gases are clearly skewed toward higher charge state. Analysis of these distributions indicates that these assymetries can be correlated with multiple ionization functions of the type shown in fig 5 [7].

Shell effects upon the capture cross sections shown in figs.3 and 4 are also visible in the equilibrium distribution as seen in fig.8 where a depression in the population of charge state 6 is observed.

EFFECTS OF DENSE MEDIA ON CHARGE CHANGING COLLISIONS

At this point it would be well to consider again the role of excited states in charge changing collisions. Although, as we shall later see, inner shell ionization is also of importance, let us confine out attention to the outer remaining shell of the ion. If only a single electron in the outer shell is excited, only radiative deexitation is possible with lifetimes in the order of 10^{-7} to 10^{-10}. If multiple excitation takes place much shorter lifetimes, $\sim 10^{-14}$ sec., associated with auto-ionization (and hence electron loss) are pos-

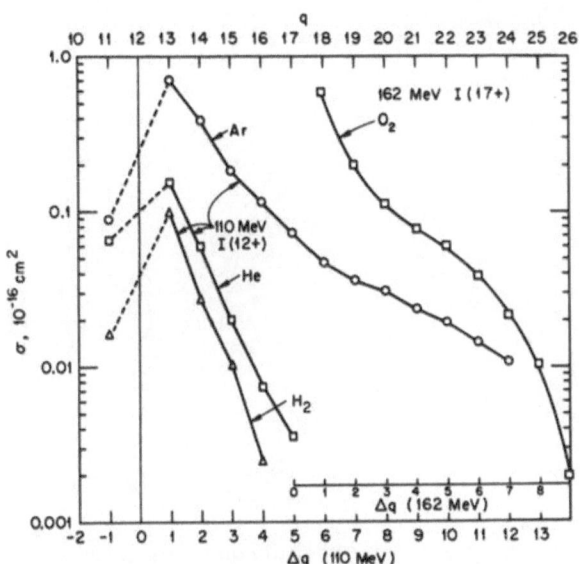

<u>Fig.5</u> : Charge capture and loss cross sections for 110 MeV I^{12+} in H_2, He and Ar and charge loss cross sections for 162 MeV I^{17+} in O_2.

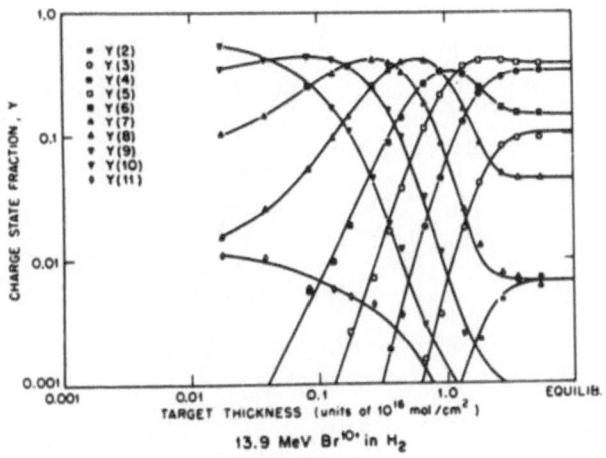

<u>Fig.6</u> : Charge fractions versus target thickness for 13.9 MeV Br^{10+} in H_2.

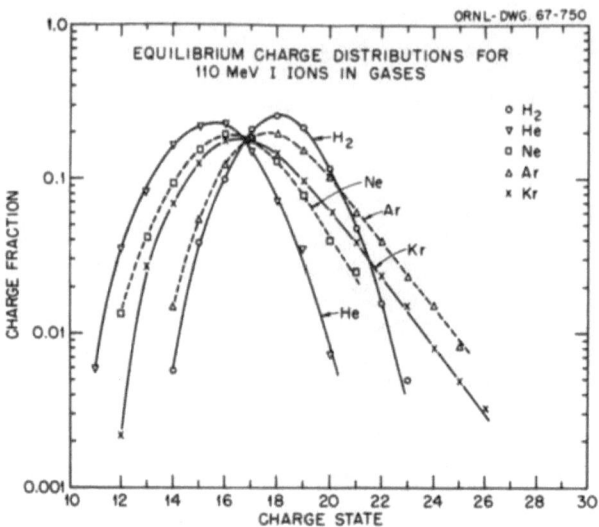

Fig.7 : Equilibrium charge distributions for 110 MeV I in various gases.

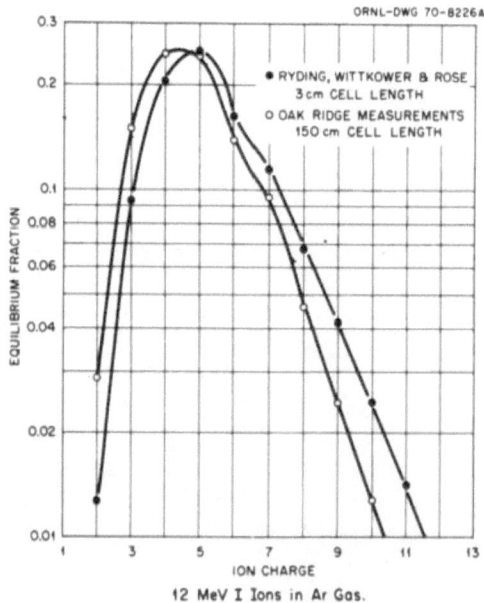

Fig.8 : Equilibrium charge distributions for 12 MeV I in Ar at two target gas densities.

sible.

The probability that an excitation caused by one collision will still be present in a subsequent collision will depend on the mean time between collision events ($\tau = 1/n\sigma v$). If we assume a cross section of 10^{-16} for a given process and consider a gas target at ~ 1 torr (5×10^{16} atoms/cm^3) we can see that for a velocity of say $v_0 = 2 \times 10^8$ cm/sec the time between collisions is 10^{-9} sec. So that even in this relatively dilute target excited states can be of importance in determining the equilibrium charge. This can be seen in fig.8. The difference between the two experiments is that in one case a long cell (150 cm) was used which assured relaxation between collisions whereas in the short cell (3 cm) pressures in the order of 1 torr were necessary to attain the same target thickness.

If one now considers the atom density in solid targets ($\sim 5 \times 10^{22}$ atoms/cm^3) the time between collisions is in the order of 10^{-14} sec. It is therefore quite clear that, if excitation cross sections are of the same magnitude as ionization cross sections, the ion will be in a constant state of excitation in the medium. The presence of this excitation will both increase the ionization cross section and decrease the capture cross section by permitting auto-ionization of a newly captured electron. Both of these effects will tend to increase the steady state ionization over that found in a gaseous medium. A glance at fig.2 amply confirms this. If we compare the most probable charges we see a value of 18+ in H_2 as against a value of 26+ for ions emerging from a thin solid carbon target. This large disparity between solids and gases especially for heavy ions was first noted by Lassen[8] who studied fission fragments emerging from solid foils. Although there is some variation between different gases (fig.7) a qualitative difference exists between solid and gas targets. This is shown in fig.9 where we have plotted the most probable charges for Br ions in many solid and gaseous targets as a function of velocity.

Although there is much data on equilibrium charge state distributions (CSD) emerging from solids very little information exists on the relevant charge changing cross sections. However, some indications of the magnitude of these cross sections can be adduced from one experiment where charge equilibrium in a random solid was not attained. In this work a beam of 100 and 140 MeV Br ions was passed through carbon foils of varying thickness (8-80 μg/cm^2)[9]. The resulting distri-

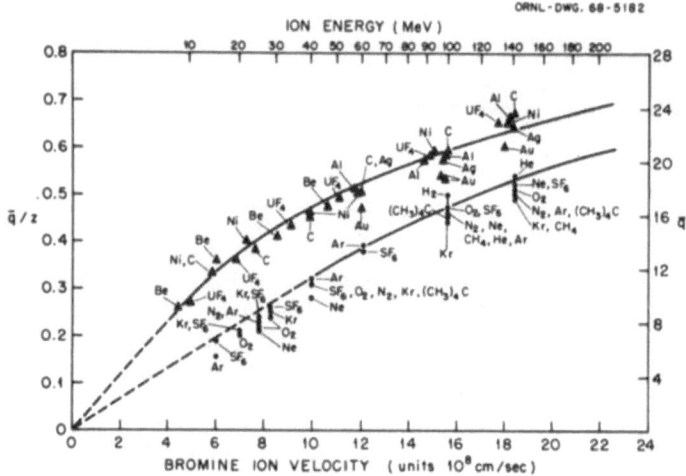

Fig.9 : Most probable charge states of Br ions at equilibrium in solid dans gas targets.

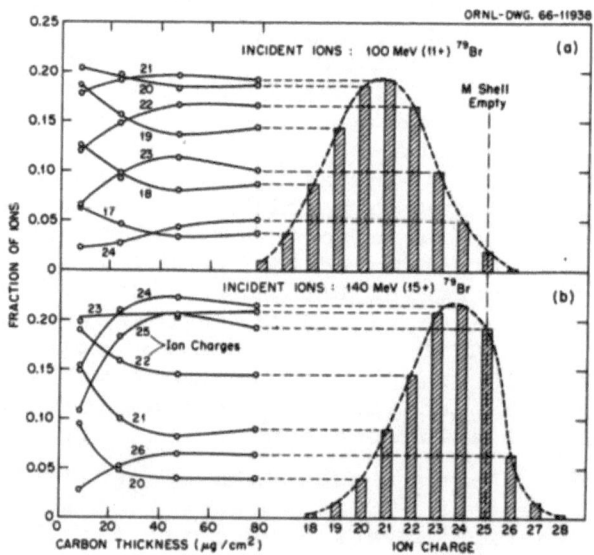

Fig.10 : (right) CSDs for 100 and 140 MeV Br ions emerging from 80 μg/cm² carbon foils. (left) CSDs as a function of carbon thickness (8-80 μg/cm²).

butions are shown in fig.10. A shift in the distribution with thickness indicates lack of equilibrium. Since thickness in the order of 40 $\mu g/cm^2$ (2 x 10^{19} atoms/cm^2) are required for equilibrium, cross sections of the order of 10^{-19} cm^2 must be involved in the equilibrium process. In general one expects that the cross sections for charge states close to equilibrium decrease with increasing velocity. However, the cross sections involved in attaining the higher charge states emerging from solids here appear to be lower than those required for attainment of the lower charge states obtained in gases at the same velocity. Another interesting feature shown in fig.10 is the sharp drop in population at charge 26+. This is a shell effect and reflects the additional difficulty of removing the first electron from the L shell of bromine due to the step increase in ionization potential.

COLLISIONS OF CHANNELED IONS

As we have seen, it is difficult to assess the effects of single collisions in solid media. Aside from the higher collision frequency which makes it difficult to consider the effects of isolated collisions, the range of impact parameters is restricted, i.e., impact parameters greater than a lattice spacing are clearly forbidden. Thus, that part of the impact parameter dependence of a charge changing collision which extends past this distance is excluded from the physical picture. Channeled ions in crystal axes or planes on the other hand also restricts lower limit of impact parameter. Thus, if we study charge distributions for channeled ions we will observe the results of collisions in a fixed range of relatively large impact parameter. Since these impact parameters are large the ion can only directly interact with valence and/or conduction electrons, all of which have velocities lower than the penetrating ion. In this case for ions with $v > v_0$ we have simulated a high density, almost static electron gas.

A relatively straightforward example of the effect of channeling is seen in fig.11 for 60 MeV iodine ions channeled in Au[10]. First, it can be noted that the equilibrium CSD for channeled ions is lower here than for random directed particles. This is due to the suppression of multiple ionizing collisions which are small impact parameter events. Second, it can be seen that equilibrium for channeled ions does not take place until the target thickness is $\sim$ 1000 Å because of the reduced cross sections for both capture and loss

136

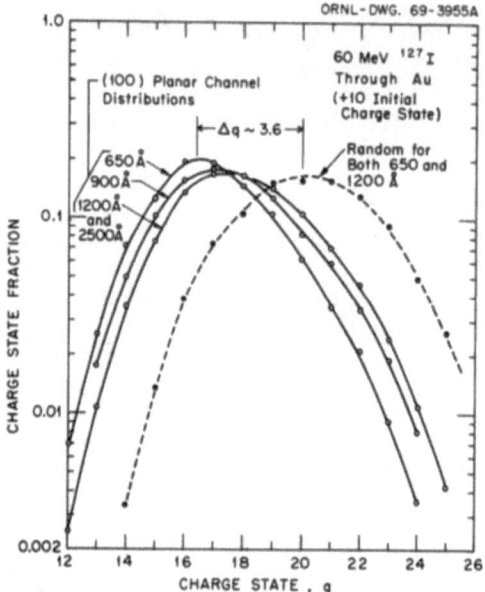

Fig.11 : CSDs of 60 MeV I ions channeled in Au as a function of
path length and the random CSD.

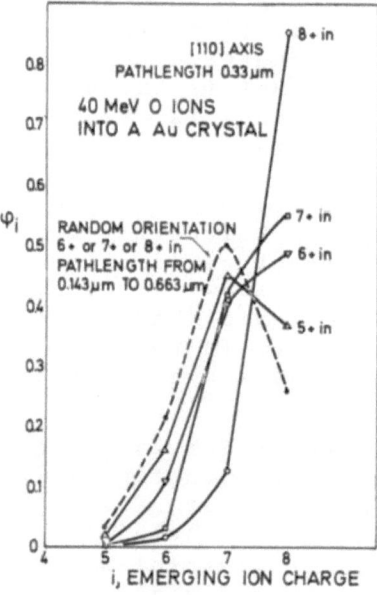

Fig.12 : CSDs obtained for 40 MeV O ions channeled in the [110]
axis of an Au crystal (0.33μm) as a function input charge state and the random CSD

events. The suppression of the loss cross section occurs because these ions are traveling at velocities
equivalent to 250 eV electrons and interact most directly with electrons of only $\sim$ 10 eV such as those
found in light gas targets.

The effect is more startling for higher velocity,
almost totally stripped ions as can be seen in fig.12[11]
Here we show the emergent CSDs for 40 MeV oxygen ions
channeled in the [110] axis of an Au crystal 0.33 μm
(3300 Å) thick as a function of the input charge state
of the oxygen ion together with the CSD obtained from
injection in a "random" direction. The random distribution (dashed curve) is representative of charge-state equilibrium since it is independent of a pathlength
(0.143 to 0.663 μm) in the crystal and independent of
input charge state (6+, 7+ or 8+). The channeled-ion
CSDs are clearly non-equilibrium cases. Since the 8+
fraction is enhanced for all initial charge states, it
is immediately obvious that the equilibrium CSD for
channeled ions will contain a higher fraction of higher
charge states than the random case. Moreover, it is
clear that some of the charge-changing cross sections
for the channeled ions in attaining equilibrium must
be quite small, i.e., if we assume that equilibrium is
attained in <u>ca</u>. 10 mean-free paths for charge exchange,
then charge-changing cross sections of $<< \sim 5 \times 10^{-18}$
cm^2 must be involved if equilibrium is not achieved in
0.5 μm with target densities of $\sim 5 \times 10^{22}$ atoms/cm^3.

These non-equilibrium effects persist at lower
energies and are observed in planar as well as axial
directions. In fig.13 data obtained for 28-MeV O^{8+} beams in the <110> axial channel, the {111} and {100}
planar channels are compared with the random CSD. The
tendency toward higher charge states diminishes with
decreased channel dimension and approaches the random
distribution.

Since the emergent CSDs for channeled ions are
clearly far removed from equilibrium, it should be possible to adduce the relevant charge-changing cross sections from variations in CSD with change in pathlength
and, in some cases, from the additional information
obtained from different input charge states. In the
case of axial channels, a wide range of pathlengths is
difficult since each point requires a different crystal.
For planar channels, a wide range of pathlengths is a-
vailable within a single crystal by simply tilting the
plane of the crystal with respect to the beam direction.
The points shown in fig.14 were obtained for channeling

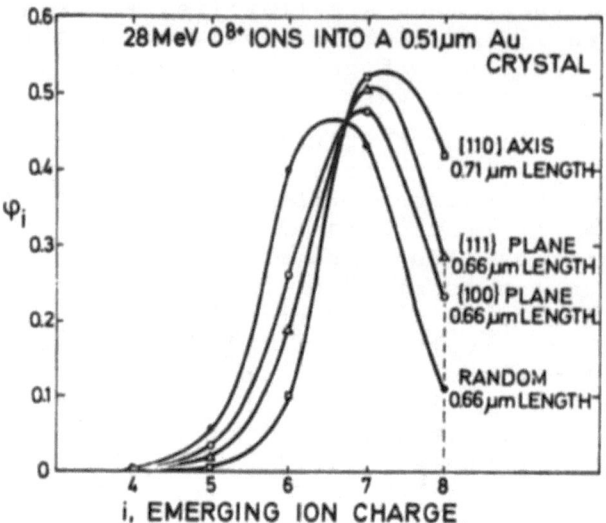

Fig.13 : CSDs for 28 MeV O ions in Au channels

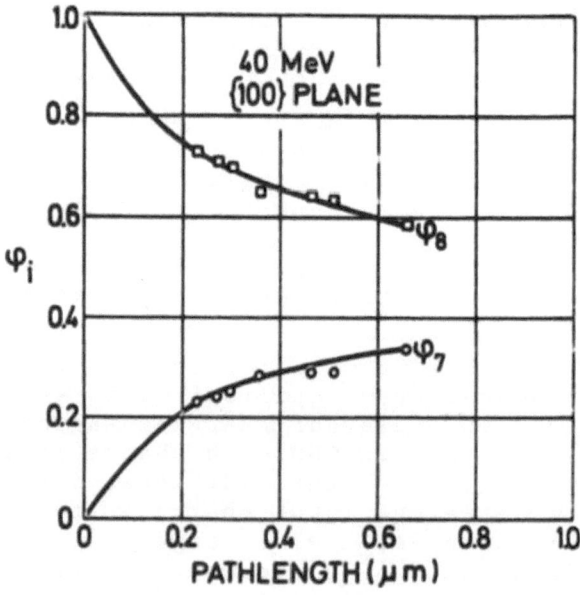

Fig.14 : Charge fraction of 8+ and 7+ emerging from a 100 plane in Au as a function of pathlength. The injected beam is 40 MeV O^{8+}.

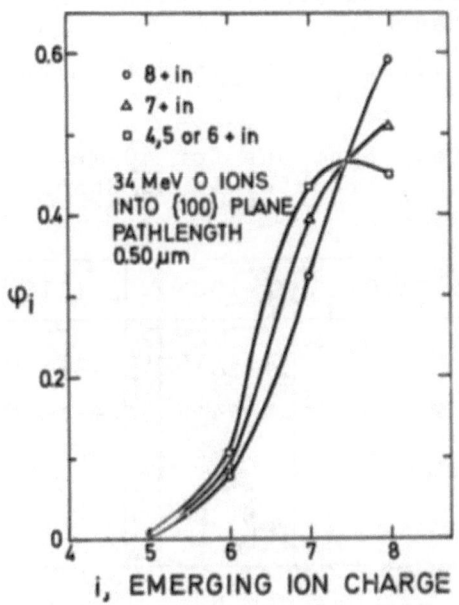

Fig.15 : Emergent CSDs for 34 MeV O injected into (100) planar Au
channels for different input charges.

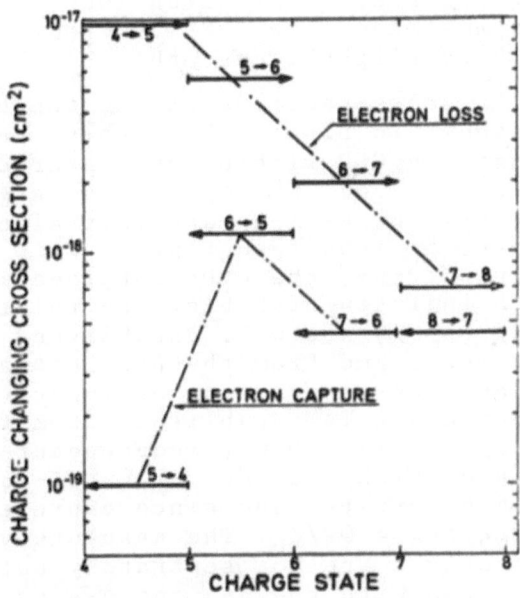

Fig.16 : Derived capture and loss cross sections for 34 MeV O in
100 planar channels in Au.

of 40-MeV O^{8+} ions in varying pathlengths along a {100} planar channel.

The cross sections obtained from these measurements are shown in Table I.

Table I. Cross Sections for 40-MeV Oxygen Ion Electron Transfer in Au Channels

Direction	<110>	<100>	{111}	{100}
σ_{87} (cm^2)	1.5×10^{-19}	1.9×10^{-19}	2.4×10^{-19}	2.5×10^{-19}
σ_{78} (cm^2)	3.4×10^{-19}	6.0×10^{-19}	2.9×10^{-19}	4.2×10^{-19}
$\sigma_{78} / \sigma_{87}$	2.3	3.2	1.2	1.7
Number of thicknesses	3	3	2	8

Data taken with 34-MeV O ions with entrance charges ranging from 4+ to 8+ for pathlengths ranging from 0.45 to 0.58 μm gave the rather remarkable result shown in fig.15, i.e., although entrance charges 7+ and 8+ gave distinctly different distributions, charges 4, 5, and 6+ gave almost identical CSDs. Moreover, the CSDs changed only very slightly over the thickness range.

The set of cross sections giving the best fit with the data are shown in fig.16. From this, the relatively small changes in CSD within the measured thickness range can be understood. The σ_{45} is so large and σ_{54} so small that all the 4+ ions are lost almost immediately. Similarly, 5+ ions feed rapidly into the 6+ state so that with 0.2 μm, the CSDs obtained with 4, 5, or 6+ would be indistinguishable. The relation and magnitude of σ_{78} to σ_{87} are well established from the 40-MeV measurements and from the preponderance of 8+ in the distributions independent of input charge state. The ratio σ_{56}/σ_{65} is established from the observation that $\psi_5/\psi_6 = 0.17 \pm 0.01$, independant of incident charge state and thickness. Hence, quasi-equilibrium between 5+ and 6+ exists, and since capture to 4+ is very small, $\sigma_{65}/\sigma_{56} \simeq \psi_5/\psi_6$. The absolute values of these cross sections are not separately established, but in combination with the σ_{78} and σ_{87} cross sections, they must meet the boundary condition that 6+ to 7+ transfer is slow. This condition also precludes a monotonically increasing capture and decreasing loss cross section.

The data clearly indicate that oxygen ions at energy from 10 to 40 MeV must travel much greater distances for charge-state equilibrium in the channels of gold crystals than in ordinary materials. They also indicate a correlation of nonequilibrium behaviour with amplitude of oscillation in a channel for the measurements at 10 and 17.9 MeV, when groups of different amplitude could be resolved. These observations imply that the charge-transfer cross sections are generally smaller in channels. The processes of electron capture and loss are both theoretically expected to decrease in a channel since capture depends principally on overlap of the ion wavefunctions with those of atoms in channel walls, and loss depends on hard collisions of ionic electrons with the electrons and nuclei of atoms located principally in the channel walls.

The principal cause of the higher charge states for oxygen channeled in gold is apparently the much reduced capture cross section from charges 7 and 8. The oxygen ion has a marked shell structure with an energy gap of $\sim$ 500 eV between the K and L shells, and the decrease in capture cross sections appears to be associated with the transition between shells. At 40 MeV, the oxygen ion has a velocity equal to that of a 1250-eV electron. The probability that an electron is captured from the medium depends strongly on the velocity of the electron to be captured, i.e., capture probability is highest for those electrons moving in orbits on the target atom at the same velocity as they will find themselves in the newly bound state. Many electrons in the Au target atom meet this criterion, but channeled ions are constrained to move at large impact parameters with respect to the atoms in the channel wall and do not come close enough to capture these electrons. The kinetic energy of the $6s^1$ and $5d^{10}$ electrons is <q9 eV, and these are the only electrons to be found in the central $\sim$ 40 percent of the channel. The channel half-spacing is 1.05 Å, and the next lower-lying electrons in Au $5p_{3/2}$, and $4f_{1/2}$ are bound with $\sim$ 250, 340 and 460 eV and have mean radii of 0.60, 0.55, and 0.28 Å, respectively.

An interesting analogy can be made between the channeled and random charge-exchange cross sections for 40-MeV O ions in He and in Ar, which are listed in Table II. In the case of He, there are no electrons with sufficient orbital velocity to meet the matching criterion, and the capture cross section is a factor of

100 lower than that in Ar, where such electrons are available for transfer. (The observed factor of $\sim$ 10

Table II. Cross Sections for 40-MeV Oxygen Ion Electron Transfer in He, Ar, and Au {100} channels

	He (gas)	Ar (gas)	Au {100}	Au random
σ_{78} (cm^2)	2.5×10^{-19}	3×10^{-18}	6.0×10^{-19}	-
σ_{87} (cm^2)	9×10^{-20}	1.5×10^{-17}	1.9×10^{-19}	-
$\sigma_{78}/\sigma_{87} =$ ψ_8/ψ_7	2.8	0.2	3.2	0.5

decrease in loss cross section for O in He compared to Ar is about that anticipated from the considerations of Bohr and Lindhard, i.e.,

$$\sigma_{1Ar}/\sigma_{1He} = Z_{Ar}^{1/3}(v/v_o)Z_{He})^2 \qquad (13)$$

where v is the ion velocity). Collisions in He may thus be compared to encounters of the channeled ion with the loosely bound electron gas in the center channel. Collisions of randomly penetrating ions are more comparable with those in Ar. (The ratio σ_{78}/σ_{87} for random penetration is given by the charge-fraction ratio ψ_8/ψ_7 at equilibrium). The higher fraction of 8+ in Au as compared with Ar is anticipated from the "condensation effect" due to the additional ionization from excited states in rapid sequential collisions.

Changes in equilibrium CSDs have also been noted for protons and helium ions[12,13] emerging from channels and it has been recently observed that the periodic electric field experienced by a channeled He ion can excite a specific state of He$^+$ when the frequency of the field variation is resonant with the frequency of the transition[14]. These results imply integrity of the electron nucleus system when passing through the crystal in apparent contradiction to the suppositions of Brandt and Sizman[15] that these states for these ions are not well defined within the crystal but instead are established at the surface.

CHARGE STATES AND STOPPING POWER

Before proceeding further I think it might be well to consider an apparent paradox in the relationship of energy loss to charge state.

Electronic stopping involves the interaction of an electrically charged moving body with the electrons of the atoms in the medium. It follows that the stopping power should be charge dependent. It is observed that heavy ions emerging from solids have much higher charge states than those emerging from gas targets. It is, however, also observed that the stopping powers in solid media are essentially identical to those in gaseous media and that the "effective charge" derived from solid stopping powers is that observed for the ions in a gas.

For example, in order to obtain a classification of stopping powers for purposes of interpolations with respect to energy and ion species, a description of stopping in terms of effective charge has been proposed. The procedure is to relate the electronic stopping of a given heavy ion to that of the proton at the same velocity through the use of a parameter γ :

$$(\frac{dE}{dx})_{Z,A,E} / (\frac{dE}{dx})_{p,E/A} = \frac{\gamma^2 Z^2}{\gamma^2_p} \qquad (14)$$

where
$$Z_{eff} = \gamma Z \qquad (15)$$

and Z_{eff} is the effective charge for stopping. These methods have been used to classify the data taken in Oak Ridge for Br, I, and U stopping powers[16]. Over most of the energy range of interest here $\gamma_p = 1$. At the lower energies an empirical function due to Booth and Grant[17] is used.

$$\gamma^2_p = (1-e^{-150E_p}) \exp(-0.835e^{-14.5E_p}) \qquad (16)$$

where E_p is proton energy in MeV. Values of proton stopping power were taken from the tabulations of Northcliffe and Schilling[18]. More than 500 data points were used to obtain a least-squares fit to the function shown as a solid line in fig.17. Although the individual points cannot be identified in this figure size, the conclusions are the same for each of the three ions. The analytical function which was derived is

$$\gamma = 1-1.034 \exp[-(v/v_0)Z^{0.688}] \qquad (17)$$

where $v_0 = 2.19 \times 10^8$ cm/sec and v is the

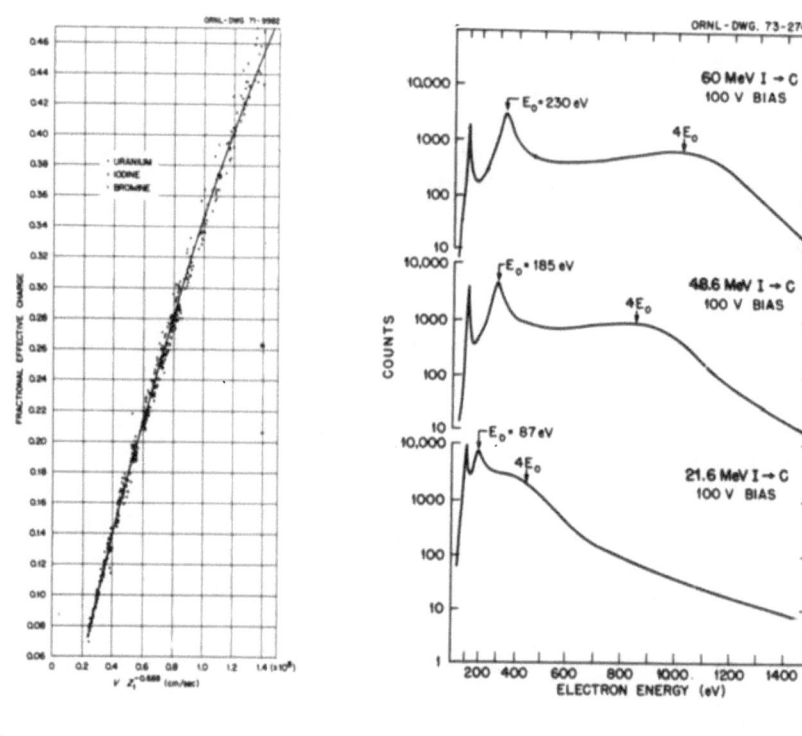

Fig. 17   Fig. 18

Fig.17 : Fractional effective charge versus $vZ^{-0.688}$ (cm/sec) for
Br, I and U ions over a wide range of velocities in several different solid targets.

Fig.18 : Energy distributions of electrons ejected in the forward
direction by I ions (60, 48.6 and 21 MeV) passing through
thin C.

ion velocity in cm/sec. If either number (1.034 or 0.688) is varied by as much as 0.002, then no value of the other number will produce an equally good fit to the data. The data for U, I, and Br, treated separately, produced the same numbers as given in Eq.(17). The function derived in Eq.(17) can be used to obtain estimates of electronic stopping power for heavy ions with Z > 35. To these values a small contribution due to nuclear stopping should be added as derived from Lindhard et al[19] to obtain the total stopping power.

It is interesting to observe that the values of effective charge for stopping happen to agree so closely with the average ionic charge emerging from gas targets for Br and I ions, and that the exponent lies so close to the 0.666 predicted from the Thomas Fermi model.

Two solutions to this disparity between charge state stopping power in solids have been suggested.The first suggests that screening by polarized electrons in the solid tends to neutralize the higher charge state in the solid. The second based on a recent agreement by Betz and Grodzins[20] suggests that the actual charge state in the solid is essentially identical to that in the gas and hence the same stopping power. They then state that many electrons are present in highly excited states and, following its emergence from the solid, the excitation is relaxed through an Auger shower which results in the higher observed charge state. However, if the electrons are sufficiently excited it might be difficult to tell the difference between them and unbound but accompanying screening electrons.

In a recent experiment we measured the energy distribution of electrons emerging in the foward direction[21]. Energy spectra of forward emitted electrons from 21.6-60 MeV I ions impinging on various targets (C, Au and Air) have been measured. Fig.18 shows the spectra obtained from carbon foil targets (the electrons have been accelerated by 100 V). The observed spectra can be qualitatively understood as follows : 1. The peak at zero corresponds to electrons knocked off the target atom by relatively gentle collisions.2. The peak at E_0 corresponds to electrons moving with zero velocity in the rest frame of the moving ion. These can come about either by capture of electrons into continuum states or by light ionizing collisions of the emergent ion. 3. The broad peak centered at $4E_0$ arises from electrons which have experienced a head-on collision with the ion (δ rays). The value $4E_0$ is obtained for electrons which are essentially at rest in the laboratory frame while the higher and lower values are caused by orbital motion towards and away from the moving ion. The spectra obtained with gold targets were similar but showed a higher zero energy peak and a broader hump which extended to somewhat higher energies. These differences may be understood from the higher electron density in Au and the availability of more electrons with higher orbital velocities. A spectrum obtained by scattering through diffuse air target ($\sim 10^{-4}$ torr) showed essentially the same features. The additional Auger electrons from ions emerging from solids anticipated by the Betz Grodzins theory are not observed.

INNER SHELL IONIZATION

With all of the evidence and controversy discussed above it is becoming apparent that we do not know much about the states of ions while they are moving in solids. There are those who would say with some justification that no state can be defined in such a medium and that all such studies are academic. However, there are some recent studies on inner-shell ionization which have begun to shed a more quantitative light on this question. In order to understand these it will be necessary to consider the physics of inner-shell ionization briefly. This subject in itself is not a digression since inner-shell ionization becomes the dominant ionization process for ions at high energy, and electron capture from inner shells of target atoms is the principal capture process at high energies.

Consider first inner-shell ionization which occurs by the collision of fast light particles in which the Coulomb interaction between particles and electrons is predominant and provides the perturbation necessary to cause excitation of the inner bound electrons to outer states. Clearly this model is only strictly valid if the probability for ionization is small enough and the speed large enough. This condition is met if the scattered wave emitted in the collision between particle of atomic number Z with velocity v and the atomic electron have small amplitude ; i.e., if $Ze^2/h \ll 1$. The theory for this type of ionization has been worked out using the "Plane Wave Born Approximation" (PWBA) by Mertzbacher and Lewis and we will only quote the result. The cross section for ionization of s subshell is given by

$$\sigma_s = \frac{8}{Z^4_s} \frac{Z}{\eta_s} a_0^2 \int_{W_0}^{\infty} dW \int_{Q_0}^{\infty} \frac{dQ}{Q^2} |FW_s(Q)|^2 \qquad (18)$$

In Eq.(18) dimensionless quantities have been used for the incident particle energy,

$$\eta_s = \frac{1}{Z_s^2}[\hbar v/e^2]^2 = (me/M)(E/Z_s^2 R_\infty) \qquad (19)$$

the energy transfer,

$$W = \frac{\varepsilon}{Z_s^2 R_\infty} \qquad (20)$$

and the square of the momentum transfer,

$$Q = \frac{a_o^2 q^2}{Z_s^2 R_\infty} \qquad (21)$$

m and M are the electron and projectile masses, a_o is the Bohr radius of hydrogen, and Z_s is the effective Z of the atomic nucleus, corrected for screening by the inner-shell electrons. The minimum value of Q is then

$$Q_0 = [\frac{a_o q_o}{Z_s}]^2 = \frac{W^2}{4\eta_s} \qquad (22)$$

For ionization, the minimum energy transfer W_0 is given by the binding energy of the s-electron as

$$W_0 = \theta_s/s^2 \qquad (23)$$

where s in the denominator is s - 1,2,3 for the K-,L-, M-shells respectively. The screening parameter θ_s is determined as the ratio between the measured binding energy I_s and the hydrogenic value.

$$\theta_s = I_s s^2/Z_s R_\infty \qquad (24)$$

Since for large values of Q_0 the lowest momentum transfers make the largest contributions to the cross section, it is apparent from equations (18) and (22) that at low incident energies the cross section obeys an approximate scaling law :

$$\theta_s Z_s^4 \sigma_s = f[\frac{\eta_s}{\theta_s^2}] \qquad (25)$$

For K-shell ionization with protons a great mass of data exists at energies ranging from 100 keV to 160 MeV on many targets. The results are in fact in

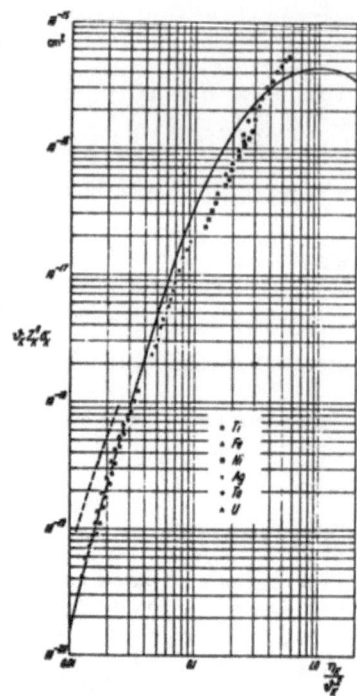

<u>Fig.19</u> : K shell ionization cross section versus proton energy
(both in reduced units). The solid curve is PWBA calculation and points indicate data from various targets.

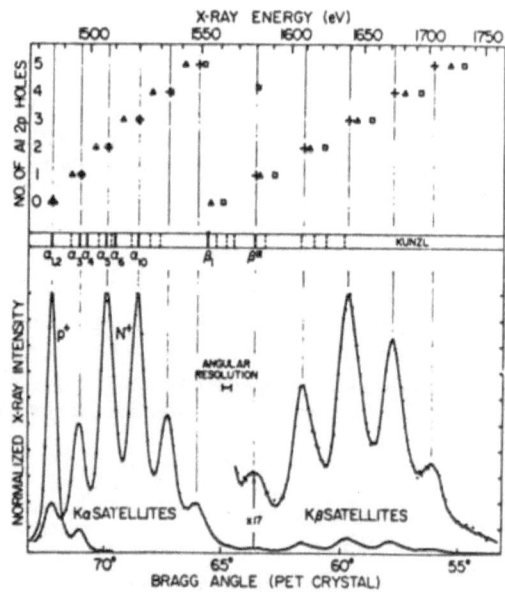

<u>Fig.20</u> : Al K-x-ray spectra induced by proton and 14 MeV N ion
bombardment showing increase in multiple L ionization.

quite good agreement with the predictions. In fig.19 a plot of some of the older data in terms of the reduced units described above is compared with theory. Newer data and some modifications in theory to take into account non-linear paths and polarization and binding effects have further reduced the discrepancies. Qualitatively speaking the cross section for Coulomb ionization first increases with velocity ($\sigma \underset{\alpha}{\sim} E^4$ at the lower velocities) reaches a maximum at a velocity about equal to the orbital electron velocity and then decreases. Thus as the proton velocity increases maxima in the ionization cross sections for the outer shell occurs first and then the next inner shell, etc. The sum of these various shell ionization cross sections weighted by the energy transfer involved in a given ionization enters directly into the stopping power.

As the Z of the ion increases the ionization cross section naturally increases as Z^2 but other effects also come into play. Recent work with heavier ions, for example, carbon, oxygen, and nitrogen, in the MeV per nucleon region, carried out in a number of laboratories has shown that, in addition to normal Coulomb ionization of, for example, K electrons in a target, multiple ionization in adjacent shells (L, M, ...) occurs during the same collision. Using x-ray detectors with resolution about 200 eV (lithium-drifted silicon solid-state detectors), these multiple ionization events are observed as a shift of the x-rays to higher energies. High-resolution studies using crystal spectrometers have confirmed this notion. In fig.20 the normal K line from aluminum produced by proton bombardment is compared with the spectrum observed with 15-MeV nitrogen-ion bombardment[23]. In the latter case a whole set of lines is produced in the Kα region. Each of the lines in this complex set comes from transitions which take place while the excited atom has, in addition to the K vacancy, L vacancies ; for example, line n°2 has a single L vacancy, n°3 a double L vacancy, etc. Thus the shift in the low-resolution measurements of the K x-ray lines can be used to estimate the degree of simultaneous excitation in the L shell. Although a great body of experimental data on this subject has come into existence in the last two years, the theory is not well understood. A recent experiment by Van der Woude and Saltmarsh[24] at ORNL has shed some light on this phenomenon. In this work, they found that the energy shifts obtained by bombardment with 1 to 10 MeV per nucleon oxygen and carbon ion bombardment could be correlated with the stopping powers for these ions in the media in question. This can be understood since the stopping powers are also directly related to the amount of ionization taking place within the media.

Even more recent work by Sellin and co-workers at ORNL has revealed a dramatic effect of particle charge state on Coulomb ionization cross section. An example is shown in fig.21 where the cross section for Ne K x-ray production created in a single collision with 80 MeV Ar ions is plotted as a function of the Ar ion charge state[25]. It is seen to vary exponentially and increase by a factor of $\sim$ 60 with increasing charge state. (Other causes for the increase in x-ray yield such as increasing fluorescence yield have been discounted by other experiments). One might well ask what the cross section is in solid Ne and would this result indicate the charge state of the Ar ion <u>within</u> the solid. Experiments of this nature are presently underway.

A recently reported result by Brandt et al[26] bears on this conjecture. They observed a dependence of the Al K x-ray cross section in solid Al targets for thicknesses ranging from 0.5 to 25 $\mu g/cm^2$ on the charge state of incident oxygen ions, in the energy range 12-68 MeV. These experiments were performed under conditions where Coulomb excitation occurs over large impact parameters compared to the Al K shell radius. Their results are summarized in fig.22. The relative Al K x-ray production yields $R(E_1)$ measured in thin-target transmission experiments are shown as a function of the target thickness D for different incident oxygen charge states n. The data converge toward thickness-independent limiting values following penetration through some 10 $\mu g/cm^2$, corresponding to several hundred atomic layers.

They account for all the observed data, fig.22, in terms of only two parameters which have values very close to theoretical estimates, namely, a dynamic screening rate λ and an electron-density enhancement constant γ. They assume that the target electrons form a Fermi gas ; that, in contrast to the situation in dilute gases, screening and collision broadening by the electron gas in a solid make the distinction unimportant between the screening of the projectile charge by bound states undergoing electron capture and loss processes, and the screening by the dynamic polarization of the target electron gas ; and that a statistical linear-response approach suffices to characterize the dynamic response of the Fermi gas to the moving projectile. In this frame of reference, the appearance of the

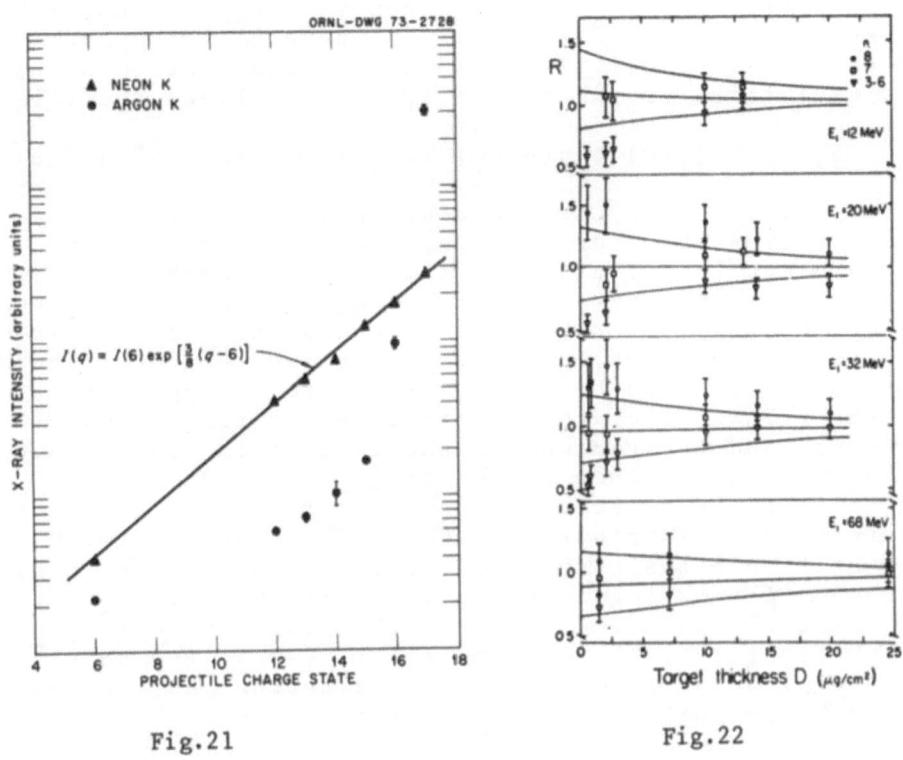

Fig.21   Fig.22

Fig.21 : Intensities of Ne K-x-rays and Ar K-x rays as a function
        of Ar charge state for single collisions of 80 MeV Ar on
        Ne.

Fig.22 : Ratio of Al K-x-ray yields obtained with oxygen ions 8+,
        7+ and 3-6+ to the thick target yield as a function of
        Al film thickness.

projectile in the target at x = 0 creates a charge imbalance at time t = 0. As the particle penetrates into the solid, the particle-target system approaches steady-state conditions at a rate λ. In linear response theory, λ is related to the inverse relaxation time of the electron gas and can be given in terms of the plasma frequency ω_p as $\lambda = (\beta/2\pi)\omega_p$, where β is a number in the order of 0.1. The approach to equilibrium along the trajectory of the particle, then, follows the form

$$q_1(x) = q_1(0) \exp(-\frac{\lambda x}{v_1}) + q_1(\infty) [1-\exp(-\frac{\lambda x}{v_1})] \quad (26)$$

where $q_1(0) = n$ is the incident charge state, $q_1(\infty)$ the steady-state charge state, and $v_1 = (2E_1/M_1)^{1/2}$ the projectile velocity. This linear approach neglects asymmetries in λ between $q_1(0) > q_1(\infty)$ and $q_1(0) < q_1(\infty)$. Moreover, they observe no differences between R measured for n - 3,4,5,6, presumably for two reasons : The outermost, loosely bound electrons are stripped off at depths much smaller than their thinnest target (0.5 $\mu g/cm^2$), and the residual L-shell electron density distribution is screened by the target electron gas. As the projectile penetrates to depths $x > v_1/\lambda$, a steady-state charge distribution establishes itself through a competition between the rate of perturbation set up by the particle moving in the solid with velocity v_1 and the rate of enhancement $\sim \lambda$ of the electron density near the particle. They estimate for $v_1 > v_F$ (v_F being the Fermi velocity of the target electron gas, on the order of the Bohr velocity v_0) that

$$q_1(\infty,E_1) = Z_1(1-\gamma v_F/v_1), \quad v_1 > v_F \quad (27)$$

where γ is a constant of the order of 1.

The authors then calculate the ratio R for the experimental conditions by fitting the two parameters of the theory. This yields values that are consistent with their approach : $\lambda = 5.5 \times 10^{14}$ sec^{-1} and $\gamma = 1$. Setting $\lambda = (\beta/2\pi)\omega_p$, where $\hbar\omega_p(Al) = 15$ eV, they obtain $\beta = 0.15$. Since $\hbar\omega_p$ is nearly the same for most solids, the equilibration rate should be insensitive to the target material. It follows that the equilibrium projectile charge state defined in this way is reached in solid targets of thickness $D \gg v_1/\lambda \simeq 4 \times 10^2 \, v_1/v_0$ Å, where

$$v_1/v_0 = [40E_1(MeV)/M_1(amu)]^{1/2} \quad (27a)$$

This concept of charge state equilibrium in terms of screening by Fermi electrons is interesting. But ex-

periments just completed by our group in Oak Ridge are
in apparent contradiction. We measured the energy loss
of 28 and 40 MeV O ions in channels of Ag as a function
of entrance and exit charge state. The charge exchange
behaviour in Ag is similar to that which we reported
in Au i.e., very low cross sections for charge exchange at high charge states. The effects of screening by
Fermi electrons should not depend upon wether an ion is
channeled or not. Yet we find that for those ions which
have not changed their charge state the energy loss is
directly proportional to the square of the ion charge
i.e. $s = kq^2$. The crystal used in this work was 0.8 µm
thick and the characteristic thickness prescribed by
eq.(27a) is .04 µm yet no indication of departure from
the $s=kq^2$ relationship was observed. This finding of
course also bears on the screening argument as it is
applied to explanation of the similarities of stopping
power in solid and gaseous media where the exit charge
states are vastly different.

We must now turn our attention to another mode of
inner shell ionization which in fact predominates for
heavy ions at low velocities.

To do this let us first consider the results of
Everhart and his coworkers and Federenko and coworkers
of about ten years ago. They had noted that for large
angle scattering of Ar ions from Ar at energies as low
as 10 keV that multiple ionization was predominant.
This, in spite of the fact that cross sections for even
single ionization by pure Coulombic interations should
be immeasurably small. Everhart et al[27] were able to
correlate the degree of ionization with the distance
of closest approach in a collision. This is arrived at
by measuring the mean charge of ions scattered into a
given angle at a given velocity. The distance of closest approach is obtained from a knowledge of the repulsive potential and scattering angle. A plot of their
results is shown in Fig.23. Here it can be seen that
there is a step function increasing from a mean charge
of ∿ 1.5 to ∿ 3.5 at a distance of ∿ 0.2 Å. A similar
experiment in which inelastic energy loss was measured
showed similar results (fig.24). Here we see that in
this region of impact parameter three different energy
losses are observed. The distance at which the onset
of these processes occures is the distance at which th
L shells of the Au-atoms begin to overlap. From these
data it was adduced that the multiple ionization was

154

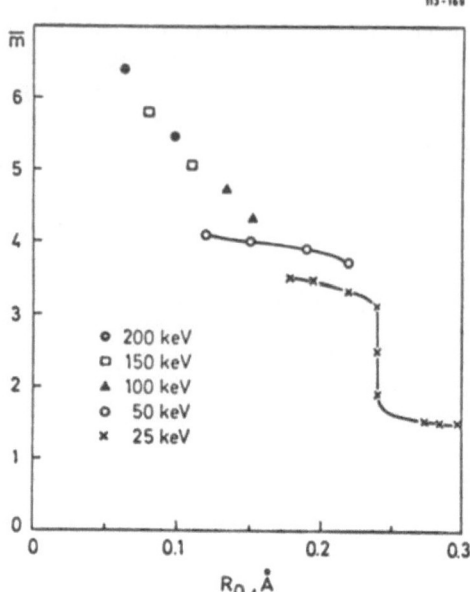

Fig.23 : Most probable charge state of Ar ions emerging from a
single collission with Ar target atoms as a function of
the distance of closest approach.

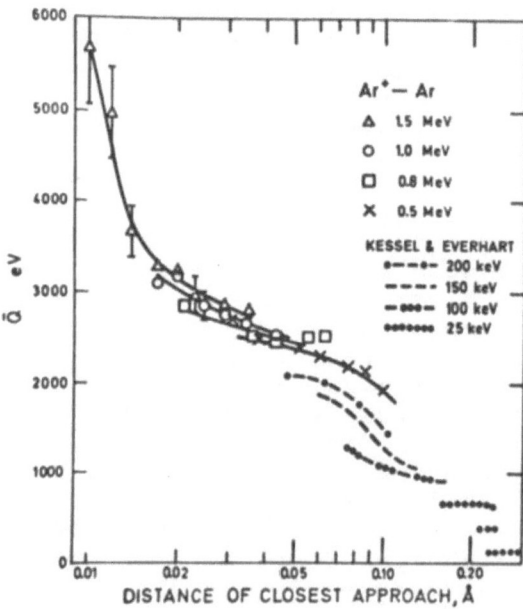

Fig.24 : Energy loss of Ar ions emerging from a single collision
with Ar target atoms as a function of distance of closest approach.

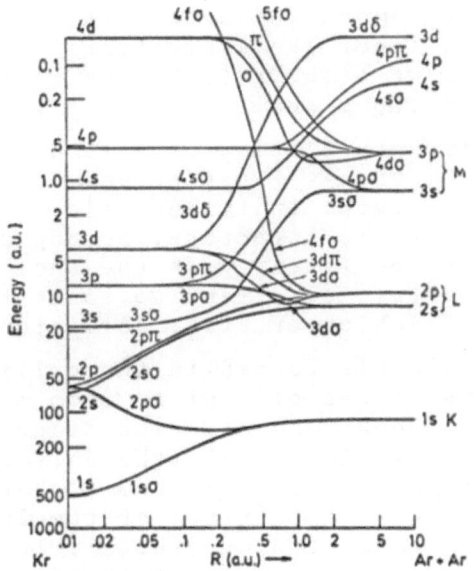

Fig.25 : Diabatic energy level diagram for the Ar-Ar system as a function of internuclear distance

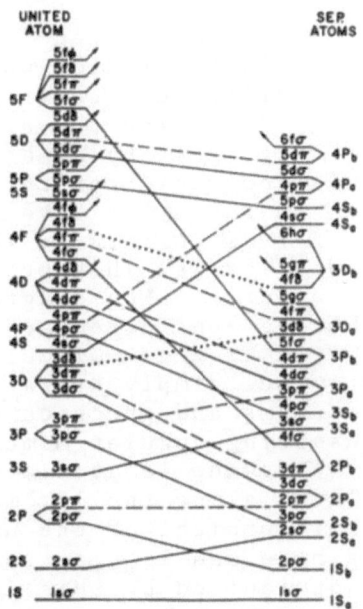

Fig.26 : Diabatic correlation diagram for two disimilar atoms ($Z_a > Z_b$).

not due to simple removal of outer shell electrons but to the removal of inner shell electrons. The three e-nergy loss values were shown to be due to the removal of zero, one, and two L shell electrons respectively.

A qualitative explanation of this data was propo-sed by Fano and Lichten[28]. They proposed a model invol-ving transitions between individual electronic orbitals in a transient diatomic structure. Strong coupling of deep lying levels consistent with the exclusion prin-ciple promotes certain electrons to ionization when the ions depart from each other after collision.

The quasi-molecule formed in the collision is des-cribable in terms of the old united atom model for e-lectron orbitals. Such a correlation diagram for the Ar-Ar system is shown in fig.25 (the united atom (i.e. r = 0) in this case is Kr). This shows that when two Ar atoms are brought together two of the 2p electrons are put into the $4f\sigma$ molecular orbital. This orbital is highly repulsive and makes many crossings. The probabi-lity of these electrons returning to their initial or-bitals after a violent collision is practically zero. Hence the observation of an essentially geometric cross section for ionization of two L electrons in such col-lisions.

More recently Barat and Lichten[29] have shown that it is really quite simple to predict which electrons will be ionized in a collision involving even two dis-similar atoms. This is done with the aid of a diabatic correlation diagram (fig. 26). By diabatic we mean here that the collision is too rapid for the electron to change the number of nodes in its wave function, i.e., that the quantity $n-\ell-1$ is conserved. This assumption implies that all levels cross. This greatly simplifies the construction of a correlation diagram. Criteria for non crossing can then be considered separately for each junction of correlation diagram molecular levels. To construct the diagram one simply draws the atomic levels of the two isolated atoms and proceeds to fill elec-trons (properly conserving angular momentum) into the available orbitals. Thus the 1s atomic orbital of the higher Z ion correlates along the $1s\sigma$ molecular orbi-tal with the 1s atomic orbital of the united atom. The-re are no crossings here and 1s electrons of the higher Z ion are clearly not ionized by this process. The 1s electrons of lower Z atom correlate along the $2s\sigma$ or-bital to the 2s of the united atom, etc. When we get to the 2p electrons of the lower Z ion we see that two of them are placed into the 4fσ orbital which leads to ionization. There are clearly many other possibilities here including rotational coupling of non-crossing levels which we will not discuss here. The main point I wish to make is that large cross sections for ionization exist which depend strongly on the individual collision pair.

An example of the size of the cross section can be seen in fig.27 where the results of Der et al[30] are shown for K ionization of C by different ion projectiles. This is compared with the universal curve for Coulomb ionization which is closely followed by protons. The discrepancy with Coulomb ionization is 5 to 6 orders of magnitude. The ion specificity is shown by the data of the Livermore group[31] in fig.28 where the cross section for Cu L x-ray production is shown for a number of different projectiles at the same velocity. Three orders of magnitude variation are observed over the periodic table. The maximum cross sections are observed in the region where the projectile inner-shell binding energies match the binding energy of the Cu L electron (i.e., K, L and M shells of the projectile are matched in binding energy with Cu L level). Fig.29 shows the results for Cu L x-rays when Cu ions are used to bombard targets of different Z. The results are similar to those of fig.28 but not identical. The only difference here is that the Cu is the moving ion in the solid target and hence is in a different state at the time of the collision than when it was the neutral target atom. Here we have a hint that x-ray spectra and yields may indicate the state of the ion while it is moving in a solid medium.

Now let us consider a case where direct information has been obtained. This work was done at ORNL with 22-48 MeV iodine ions in Se, Br_2 and Kr targets[32] These are adjacent elements in the periodic table and the principal difference is that Se is a solid while Br and Kr are gases. From observations of the I L x-ray spectra (energy shift, etc.) it could be determined that the state of the I ion following an L shell ionization event was the same for all targets. This is not unexpected because an atypically violent collision is necessary to create such a vacancy. Nonetheless, the cross section for L shell ionization was shown to be larger by a factor of 10 in Se as compared to Kr as shown in fig.30 which shows absolute yields of x-rays observed with a proportional counter. The principal

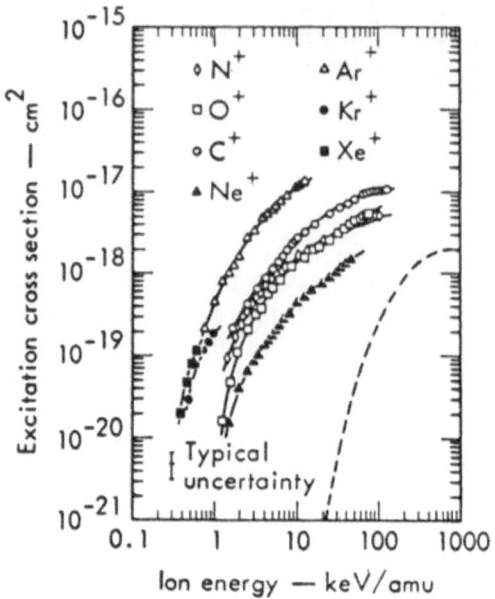

<u>Fig.27</u> : Excitation cross section for C K-x-rays by various ion
projectiles as a function of ion energy. The dashed curve gives the PWBA prediction.

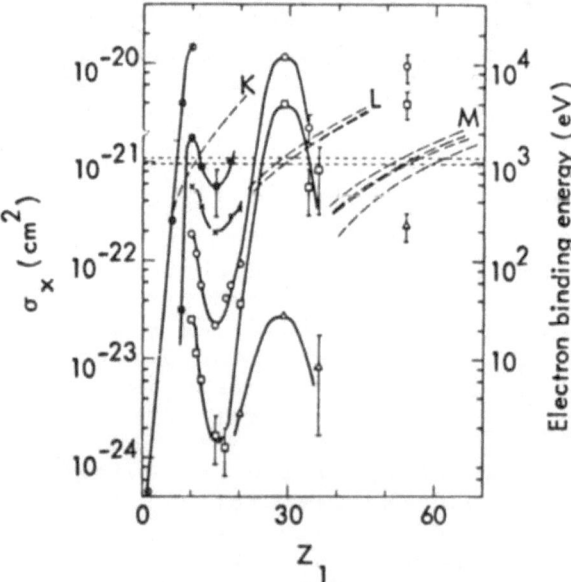

<u>Fig.28</u> : Cu L-x ray production cross section in a thick Cu target
for incident ions of varying Z_1 at a fixed number of keV/amu ranging from 1.0, 3.0, 5.0, 10 and 50 keV/amu. The lowest cross section corresponds to the lowest energy.

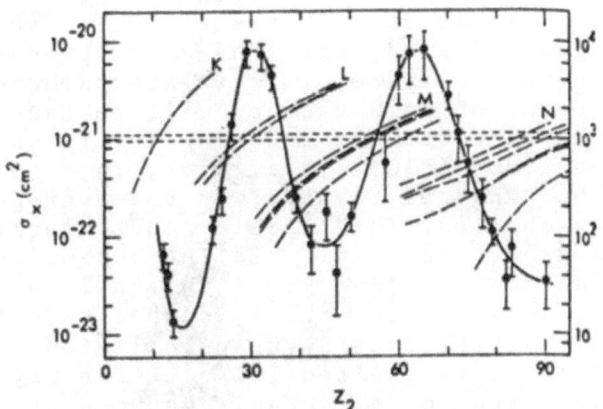

Fig.29 : Cu L-x-ray production cross section for Cu ions at 160
keV (2.5 keV(amu)) in targets of varying Z_2.

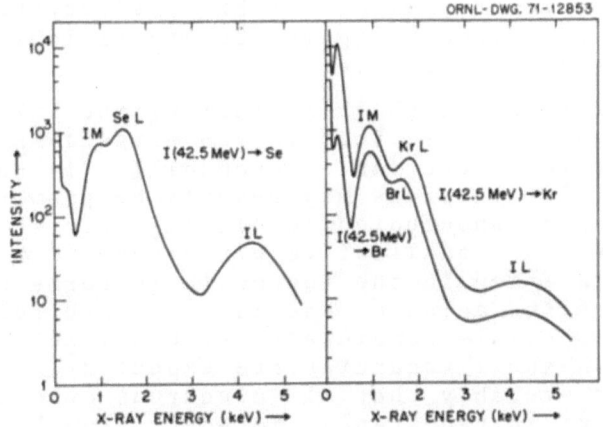

Fig.30 : Proportional counter x-ray spectra from collisions of
42.5 MeV I ions in Se (solid) and Kr and Br_2 (gas)

difference in the x-ray production in solid and gaseous
targets appears to arise from the state of the ion pri-
or to the collision. The high collision frequency in a
solid can lead to considerable steady-state ionic excitation in the M shell of I. For example, in the ion e-
nergy range investigated, the I M-vacancy production
cross section in the system I-Se is approximately 10^{-17}
cm^2. The lifetime of an M vacancy will be about 5×10^{-14}
sec, while the time between exciting collisions in the
solid will be about 5×10^{-15} sec. Thus, in a solid the
I ion would be expected to have, on the average, several M-shell vacancies. This should render possible an
I L-electron promotion, e.g., via the $3d\sigma$ molecular level, to an unoccupied bound state in the I M-shell
(fig.31). In a gas target, the I ion has time to de-
excite M-shell excitations between collisions. In this
case, a collisionally excited I L electron may have to
be transferred high up to a weakly bound state or directly into the continuum, resulting in a small cross
section. The difference in Se and Kr L x-ray yield is
also attributable to a difference in collisional vacancy production, in analogy to the I L-shell excitation.
The promotion of Se 2p electrons to an unoccupied I M
level is possible, e.g., via $4f\sigma$ (fig.31). In agreement
with the observations, the gas-solid effect on the Se
and Kr L excitation yield is expected to be smaller
than in I, since, at somewhat smaller internuclear separations, the Se as well as the Kr 2p electrons can be
excited by rotational coupling of $4f\sigma$ to ionized outer
shells in the I projectile.

In contrast to the L radiation, the I M radiation
yield is less in solid than in gas targets. A difference in collisional excitation probability is not expected since I 3d electrons may readily be promoted to a
large number of unoccupied levels in upper shells. There is, however, a qualitative difference in M and L
vacancy relaxation in the two media in terms of the
structure of the adjacent shells. For both solid and
gas, the O shell is completely stripped. In a gas target 4 to 5 N-shell vacancies are expected[7], while in a
solid it is probably that all electrons outside the M
shell are in excited states. Moreover, in the solid the
ion is always moving through regions of high electron
density. This milieu may tend to enhance decay via less
selective nonradiative transitions and, thus, cause an
effective suppression of the fluorescence yield. This
effect was also observed for I ions in targets of Xe
and Te ; at 22.5 MeV, the collisional M x-ray yield was
found to be about a factor of 6 larger in the system

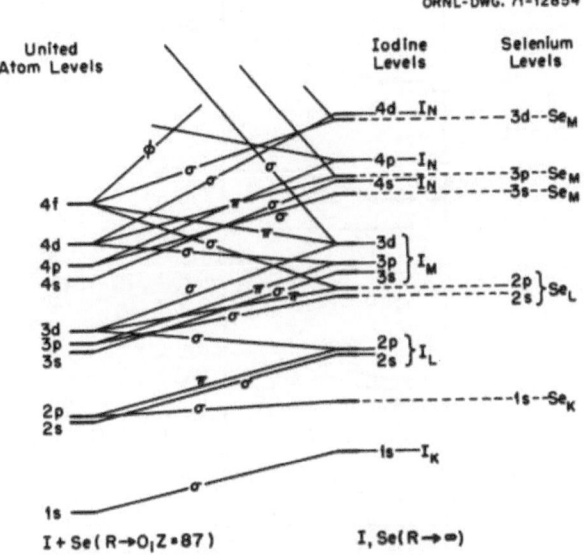

ORNL-DWG. 71-12854

Fig.31 : Diabatic correlation diagram for the I-Se system.

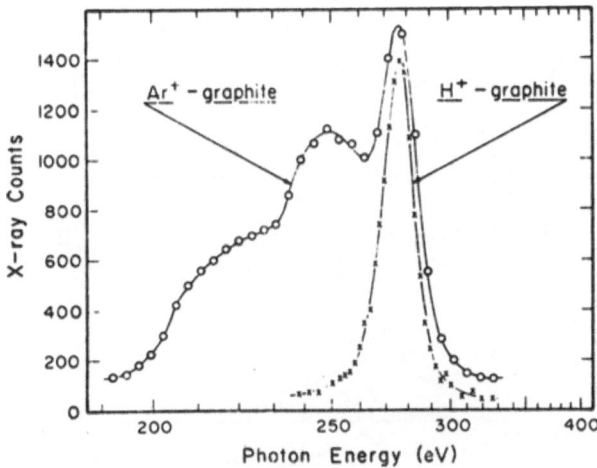

Fig.32 : X-ray spectrum for Ar on C compared with proton induced
C-K x-ray line.

162

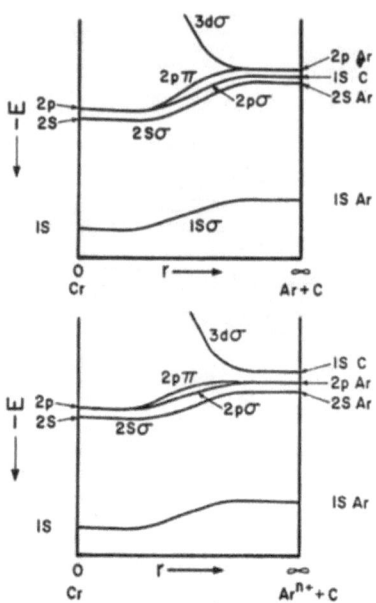

Fig.33 : Correlation diagrams for the Ar-C system. Lower diagram pertains to a highly ionized Ar ion in combination with neutral C atoms.

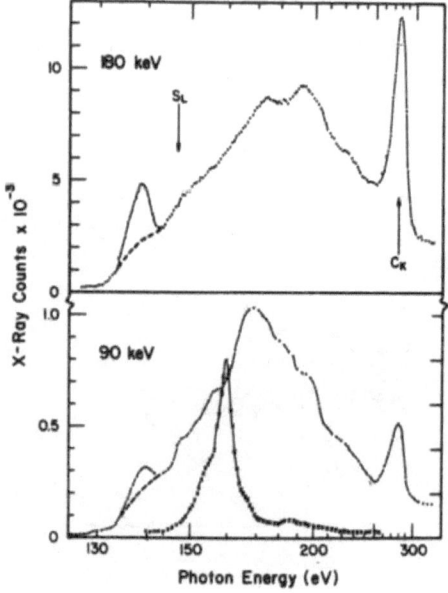

Fig.34 : X-ray spectra for 180 and 90 keV S ions on C (solid) and CH_4 (gas). The sharper spectrum in the bottom portion is the spectrum obtained with CH_4 target.

I-Xe than in the system I-Te. At 48 MeV, this factor
was 4.

Another approach to this problem has been taken
by Der et al[33] who measured x-ray spectra for Ar-C collisions in gas (CH_4) and solid (graphite) targets for
collision energies of 30 and 90 keV. In the gas targets
with either carbon or argon as the target only Ar L
shell vacancies are produced in accordance with the Fa-
no-Lichten model using the atomic levels of Ar and C.
For Ar ions in graphite, however, a high yield of carbon K x-rays is observed (fig.32). The authors attribute the increase in C-K x-ray cross section to a high
degree of ionization in the Ar. If the Ar is highly ionized the binding energies of the remaining electrons
is increased and, as is shown in fig.33, the Ar L electrons become more tightly bound than carbon K electrons.
Hence molecular promotion would act to ionize the C-K
electrons. Another demonstration of this effect has
been made in the $S^+ + C$ system. Shown in fig.34 are the
x-ray spectra observed on bombarding C and CH_4 with 90
and 180 MeV S ions[34]. In the lower portion of the figure, the sharper spectrum is that obtained from CH_4
at 90 keV S bombarding energy. The normal S L-x ray lies
at 148 eV. The sharp peak appears at $\sim$ 160 eV and corresponds to a doubly ionized S atom ($3s^23p^2$) with an L
vacancy. In the graphite target much higher x-ray energies are observed and attributed to even more highly
ionized states. In fact, the x-ray energies above 200
eV imply the presence of 4-5 L shell vacancies at steady-state. (Note the fraction of these ions is not as
large as indicated in the figure because of the sharp
increase in fluorescence yield with increasing x-ray
energy). The line at 280 eV is the C K-x ray. At 180
keV steady state inner shell ionization is even more
pronounced.

Indeed, if one considers the quasi-molecular promotion cross section and the lifetime of these states
it follows that there should be appreaciable steady
state concentration of inner shell vacancies. Evidence
for these was first presented by Saris et al[35] who postulated that anamolously high energy x-rays occurred
because of the relaxation of vacancies during the collision period. For example, consider the 2pπ orbital
in fig.25 from an Ar-Ar collision. If the atoms come
into the collision with a vacancy in the 2p orbital
which passes along the 2pπ orbital and the orbital is
relaxed at essentially zero internuclear distance the
resultant x-ray will be characteristic of Kr and not

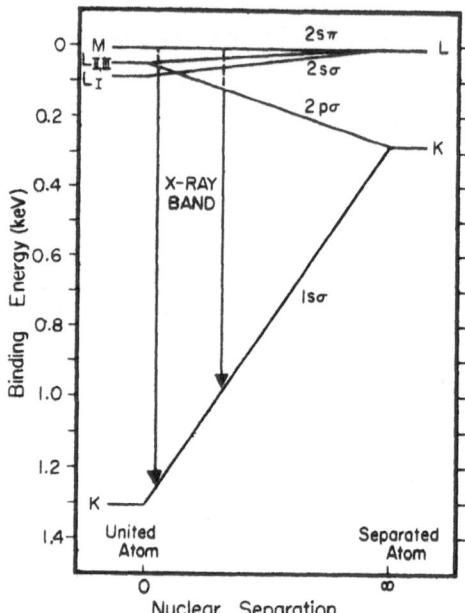

<u>Fig. 35</u> : Correlation diagram for, e.g., the C-C system showing x-ray energy band expected for 1sσ relaxation during a close encounter.

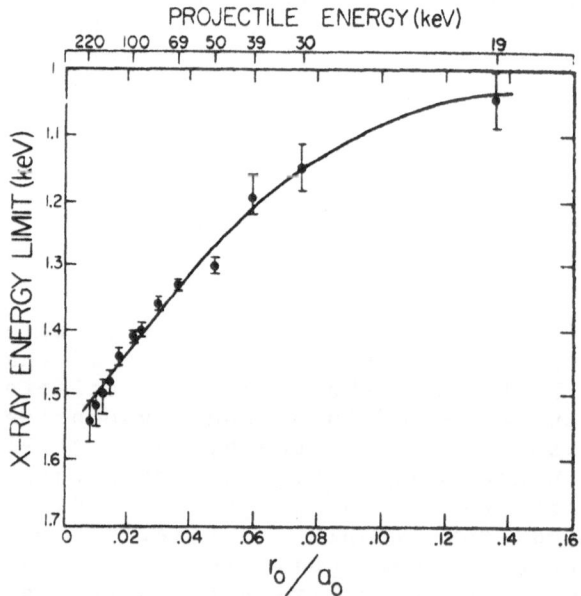

<u>Fig.36</u> : X-ray energy band limit in C-C collisions as a function of minimum distance of closest approach.

Ar. At intermediate distances intermediate energy x-rays will be emitted. Such x-rays were in fact observed as a high energy tail on the Ar x-ray spectrum when Ar collided with Ar imbedded in an Si target. This phenomenon has been further investigated by Macdonald et al[36] who reported the yield and high energy limit of a K x-ray band emitted by the transient C-C system in collisions of 19-24 keV carbon ions in graphite. The transitions in the transient state are pictured in fig. 35. As the collision energy grows, the distance of closest approach grows smaller and the maximum energy of the x-ray band grows larger as is shown in fig. 36. Although the total picture of these events may not really be as clear-cut as seems indicated in the above, it is highly probable that steady state inner shell ionization is present in relatively low velocity ions ($v < v_i$) while penetrating solids.

All of this evidence is convincing and, at the same time, confusing since the equilibrium charge predicted from this is quite high. Yet at this velocity the equilibrium charge could be between zero and one even in the gas phase. This in contrast to Betz and Grodzins'[20] prediction that charge states are low in solids and that higher charges are attained upon leaving the surface. In summary then, it can be stated that much information on the states of energetic ions in solids exists and is presently being obtained by a number of groups. However, a truly clear picture remains to be developed.

REFERENCES

1. Mapleton R.A., "Theory of Charge Exchange" Wiley-Interscience, 1972 .

2. Bohr N. and Lindhard J., Kgl.Dansk.Vid, Mat.-Fys. Medd. 28, n°7, 1954 .

3. Betz H.D., Rev.Mod.Phys., 44, 465, 1972

4. Bell G.I., Phys.Rev., 90, 548, 1953

5. Bethe H. and Saltpeter E.E., in "Quantum Mechanics of One and Two Electron Systems" Springer Verlag Berlin, 1957 , pp.320-322

6. Datz S., Lutz H.O., Bridwell L.B., Moak C.D., Betz H.D., and Ellsworth L.D., Phys.Rev., A2, 430, 1970

7. Moak C.D., Lutz H.O., Bridwell L.B., Northcliffe L.C. and Datz S., Phys.Rev. 176, 427, 1968

8. Lassen N.O., Kgl.Dansk.Vid., Mat.-Fys. Medd. 26, n°5 and n°12, 1951

9. Moak C.D., Lutz H.O., Bridwell L.B., Northcliffe L.C. and Datz S., Phys.Rev.Lett. 17, 41, 1967

10. Lutz H.O., Datz S., Moak C.D. and Noggle T.S., Phys.Rev.Lett. 33A, 309 1970

11. Datz S., Martin F.W., Moak C.D., Appleton B.R. and Bridwell L.B., Radiation Effects 12, 163, (1972)

12. Kaminsky M., Proc.Int.Conf.on Mass Spectroscopy Kyoto, Japan, Sept.1969

13. Andersen T., Datz S., Hvelplund P. and Sorensen G., Phys.Lett. 33A, 121, 1970

14. Okorokov V.V., Tolchenkov V.L., Khizhnyakov I.S., Cheblukov Y.U., Lapitskii Y.Y., Iferov G.A., and Zhukova Y.I., JETP Lett. 16, 415, (1972)

15. Brandt W. and Sizmann R. (private communication)

16. Brown M. and Moak C.D., Phys.Rev. B6, 90, 1972

17. Booth W. and Grant I.S., Nucl.Phys. 63, 481, 1965

18. Northcliffe L.C. and Schilling R.F., Nucl.Data, Sect.A7, 233 1970

19. Lindhard J., Nielsen V. and Scharff M., Kgl.Dansk. Vid., Mat.-Fys.Medd. 36, 10, 1968

20. Betz H. and Grodzins L., Phys.Rev.Lett. 25, 211, 1970

21. Datz S., Appleton B.R., Biggerstaff J.R., Menendez M.G. and Moak C.D., Bull.Am.Phys.Soc. 18, 662, 1973

22. Merzbacher E. and Lewis H.W., Handebuch der Physik (Springer Verlag, Berlin) 34, 166, 1958

23. Knudson A.R., Nagel D.V., Burkhalter P.G. and Dunning K.L., Phys.Rev.Lett., 26, 1149, 1971

24. Saltmarsh M.J., van der Woude A. and Ludemann C.A., Proc.Int.Conf.on Inner Shell Ionization Phenomena USAEC Conf. 720404, Oak Ridge, Tenn., 1973

25. Mowat J.R., Sellin I.A., Pegg D.J., Peterson R.S., Brown M.D. and Macdonald J.R., Phys.Rev.Lett. 30, 1289, 1973

26. Brandt W., Laubert R., Mourino M. and Schwarzchild A., Phys.Rev.Lett. 30, 358, 1973

27. Everhart E. and Kessel Q.C., Phys.Rev. 146, 27 1966

28. Fano U. and Lichten W., Phys.Rev.Lett. 14, 627, 1965

29. Barat M. and Lichten W., Phys.Rev. A6, 211 1972

30. Der R.C., Fortner R.J., Kavanagh T.M. and Khan J.M., Phys.Rev., A4, 556, 1971

31. Kavanagh T.M., Cunningham M.E., Der R.C., Fortner R.J., Khan J.M., Zaharis E.J. and Garcia J.D., Phys. Rev.Letters 25, 1473 1970

32. Lutz H.O., Stein H.J., Datz S. and Moak C.D., Phys. Rev.Lett., 28, 8 1972

33. Der R.C., Fortner R.J., Kavanaugh R.J. and Garcia J.D., Phys.Rev.Lett., 27, 1631 1971

34. Fortner R.J. (Private communication)

35. Saris F.W., van der Weg W.F., Tawara H. and Laubert R., Phys.Rev.Lett. 28, 717 1972

36. Macdonald J.R., Brown M.D. and Chiao T., Phys.Rev. Lett. 30, 471, 1973

Part 2

Radiation Damage Processes

IONIZATION DAMAGE PROCESSES IN INORGANIC MATERIALS

M.N. KABLER

Naval Research Laboratory
Washington, D.C. 20375

I. INTRODUCTION

Inorganic materials display essentially the entire range of radiation damage phenomena, from simple electronic charge redistribution to radiolysis to knock-on displacement[1]. Although this diversity can lead to extremely complex radiation effects, particularly at higher temperatures where lattice defects as well as electrons and holes are mobile, it is counterbalanced to a large extent by the availability of a wider range of workable experimental approaches and techniques. One of the most useful of these approaches has been the simple act of choosing to work on materials which are chemically and structurally simple and which can be prepared in some degree of purity. It is no accident that the alkali halides have received more attention than any other class of materials in this category, although in fact these deceptivly simple crystals harbor a most complicated combination of damage phenomena.

We shall herein deal almost exclusively with halide compounds, since these offer the most highly developed example of radiolysis in inorganic materials. The other form of ionization damage, charge redistribution among preexisting defects, is conceptually straight-forward and might be illustrated by any number of non-metallic materials. We have choosen as an example the scintillation process in doped alkali halides. Scintillation and radiolysis in alkali halides are related in several interesting ways, one being the fact that in both areas the experimental technique of real-time optical absorption spectroscopy has recently produced significant advances.

The alkaline earth oxides have become a second major area of damage research in inorganic crystals. We shall not have occasion to discuss these here, since the primary intrinsic damage process is collisional and this aspect will be covered in other lectures. True enough, many hole centers in crystalline oxides are created by electron redistributions among existing vacancies, particularly hole centers involving alkaline-earth-ion vacancies, but these effects are qualitatively similar to those in other materials. The companion lecture of D.L.Griscom on charge redistribution among network defects in amorphous oxides concerns a closely correlated area of current research.

Even within its limited scope, this presentation is necessarily selective in its coverage, and a number of other important and interesting topics have had to be omitted or noted only briefly. An attempt has been made to emphasize recent and timely results concerning the simplest, most fundamental processes.

II. ELECTRON-HOLE REDISTRIBUTION IN DOPED HALIDES : THE SCINTILLATION PROCESS

A. - BACKGROUND

All materials are inpure, and it is rarely possible to escape awareness of some effect of electron redistribution among impurities or preexisting defects. The effects are, of course, more easily identified in doped crystals. A straightforward illustration might be ruby, $Al_2O_3 Cr$, where a fraction of the Cr^{3+} ion is converted to Cr^{4+} and Cr^{2+} by hole capture and electron capture during irradiation. Other transition-metal ions produce similar effects. Entirely analogous processes can of course occur in semiconductors as a result of carrier redistribution among donors and acceptors.

The standard alkali iodide scintillator, for example NaITl, serves very well as a paradigm for this simple class of damage processes[2]. The activator ion produces luminescence in response to ionizing radiation, and for radiation of low stopping power (high energy electrons for example) the scintillation efficiency is such that nearly every electron-hole pair created produces a luminescent photon[3]. Since the activator concentration is generally about 0.1 mole %, most of the incident energy is absorbed in the undisturbed crystal and not at lattice ions adjacent to the activators. There are several mechanisms whereby this absorbed energy might be transferred to the activator ions.

Transfer by intrinsic luminescence is excluded because, at the normal operating temperature, 300K, intrinsic luminescence is negligible. This leaves two competing primary mechanisms :
1) the capture of excitons by the activator ions, and
2) ambipolar diffusion of electrons and holes and consecutive capture by the activator.

Early results were generally interpreted in terms of exciton diffusion, although the evidence was largely circumstantial. We will not describe that work but will instead discuss a recent set of experiments which established the contrary conclusion, namely, that almost all of the energy transport occurs by ambipolar diffusion of electrons and holes[4,5]. This conclusion holds for KI:Tl and NaI:Tl near room temperature, and exciton transfer at lower temperatures has not been ruled out. We shall, in fact, have occasion to mention data in support of the latter possibility.

It is useful to catalog and briefly describe the defects and charge states of concern. Thallium enters alkali halide crystals substitutionally with a normal valence of +1. It possesses a distinctive sequence of optical absorption bands, and at 300K gives a broad emission band around 3 eV. Scintillation also occurs in this energy range and is attributed to the Tl^+ ion. However, the question as to just which states of the Tl^+ are involved is complex. Both Tl^0 and Tl^{2+} are stable in alkali halides, at least through some temperature range. Both valence states have characteristic optical absorption bands by which their presence may be identified. They are readily formed through capture of an electron or a hole by a Tl^+ ion.

The self-trapped hole, or V_k center, is one of the most pervasive of radiation produced defects in halide lattices[6]. Free holes can evidently exist only for extremely short times, perhaps no more than 10^{-12} sec. The self-trapping is an intrinsic effect whose driving force is energy gained in the formation of a X_2^- molecular ion from two adjacent halide ions. The bonding involves the p valence orbitals. The hole occupies one of the two highest antibonding σ_u orbitals, thus providing a net bonding which decreases the spacing of the two halide ions by 20 - 45 % relative to the perfect lattice. The V_k center has been thoroughly investigated by means of EPR and optical spectroscopy, and the nature of the principal electronic states is well-known. Efforts at computing its properties have enjoyed modest success[7]. Figure 1 shows the morphology of this center in the NaCl structure. The EPR and optical spectra are generally anisotropic and can be characterized to good approximation in terms of <110> axial symmetry, even though the actual point group is D_{2h}. At a characteristic temperature, about 90K for KI, the V_k center becomes unstable and diffuses through the lattice[8]. The motion involves rotation of the V_k axis through an angle of 60°.

B. - THE SCINTILLATION MECHANISM : TIME-RESOLVED OPTICAL SPECTROSCOPY

Knowing characteristic optical spectra for all the states which might logically be involved (excluding the exciton for the moment), it becomes apparent that time-resolved optical experiments can identify directly the scintillation mechanism. Such an experiment has recently been carried out by Dietrich, Purdy, Murray and Williams[5], and we shall briefly describe their results. They were chiefly concerned with KI, and set out to measure the evolution of optical absorption and emission following pulsed irradiation at various temperatures. Excitation was obtained from a field emission source with provided a 5 nsec electron pulse of roughly 500 keV average energy. A geometry employing total internal reflection was used to maximize the optical paths through the irradiated region of the crystal, thus lowering the required excitation intensity.

Observation of the V_k centers themselves proved difficult because of various overlapping absorption bands due to other defects. However, the desired information on hole capture was readily obtained by monitoring a distinctive, well-resolved absorption band of Tl^{2+}, which falls at 306 nm in KI:Tl. Figure 2 shows typical data on this band at 298 K in a crystal containing 0.12 mole % Tl. The initial rise is due to hole capture at Tl^+, while the millisecond decay is due to the annihilation of Tl^{2+} by electrons released thermally from other trapping sites, primarily Tl^0. There is also a very rapid rise, not evident in fig.2, which is due to immediate trapping of those holes initially created within about 15Å of a Tl^+ ion, approximately 10 % of the total. Data such as those of fig.2, in particular the Tl^{2+} build up shown in the inset, are in good agreement with diffusion-limited reaction theory. The diffusion times thus obtained are consistent

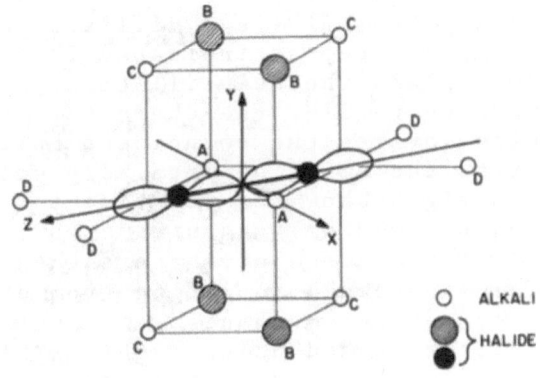

<u>Fig.1</u> : Self-trapping morphology in the NaCl structure.

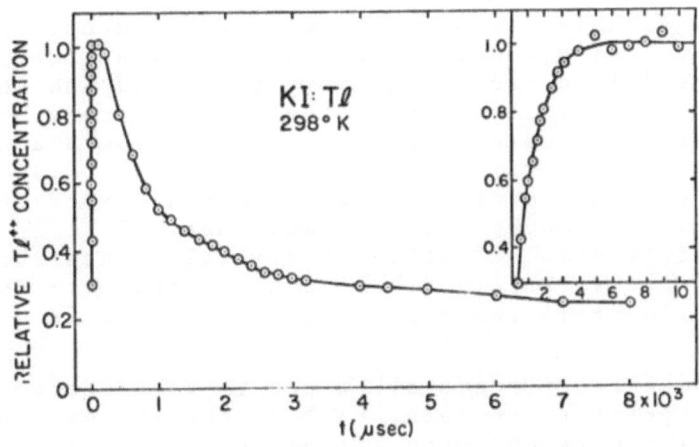

<u>Fig.2</u> : Relative Tl^{+2} concentrations vs time following pulsed ir-
radiation of KI:Tl (from ref.5).

with V_k-center diffusion times extrapolated from separate low-temperature measurements, carried out under circumstances more favorable for the unrestricted observation of V_k-center motion[4].

Since the annihilation of Tl^{2+} by electrons released thermally from Tl^0 is a millisecond process around room temperature, as indicated in fig.2, it is necessary to explain the scintillation process in terms of hole diffusion to Tl^0, the characteristic time being that which governs the coincident buildup of Tl^{2+}. This model fits the emission data very well. The circle points in fig.3 show the relative intensity of the scintillation (at 430 nm) measured as a function of time at 236K. The dotted curve represents that fraction of the luminescence resulting from direct excitation of Tl^+ by various mechanisms including immediate capture of electrons and holes ; this lifetime was obtained from photoexcitation data. The dashed curve represents the luminescent intensity calculated from the hole diffusion rate measured by the transient absorption. The sum of these curves, the solid line in fig.3, obviously gives an excellent fit to the data points. The agreement at room temperature is similar, except that here a long, weak tail appears in the luminescence due to thermal release of electrons from Tl^0 and subsequent radiative recombination at Tl^{2+} ions. This tail coincides with the long term Tl^{2+} decay shown in fig.2.

Comparable measurements were carried out on NaI: Tl and the same conclusions were drawn. At room temperature hole diffusion is very rapid and the decay time is essentially that of the excited state of Tl^+, 0.25 µsec. The weak component arising from electron release at Tl^0 exhibits a time constant of 0.35 µsec and is thus not distinguished in normal experiments. These times account for the fact that NaI:Tl is the most widely used scintillator crystal, since about 90 % of the total light has been emitted at 1 µsec whereas in KI:Tl the comparable percentage is under 25 %.

These experiments allow one to conclude that near room temperature at least 95 % of the energy transport in KI:Tl occurs by ambipolar diffusion of electrons and holes. It is perhaps useful to add that the build-up of Tl^{2+}, as in fig.2, cannot be due to exciton diffusion to Tl^+ followed by an Auger process which ejects an electron. The lifetime of self-trapped excitons is, as we shall see, orders-of-magnitude smaller than re-

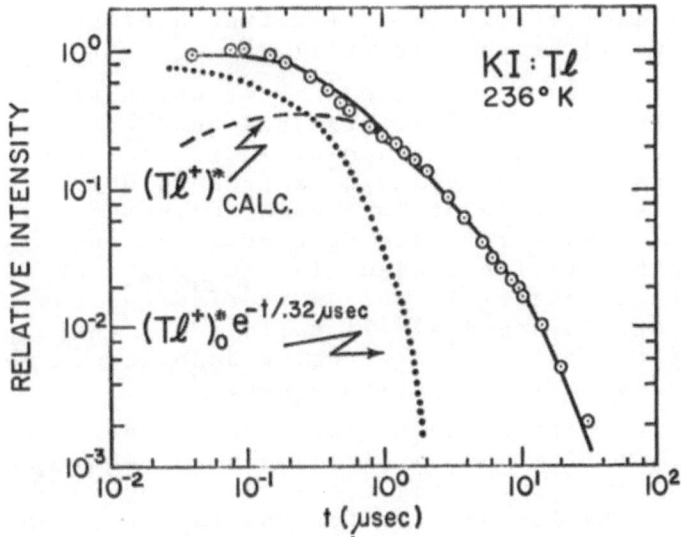

Fig.3 : Relative luminescence intensity of KI:Tl at 236K. The initial sharp rise was not recorded. Broken curve : calculated (Tl⁺)* concentration normalized to data at 4 μsec. Dotted curve : calculated (Tl⁺)* based on exponential decay from prompt excitation. Solid curve : Sum of two calculated curves normalized to data at 0.2 μsec. Circles are experimental points. (from ref.5)

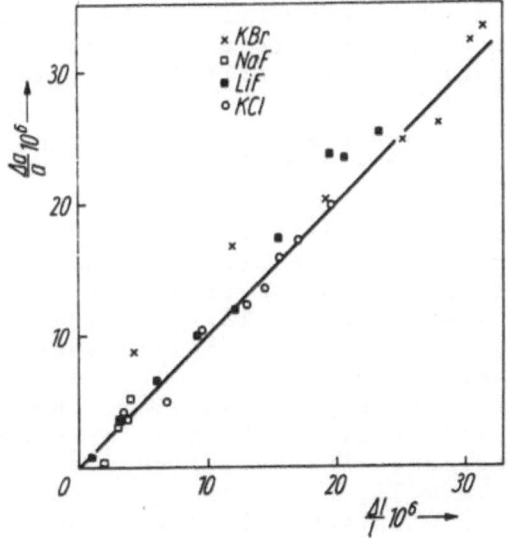

Fig.4 : Relative lattice parameter and length changes for alkali halides x-ray irradiated at room temperature. (from ref.10).

quired ; and the lifetime of a free exciton, if such exists, would be shorter still.

At temperatures below that at which the self-trapped hole can diffuse, trapped charge builds up to a saturation level governed by intrinsic recombination of electrons and V_k centers, a process which can be an efficient source of intrinsic luminescence. Recombination of electrons and V_k centers through an alternative nonradiative channel is the origin of radiolysis in these materials and is a principal subject of this paper. In doped alkali halides all these processes compete for the energy being deposited, and the balance is strongly temperature dependent.

It is of interest to note some experimental evidence indicating that excitons may be mobile for a brief interval after their creation. Tomura et al[9] have obtained emission spectra for RbI and RbI:Tl at 5K for two different ultraviolet excitation wavelengths, one just above the band gap and the other in the high-energy tail of the first exciton absorption band. Of the four possible combinations, they find that only in the case of the Tl-doped crystal with excitation in the exciton band is the Tl^+ emission strong compared to the intrinsic luminescence. The excitation which presumably produces free electrons actually produces much less Tl^+ emission, which is consistent with the holes being immobile. The data also suggest that the exciton mobility decreases with temperature due to scattering by phonons and thus would become negligible at liquid nitrogen temperature or above.

III. IONIZATION DAMAGE PHENOMENA IN HALIDES

A. - BACKGROUND

It is well-known that x-ray or particle irradiation will color most simple halide crystals at temperatures in the vicinity of room temperature and below. The mechanisms whereby the coloration is produced have been pursued by many workers over the years, and the nature of the defects which cause the coloration and related effects such as paramagnetism, changes in lattice parameter and density, changes in mechanical properties, and decreases in thermal conductivity are generally known. We will, of course, be concerned with these defects but will generally avoid mention of the methods whereby their identities were revealed. The important fact is that, in pure crystals at low temperatures where defect aggregation is negligible, the products are Frenkel defects in the halide sublattice : F-center interstitial-atom pairs (F-H pairs) for which the charge of each constituent is zero, or F^+ center interstitial ion pairs (F^+-I pairs). Other possibilities such as Schottky defects or Frenkel defects in the cation sublattice have been considered. One cannot state categorically that these defects are never produced by radiation, yet there is little evidence for their production.

To illustrate the predominance of Frenkel defects, one might choose a class of experiments involving the simultaneous measurement of lattice parameter and crystal volume during the course of irradiation. If Schottky defects are produced the crystal will expand as vacancies appear and the corresponding ions are added to dislocations or surfaces ; however, the lattice parameter will change only very slightly as a result of the small dilation around the vacancies. On the other hand, if Frenkel defects are created the lattice parameter and volume changes will be identical since no new lattice sites are being created. Measurements in several alkali halides have shown that if Schottky defects are produced by radiation they can be at most a small fraction of the Frenkel defects produced[10]. The data of fig.4 for x-ray irradiation at room temperature illustrate this fact, since they establish that the relative change in lattice parameter a is equal to the relative change in overall length ℓ.

It is possible to view past progress in defining the production mechanism largely in terms of experimental advances toward determining the upper limit to the intrinsic production efficiency. A number of plausible models have been proposed over the years, and most of these have continued to be tenable as long as the recognized maximum production efficiency remained relatively low. Energetic particles certainly produce damage by direct collisions or knock-ons, although at low efficiency. However, the efficiency is actually quite high, even for low-energy electrons which are unable to produce displacements. Thus an ionization process is clearly in operation[11].

A mechanism due to Varley, which involved multiple ionization of a single halide ion and its consequent ejection as a result of electrostatic forces, was a leading contender[12]. But its involvement became less likely in the wake of efficiency measurements which showed that in KBr at 5K one vacancy is produced for each 240 eV absorbed from a beam of 50 kV x-rays[13]. Mechanisms involving ionization of two adjacent halide ions, such as those of Klick[14] and Williams[15], proved more difficult to assess in terms of efficiency. However, these mechanisms were also forced to assume a subsidiary role as data accumulated in support of a single ionization mechanism. Although early work had indicated that ultraviolet photons could produce coloration in alkali iodides, a key event was the demonstration by Pooley[16] and others[17] that ultraviolet radiation could create higher concentrations of F centers than could be accounted for by extrinsic processes. Ultraviolet light has since been shown to produce F centers in several alkali halides and, although details such as excitation spectra remain subject to speculation, these measurements make it clear that a single ionization mechanism of relatively high efficiency is in operation.

The rates at which particular color centers are produced or destroyed by radiation vary strongly and often in complicated ways with parameters such as impurity concentration, dislocation content, temperature, radiation quality and intensity, and of course the particular crystal. This variability has added to the difficulty of pinning down the damage process. We shall emphasize the basic processes, which are generally best isolated at low temperatures, and be concerned primarily with how the single ionization mechanism may operate. It should perhaps be noted that many of the color centers produced by high energy radiation can also be produced by chemical methods, that is, by additive or electrolytic coloration. These have allowed the preparation of crystals containing vacancies but not interstitials, and vice versa, and have thereby substantially aided the identification process.

B. - THE EXCITONIC MECHANISM

The idea of a single ionization mechanism seems to have evolved over a period of several years, and the general model currently in favor evidently occurred to several groups independently, including those of Hersh at Zenith and Lushchik at Tartu. Pooley at Harwell has proved to be the most effective exponent of the model, and it is thus frequently referred to as the Pooley mechanism. We shall outline this mechanism here and continue in a later section with a more detailed analysis.

It had been shown that metastable states involving a self-trapped hole and a bound electron were responsible for initiation of intrinsic recombination luminescence[18]. These states are generally termed self-trapped exciton (STE) states. The hole is shared by two adjacent halide ions, which are thereby bound covalently in the same way that certain excited states of a diatomic rare gas molecule are bound. The important aspect for Frenkel defect production is that these STE states are well above the middle of the band gap. Therefore, if the system can somehow make a nonradiative transition to the ground state, a large fraction of the band gap energy becomes available as kinetic energy of the two ions, each now in its ground state[19]. This kinetic energy might be sufficient to initiate a replacement collision sequence along a close-packed line of halide ions, that is along a <110> direction in the NaCl-type structure. Pooley's potential curves, a calculated ground state and a schematic luminescent state, are illustrated in fig.5 for KI. These illustrate how the energies of the luminescent and ground states depend on the relative separation r/r_0 of the two iodide ions. E_{fi}^o is the energy available for defect production once the system has surmounted the barrier Δ_4. In KI the luminescent efficiency is high and the F-center production efficiency low at low temperatures, but as the temperature increases to the vicinity of 100K the situation becomes reversed. This behavior was considered a prime factor in favor of the model, being consistent with thermally activated transitions from the metastable STE state to the ground state with consequent F-center production at the expense of the luminescence. Pooley also investigated the energetics of the replacement collision sequence with a rough computational model which produced encouraging but inconclusive results.

Independant of the details of this model, one would first wish for direct evidence that a self-trapped hole is involved. Such evidence was eventually provided in the following way : Suppose a crystal doped lightly with an electron trapping impurity such as Tl^+ is irradiated at low temperature, thereby creating stable self-trapped holes and electrons trapped as Tl^0. If the temperatures are chosen properly, then according to the excitonic mechanism one should be able subsequently to excite the electrons with red light, cause them to recombine with the self-trapped holes, and thereby produce F centers and their complementary interstitials which are H centers. This experiment was attempted by at least two groups, at ORNL and NRL, using optical detection techniques. These experiments failed, largely because of difficulties in counting

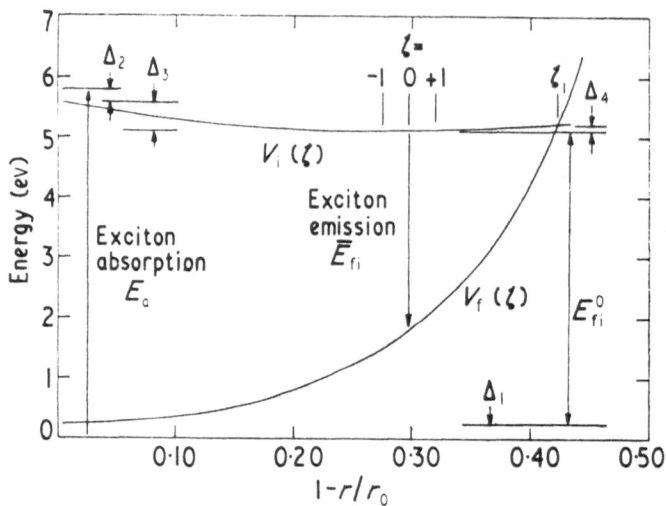

Fig.5 : Energies of electron-hole pair states in KI as a function of the relative separation r/r$_0$ of the two iodide ions. (from ref. 19).

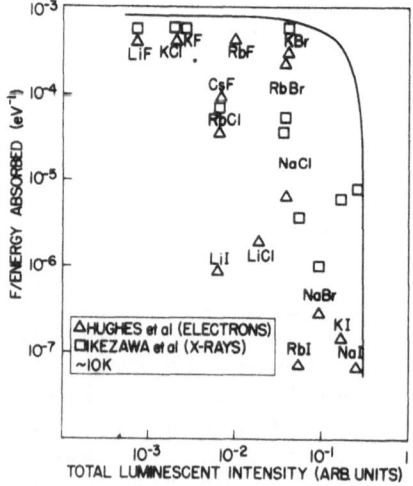

Fig.6 : F-center production efficiency vs luminescent intensity near 10K. See text for sources of data.

ionized F centers over a background of heavy metal absorption.

Keller and Patten succeeded in obtaining the desired correlation using EPR techniques to detect the production of H centers rather than vacancies[20]. When the recombination was produced by ionizing Tl^0 with red light, the well-known self-trapped hole resonance was observed to decrease and an H-center resonance to appear and grow. Thus the production of interstitials as a direct result of the recombination of an electron with a self-trapped hole was directly demonstrated. $ClBr^-$-type self-trapped holes were found to be roughly an order of magnitude more effective in producing interstitials than was the intrinsic Cl_2^- center. This is consistent with the Pooley model in that a larger fraction of the available kinetic energy will go to the lighter ion, thereby aiding initiation of the <110> collision sequence. The absolute quantum efficiency, that is, the number of H centers produced per V_k center destroyed, was roughly 0.4 for $ClBr^-$ centers and roughly 0.02 - 0.05 for Cl_2^- centers at 25K. If one assumes that 25 eV are required to make each electron-hole pair, then the latter figures imply F-center production efficiencies of 500-1250 eV per F center. This may be compared with the 2400 eV per F center measured by Ritz[13] for 50 kV x-ray irradiation at 5K. Considering inherent differences in the measurements (e.g. the luminescence is quenched at 25K but not at 5K) the agreement is acceptable.

These experiments thus indicate that the efficiency of the excitonic mechanism is adequate to account for steady-state F-center production in KCl. There exists substantial indirect evidence, much of which we must leave unmentioned, to support the premise that a mechanism of this type is common to most if not all simple halide crystals. We shall proceed on this basis, keeping in mind that in any particular case some other mechanism, such as simultaneous ionization of two adjacent halide ions, may contribute significantly.

Let us now consider the excitonic mechanism from a broader view-point. There are at least four distinct channels through which electrons and holes created by ionizing radiation can annihilate or stabilize : 1) recombine radiatively, 2) recombine nonradiatively producing only phonons, 3) recombine nonradiatively producing Frenkel defects, 4) become trapped at impurities or other imperfections, if such are available. This suggests the following relationship for pure crystals :

$$N_F + I_{RL} + W = \text{constant}, \qquad (1)$$

where the respective terms are the number of Frenkel defects, the number of recombination photons, and the heat generated per unit energy absorbed in the crystal. A relationship of this type has been shown to hold approximately for variations due to temperature in several materials, particularly KI, as mentioned above[21]. However, if N_F is high at low temperatures the complementary relationship between N_F and I_{RL} is generally not apparent experimentally[19]. On the other hand, suppose one considers a constant temperature and compares different crystals. Figure 6 plots the number of F centers produced per electron volt absorbed versus I_{RL} for a number of alkali halides near liquid helium temperature. The I_{RL} values are those obtained by Pooley and Runciman using x-ray excitation[21]. They include all observed emission bands, some probably of extrinsic origin (these nevertheless should be counted in the energy balance). The square points represent F centers created by x-ray irradiations at 10K,[22] and these are in reasonable agreement with the earlier data of Rabin and Klick[23] and Ritz[13]. The triangle points are those of Hughes et al using 400 kV electron irradiation[24]. The higher intensity of the latter source or differences in impurity content of the materials used may account to some extent for the disagreement apparent for the iodides. The curve is a plot of Eq. (1) with W assumed constant. The placement of this curve is somewhat arbitrary, the objective being to delineate the upper bound on I_{RL} and N_F such that losses due to W will cause the points to fall to the lower left of the curve by varying amounts. It is important to note that F^+ centers are also produced by these irradiations, and their concentrations exceed those of the F centers by factors in the 5-10 range, at least for KCl and KBr. To take this into account one might wish to move the curve upward by perhaps an order of magnitude.

The heat dissipation W will surely vary from crystal to crystal, causing the points to scatter somewhat. Experimental difficulties involved in measuring the luminescence intensity, for example translucent surfaces resulting from the hygroscopic nature of LiI and LiCl, will have a similar effect. Nevertheless, when all factors are considered, it is quite possible to identify a trend in the data of fig.6 consistent with

Eq.(1). Thus competition between luminescence and defect production would appear to be a general characteristic.

C. - THE SELF-TRAPPED EXCITON

On the basis of the data discussed thus far, it is clear that electronic states of the STE play a crucial role in initiating Frenkel defect production as well as recombination luminescence. It will therefore be useful to describe these states in somewhat more detail. The properties of the luminescence have thus far been the primary indicators of the nature of the STE, although measurements of excited-state absorption spectra[25] and, quite recently, of EPR-optical double resonance in the triplet states[26] have furnished important data. Those properties of the STE which appear most relevant to the defect production mechanism will be emphasized in the following description.

The STE is a bound electron-hole pair which is localized in the perfect lattice due to intrinsic ionic relaxation. The self-trapping mechanism is analogous to that which causes hole self-trapping ; a covalent bond is formed between two adjacent halide ions[6]. When an electron has been trapped to form a STE, the hole remains relatively unperturbed and the self-trapping configuration is preserved. Thus the structure in fig.1 also applies to the STE.

Luminescence from the lowest triplet state of the STE is observed in many, if not all, alkali halides[27]. Roughly half the alkali halides also exhibit luminescence from a singlet state. The photon associated with the single luminescence is roughly 1-2 eV greater in energy than that from the triplet, and it is likely that the singlet state is higher than the triplet by about the energy. As a result of the strong relaxation associated with the self-trapping, the luminescence can occur at relatively low energies ; for example, the triplet transition occurs at about 2.3 eV in KCl, KBr, and RbI, giving Stokes shifts of 5.4, 4.5, and 3.4 eV, respectively.

The lifetimes of the singlet transitions are in the 2-10 nsec range, as expected for allowed transitions to the singlet ground state[27,28]. The lifetimes of the triplet transitions vary strongly from crystal to crystal, from milliseconds in the chlorides to microseconds in iodides. The dependence of the lifetime on the identity of the halide is consistent with the multiplicity forbiddenness being broken by spin-orbit coupling involving the hole orbitals. The data also suggest that the overlap of electron and hole orbitals is somewhat less for the triplet state than for the singlet.

Under an applied magnetic field, the triplet luminescence becomes circularly polarized due to mixing of nonequivalent orbital components into the three unequally populated triplet levels[29]. The magnitude of polarization, as well as the luminescent intensity, is affected by changes in population of these levels. This effect has provided a means for carrying out EPR experiments on the triplet state. Microware resonances have recently been detected by observing changes in the luminescence as the applied field is swept through the resonance[26]. Resolved hyperfine structure is evident in some of the transitions for KBr and KCl. Analyses of the hyperfine splittinges and the fine structure splittings (at zero field, the usual D term) indicate that the electron amplitude on the two halogens is relatively small for KBr and KCl but larger for CsBr, in accord with the trend predicted by comparison of lifetimes and Stokes shifts. The symmetries of the EPR spectra are consistent with structure shown in fig.1. In addition, from the fine structure splittings in the triplet state it has proved possible to deduce theoretically the electron-hole exchange parameter.

For present considerations the most useful data on the STE have come from transient optical absorption experiments. If the crystal is excited with a fast pulse of ionizing radiation and observed immediately in times short compared to the STE lifetime, new absorption bands similar to various color center bands are evident[25]. On the basis of their lifetimes these bands can be unambiguously correlated with the STE. Because of the experimental convenience associated with their longer lifetimes, only the triplet STE states have been investigated thus far. Figures 7 and 8 show time-resolved absorption spectra originating in the lowest triplet states of the STE in alkali chlorides and bromides. The crystal temperatures were $9 \pm 3K$. The original data were separated into individual exponential decay components, and that part of each spectrum corresponding to a given decay time is indicated. In each case the solid line accompanied by experimental points represents the component of absorption which shares the decay time of the triplet luminescence. The stable bands indicated by the dashed curves arise from F centers and, for KCl and KBr, from H centers as well. The strong transient F-and H-center bands are, of course,

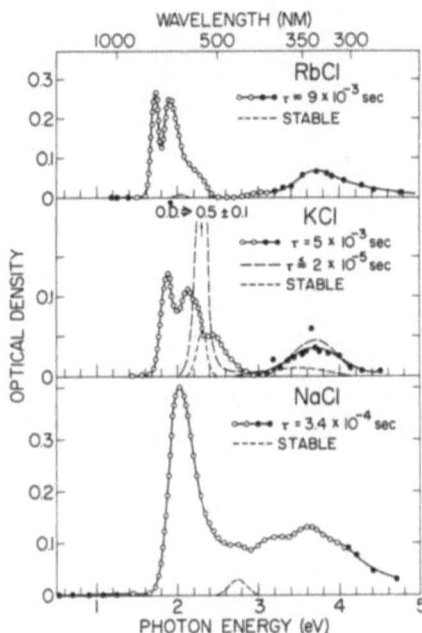

Fig.7 : Time resolved absorption spectra for chlorides at 9 ± 3K. The solid curves and data points indicate that part of the initial absorption which decays with the lifetime of the lowest triplet STE state. Dashed curves indicate that part of the initial absorption which is due to F and hole centers. Lifetimes are noted. (from ref. 25).

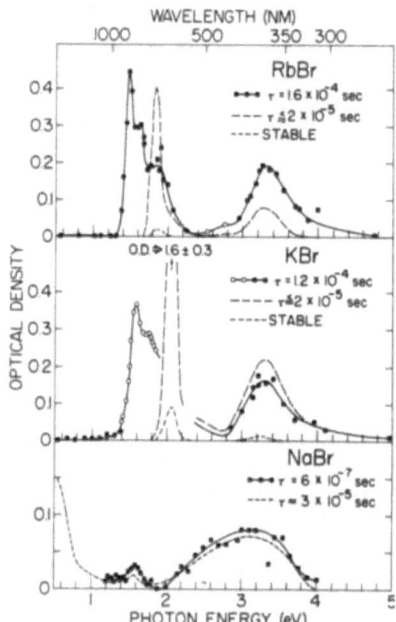

<u>Fig.8</u> : Time resolved absorption spectra for bromides at 9 ± 3K.
The solid curves and data points indicate that part of the initial
absorption which decays with the lifetime of the lowest triplet
STE state. Dashed curves for RbBr and KBr indicate that part of
the initial absorption which is due to F and hole centers. Dashed
curve for NaBr is attributed to a second component of the STE.
Lifetimes are noted (from ref. 25).

of prime interest in the present context and will be discussed presently. The solid points are data for fixed wavelength, while the open points represent data taken in a scanning mode. The curves take account of all accumulated data. All curves shown for a given crystal, when added together, accurately represent the total absorption immediately following the first 5 ns pulse of 500 kV electrons incident upont the sample. NaCl is a minor exception, since the data for this case represent the fifth pulse. The experimental apparatus was the same as that used for the real-time scintillation measurements described in Section IIB.

It has proved possible to assign most of the bands in figs.7 and 8, as well as those for the corresponding alkali iodides, to specific transitions with considerable confidence. The absorption in the 3-4 eV range is attributed to hole transitions within the STE : if the hole is perturbed only slightly by the electron these transitions should resemble those of the V_k center. This resemblance is quite apparent. The absorption in the 1-3 eV range (except for NaBr) is due to excitations involving the electron. One can argue that these bands bear strong similarity to the lower energy transitions of F and M centers, and it is not unreasonable to expect the potential which the electron in the STE sees to be comparable to that in the F or M center. The STE is, of course, the generic equivalent of an X_2 molecular ion superimposed upon an M^+ center. By inspection of these bands it is possible to make a rough estimate of the optical binding energy, E_i of the electron and hole. The numbers one obtains are roughly 2.3 eV for RbCl, 2.6 eV for KCl, 2.8 eV for NaCl, 2.2 eV for RbBr, and 2.3 ev for KBr. NaBr represents the case of minimum ionic relaxation relative to the perfect lattice, as indicated also by the relatively small Stokes shift and short lifetime. The electron transitions are thought to lie below 1 eV in fig.8, giving $E_i \leqslant 1$ eV. We note that althought the 3×10^{-5} sec decay component represented by the dashed lines has not yet been identified in the luminescence, it can nevertheless be tentatively attributed to the STE and may well originate in the A_u level of the lowest triplet state. The spectra for NaI are similar to those for NaBr, and $E_i \leqslant 1$ eV also for NaI.

These data have led to the following conclusions concerning the triplet states of the STE : Potassium and rubidium halides generally represent a limit where the X_2^- core is largely uncoupled from the lattice ;

the electron and hole are substantially independant of
each other and give rise to spectral characteristics
similar to those of analogous perturbed color centers,
specifically the M center and the.V_k center. Consequently the E_i values and Stokes shifts are relatively
large and the luminescent lifetimes relatively long.
The opposite extreme in the STE behavior is represented
by NaBr and NaI. For these the formation of the distinctive X_2^- core is still the driving force for self-trapping, but the general character of the electron is
perhaps better described in terms of a perturbation on
the behavior of the free exciton rather than as an analog to a perturbed color center. This picture can be
regarded as at least superficially consistent with ion
size considerations. That is, the farther apart the
halide ions in the undisturbed lattice, the greater the
relaxation in forming the X_2^- core and therfore the more
localized the hole and the more isolated the X_2^-
from the electron. The effective potential the electron sees in the region of the alkali ions is evidently lower than that near the X_2^- core, and in the case
of RbCl, for example, the relative volume of the low-
potential region is larger than for NaBr.

The E_i derived from the spectra of figs.7 and 8
are extremely useful in determining the energetics of
electron-hole recombination. It is simplest to consider this in terms of a configuration coordinate diagram showing total energy as a function of a single
mode of lattice distortion. This can be done for the
STE, at least schematically, but it is difficult to
explain E_i in this way. It has been shown to be advan-
tageous to consider the next alternative, two distinct
modes[25]. The first mode is primarily associated with
the internuclear separation of the X_2^- ; this separation r, or more precisely the covalent bonding associated with it, is the primary source of both the self-
trapping energy and the ground state energy variation.
The second mode, d, probably also of A_g symmetry, com-
prises the displacements of all other ions. It is comparable to the breathing mode which is mainly responsible for the Stokes shift and broadening of spectra for
M or F centers. For the purpose of discussion these
modes can be considered approximately independent,
subject to a more accurate characterization. Figure 9
illustrates the important potential surfaces for electron trapping and recombination. The curves are semi-
quantitative representations of KCl, although they indicate the general behavior for all strongly relaxed
STE's.

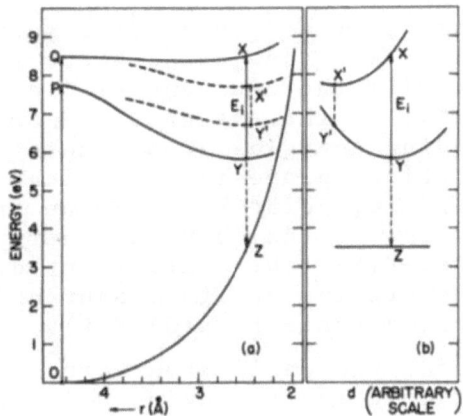

Fig.9 : Potential curves for STE states in KCl. Coordinate r represents the approximate internuclear spacing of the Cl_2^- core, and d represents schematically the displacements of other ions. The determination of these curves is described in the text.(from ref.25).

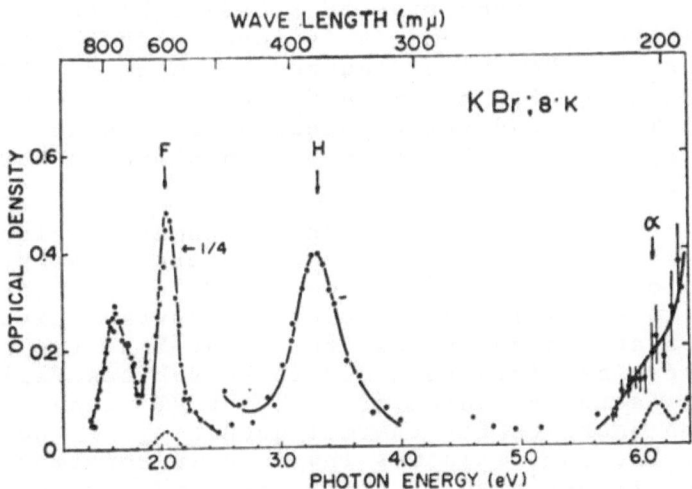

Fig.10 : Transient (solid curve) and stable (broken curve) total absorption produced in KBr by an electron pulse. The F-band region has been reduced to 1/4. The absolute optical densities for the transient and stable spectra are not directly comparable. The band designated H includes STE absorption also (from ref.33).

In the configuration space spanned by coordinates r and d, the minima of the V_k potential surface, the triplet STE surface, and the ground state surface will not coincide, nor will they in general lie in a single vertical plane. This is the essential reason for the difference in binding energies. Coordinate d is taken to designate a vertical plane intersecting the minima of both the V_k surface, point X', and the triplet STE surface, point Y. The optical binding energy of the triplet STE is E(X)-E(Y), the luminescent quantum is E(Y)-E(Z), and the binding energy measured by transitions from the ground state is E(Q)-E(P).

The energies at the designated points in fig.9 have been estimated in the following way : the V_k minimum X' is below point W by an energy which is 1.5 times the thermal reorientation energy of the V_k center, in accord with a prior estimate[30]. The points Y',X, and Y have been located so that the potential curves corresponding to coordinate d resemble those for the F center, the relaxation being shared equally by the V_k and STE curves. The ground state potential curve OZ was calculated in the conventional way using Born potential and taking r as the internuclear separation of the Cl_2^-. The value of r corresponding to the point Y was determined by requiring that E(Y)-E(Z) be the correct luminescent quantum. It is notable that r(Y) is almost exactly the value required to optimize theoretical agreement with measured V_k-center hyperfine parameters. The ground state potential should not vary strongly with d ; it has simply been sketched horizontal in fig.9b.

The curves other than OZ are schematic, although it is possible to estimate the curvature at point Y on the basis of the emission band-width. This in turn allows one to estimate the barrier for nonradiative, thermally activated transitions from point Y to the ground state. These barriers turn out to be substantially larger (a factor of three or more for KI) than measured activation energies for thermal quenching of the luminescence. The uncertainties involved in constructing the curves illustrated by fig.9 are almost certainly not large enough to accommodate triplet STE curves as shallow as that of fig.5 or barriers as small as measured luminescence quenching energies. However, a nonradiative channel of this type is essential since the depth of point Y would appear to preclude the possibility of ionization or diffusion being the cause of thermal quenching.

After describing transient defect production results, we shall investigate how the properties of STE
states can be related to the damage process and shall
have occasion to suggest an alternative path whereby
these states can be depopulated.

D. - TIME-RESOLVED STUDIES OF DEFECT PRODUCTION

In general, the production of vacancies and interstitials by an ionization mechanism will involve the
following five steps or processes:

(1) Absorption of radiation and creation of elec-
    trons and holes

(2) Population of those electron-hole pair states
    which initiate defect production

(3) Production of F-H and/or F^+-I pairs

(4) Vacancy-interstitial recombination events
    which destroy a fraction of the damage initi-
    ally produced

(5) Charge rearrangements, for example $F \rightarrow F^+ + e$
    and $H + e \rightarrow I$.

The usual experiments integrate the radiation dose
over time and thus are sensitive only to the end result
of the above sequence of events. However, each event
is distinct, and time-resolved experiments can in principle isolate one or all of the steps.

The pioneering research of Schulman and Boag[31],
and later Ueta[32], was able to demonstrate of effects
of step (4) and, in some cases, step (5) at temperatures from 300K down to 80K. Interstitial aggregation and
trapping were also observed. Subsequent work by Sendai
group and at NRL has taken advantage of lower temperatures to provide some information about each step in
the sequence.

Let us first consider the nature of the primary
defects formed by pulsed radiation and their formation
efficiencies. Returning attention to figs.7 and 8, re-
call our previous remark that the initial absorption
spectrum for a given crystal is the sum of all the
spectra shown for that crystal. For KCl, RbBr and KBr,
the spectra shown by the dashed curves are due almost
entirely to F and H centers, and the initial integrated absorption strengths due to these centers are comparable to those of the STE triplet states. Since the
oscillator strengths for the corresponding color-cen-
ter and STE absorptions are thought to differ by no

more than a factor of two, it is clear that on a nanosecond time scale the radiation pulse produces roughly
as many F centers as triplet STE's. Note that, except
for NaCl, the data of figs.7 and 8 represent the results
of the initial electron pulse into a previously unirradiated sample. Figure 10 shows comparable data of Kondo
et al for KBr using similar methods. The principal
difference is that in fig.10 the dashed curves represent approximately one-tenth of the stable absorption
after ten electron pulses. Also, the band labeled H
includes a substantial contribution from the triplet
STE. This figure indicates that the conversion of F
centers to F^+ (or α) does not account for the great
loss of F centers after the pulse (1/20 of those in
fig.8 are stable, 1/50 in fig.10). The F^+ band at 202nm
rises in 10 nsec or less, and its transient increase or
decrease thereafter is well under 50 % per pulse, according to both the Sendai and NRL results. Thus most
of the transient F and H centers evidently annihilate
through mutual recombination governed by decay times
in the 10-100 μsec range at these temperatures.

The transient F-center production efficiency for
KCl, KBr, and RbBr will clearly be quite high relative
to that measured by steady-state experiments. On the
basis of direct dosimetry the spectra of figs.7 and 8
correspond to production efficiencies of 150 eV/F center for KCl and 45 eV/F center for KBr, within an accuracy of roughly a factor of two. For KBr, Kondo et al
have obtained a similar estimate using indirect dosimetry[33]. They compared the transient and stable F-center
concentrations derived from fig.10 and assumed the creation efficiency for stable centers to be the same as
that measured in prior steady-state experiments. The
latter value is 4,700 eV per F center for electrons of
comparable stopping power[13], and since only 2 % of the
F centers in fig.10 are stable one obtains roughly 100
eV per F center when transients are included. Since this
number depends on steady-state dosimetry data at higher
overall doses and much lower dose rates, its reliability is somewhat restricted. Also, the effects of the F^+
center and possible long-term $F \rightarrow F^+$ conversion are difficult to assess.

Despite the uncertainties in these preliminary measurements, it would seem that in KCl and KBr the vacancy production efficiencies under pulsed conditions
are in the range 50-150 eV per vacancy-interstitial
pair. Note that the relatively low F^+-center concentrations observed in KBr are too small to lower these num-

bers substantially.

For KBr both the F and F$^+$ absorption bands rise
in times comparable with the width of the electron
pulse, roughly 5-10 ns. It was suggested that F-H pairs
are created exclusivly in the initial event and that
F-I pairs form subsequently as a result of electron
transfer from an F center to its nearby interstitial[34]
This process in not unreasonable, since theoretical estimates indicate that the energy of a stable F$^+$-I pair
is lower than that of an F-H pair[35,6]. However, the
KBr experiments (both at Sendai and NRL) indicate that
such a process must go to completion on a nanosecond
or faster time scale. It appears less likely that the
required number of F$^+$-I pairs could be converted to
F-H pairs under the circumstances. The pulse experiments thus tend to indicate that F-H pairs are the primary defect created, and we shall henceforth assume
this to be the case.

E. - STE STATES AND DEFECT PRODUCTION

On another point the pulsed irradiation data are
conclusive : for NaCl, KCl, KBr, and RbBr the lowest
triplet STE state is not the state which initiates F-
H pair production at low temperatures. If this were
the case, the rise time of the F and H bands would coincide with the decay time of the triplet STE luminescence, since both would originate in a common state.
No such correspondance is observed. Note that his conclusion is not in contradiction with the temperature
correlation between F-center production and luminescence for KI and some other alkali halides described
in Section IIIB.

Beyond this, the question as to the identity of
the initial state for the vacancy production process
in these easily colored crystals remains difficult to
answer. The next higher state is a singlet just above
the lowest triplet, separated from it by an exchange
energy in the range 1-100 meV[36]. This state, which should
have a lifetime in the 0.1 μsec range, has not been i-
dentified experimentally; it is clearly not that singlet
state which initiates fast intrinsic luminescence in
several of the crystals. The triplet STE absorption
spectra, coupled with the observed Stokes shifts for
singlet luminescence, indicate that all other electronic states of the STE are at least 1 eV above the lowest triplet state. These higher states have some advantage with regard to the energetics of the mechanism.
If, for example, the lowest triplet state did initiate

F-center production, the available energy for KCl would be less than 6 eV according to fig.9. In fact, the experiments still allow speculation on the basis of almost the entire band gap, roughly 8 eV.

In KI and, we suspect, the other crystals which color poorly at low temperature, the lowest triplet STE states are not yet ruled out as the origin of defect production in the temperature range where this process grows at the expense of the recombination luminescence. This point will be clarified by transient absorption experiments in progress[37].

Suppose one assumes, then, that in all cases defect production originates from nonradiative decay over a barrier to the ground state, a process which can be described in terms of figs.5 and 9 except that for the easily colored crystals the initial state is not the lowest triplet but some higher state, triplet or singlet. Several significant problems arise in connection with this hypothesis, no single one of which provides a firm basis for rejecting the model. There is first the problem that such a nonradiative crossover shares the available kinetic energy equally between the two halide ions between which the bond has been broken. Thus in KCl, for example, there is at most about 4 eV available to initiate a given interstitial creation event. Schulze and Hardy have recently calculated the energy of formation of an isolated halide-ion-vacancy-interstitial-ion pair, the value being 3.73eV[38]. Furthermore, using a Born-Haber cycle with appropriate theoretical energies, it can be shown that the energy to form an F-H pair is probably 2-3 eV greater than this. Even ignoring all dissipative processes, it is thus very improbable that the exciton energy can produce well-separated defects if shared equally between the two halide ions. Smoluchowski et al have dealt with this matter and have argued that zero-point vibrations provide some degree of asymmetry[39]. However, the transient efficiency measurements previously described indicate that better than one recombination event in five produces an F center, and thus the argument is borderline, particularly if dissipation is present.

It has been argued that focused collisions along a close-packed line of like ions is most favorable for the separation for the vacancy and interstitial. Various theoretical estimates have been made of the energy required for such a process, including dissipation[40] The results appear to indicate that it is highly unlikely that a displacement can be produced with less than

10 eV, the end product being an F^+-I pair. An alternative possibility has been conjectured by several investigators, this being that in the replacement collision sequence the electron remains in the vacancy and the interstitial emerges and propagates as an H center. This idea obviously has desirable aspects, and calculations indicate that under the most favorable circumstances a stable F-H pair can be produced with as little as 5 eV in KCl[39]. A substantial asymmetry in the energy partition is still required. There is also an apparent difficulty with the electronic states ; it is hard to visualize a mechanism for reionizing a halide ion once the crossover to the ground state (two p^6 ions) has been made. This is a basic problem if F-H pairs and not F^+ -I pairs are to be the primary product.

Finally, it has been noted in Section IIIC that the measured optical binding energies of STE's lead to barriers which are in all cases significantly larger than the barriers one would infer from the measured temperature dependences of either luminescence or F-center production.

In view of these difficulties it is reasonable to ask whether there may be other possible nonradiative decay channels for the STE rather than the simple ground state crossover. The original excitonic model of Hersh assumed that dissociation of the X_2^{-2} occured via a hypothetical state which intersected the STE states near their minima[41]. However, the existence of such a state is inconsistent with current knowledge of the STE.

An alternative hypothesis can be constructed which would appear to have significant merit in this situation. The configuration coordinate diagrams of fig.9 supposedly take account of the two most important even modes affecting the STE. Let us consider the response of some arbitrary STE state lying between points X and Y to an odd-parity mode involving translation of the X_2^- as a whole parallel to its axis. Driven to extreme this mode can clearly become an F-H pair, thereby lowering the potential of the system. The crucial question is whether or not there exists a prohibitively high potential barrier to this type of evolution. One would on first thought expect the two nearest alkali ions to provide a significant barrier ; there is in fact no indication that the V_k center alone ever penetrates (thereby forming an F^+-H pair). However, the presence of the electron can change matters, in that it may serve to lessen this barrier for certain states. We have indicated that the largest optical binding energies of the triplet STE state occur for those crystals in which the electron can best avoid encountering the X_2^- core[25]. In these cases the electron already exhibits certain properties of an F electron. If the electron responds adiabatically to the motion of the X_2^-, in such a way as to occupy the incipient vacancy, it is conceivable that the energy thus gained might counterbalance the alkali-ion barrier, allowing the X_2^- to slip through. Once begun, the motion would be aided by overlap of the hole onto the adjacent halide ion forming an X_3^{-2} -like species. The barrier against subsequent H-center motion is of course quite small. In this model all the energy of the STE state would be available (not one-half), the collisional aspect would be minimized, and motion of all ions might remain roughly coherent as implied in a configuration coordinate representation.

As mentioned in Section IIIB, the data of Keller and Patten on H-center production indicate that the process is more effective for recombination with a $BrCl^-$ -type hole than with a Cl_2^- hole[20]. This was consistent with the ground state crossover model, since the kinetic energy partition is asymmetric for the mixed V_k -center recombination. However, the barrier to translation of the XY^- core also becomes asymmetric, and for motion in one direction it should become lower than that for the X_2^-. Thus the X_2^- translation model can also agree with these data.

Schematic potential curves which might characterize this production process in KCl are sketched in fig. 11. The configuration coordinate z represents approximately the location of the center of the Cl_2^- as it translates parallel to its axis along a <110> direction. This coordinate is orthogonal to r and d in fig.9. The process originates at U, which might occur just below X' in fig.9. Point G represents an H center formed at the first interstitial position, which may or may not be stable. Point F^+ -H is the next position, which is probably stable. The energy scale is the same as that of fig.9, zero being the perfect lattice. The lower curve represents the ground state, point F +I being a stable F^+-I pair. The energy at point F+I is the Frenkel pair formation energy calculated by Schulze and Hardy[38]. However, this point should in principle be somewhat lower than shown because of the Coulomb attraction of the close pair. The location of point F+H relative to point F^++I represents a best estimate of the

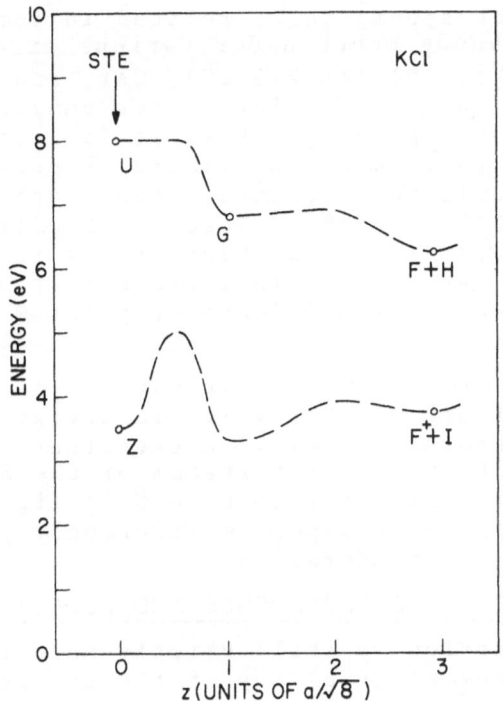

Fig.11 : Schematic potential curves describing possible defect
production in KCl by relaxation of some arbitrary STE state via
an odd-parity local mode z.

formation energy of an F-H pair relative to that
of an F^+-I pair. The actual shapes of the curves are
highly schematic. The upper curve has been drawn without
a barrier between points U and G, although a slight
barrier might not conflict with experiment. Note the
effect of transitions from F+H to F^+I.

The potentials sketched in fig.11 might conceiva-
bly explain the existence of certain additional emission bands which appear to be related in some way to
the STE.These bands occur under various circumstances
in KI(at 3.0 eV), RbI (at 2.3 eV), CsI, and KBr[42]. They
appear to originate in triplet states comparable to
those of the corresponding STE's. It is tempting to
speculate that these bands originate at point G in
fig.11. They could be described, then, either as the
radiative annihilation of a closest F-H pair or as a
perturbed STE transition. Although there is at the moment no better way to explain these emission bands,
the evidence that they originate at point G is certainly not compelling.

The principle advantages of this model are that it
makes efficient use of the available energy and produces F and H centers directly. On the other hand, it is
not obvious that there exist states of the STE with
the required low barrier from U to G in fig.11. On balance, the model would appear sufficiently plausible
to merit further consideration.

F. - TEMPERATURE DEPENDENCE AND DEFECT STABILITY

In this section we shall briefly note effects related to the thermal stability of the centers we have
been discussing. It is clear that all processes following the creation of electrons and holes can depend
upon temperature, and it is often difficult to determine which process is primarily responsible for a given temperature dependence. However, substantial progress has been made with these problems.

The importance of temperature in determining the
relative probabilities of radiative and nonradiative
decay of the STE has already been described. In general,
the inverse correlation is apparent only for crystals
which color poorly at low temperature, the prime example being KI. Other than this, the vacancy-interstitial
production event itself has not shown a strong temperature dependence, and coloration rates at 300K are
generally comparable to those at lower temperatures,
at least in pure crystals.

Neither EPR nor optical spectra show perturbation of H centers by F's and vice versa. Over the wide distribution of interstitial sites populated by the replacement sequence, the far pairs appear to remain F and H centers while the close pairs become F and I. The electron transfer process $F+H\rightarrow F^++I$ is more probable the smaller the distance. In typical alkali halides the interstitial ion is the least stable defect. F^+-I pairs become thermally unstable in the KBr between 10 and 20K[43] and in KCl below about 35K[44]. H centers are somewhat more stable and decay around 50K in both these crystals. Next, the V_k center becomes mobile at around 140K in KBr and 170K in KCl[45]. Vacancies are the last defects to move, the F^+ center becoming unstable in the 200-240K range and the F center being stable above room temperature in both materials[1]. When alkali halides are irradiated at low temperatures and warmed, one expects to lose centers at temperatures where either the center itself or its complement becomes thermally unstable. On the other hand, both vacancies and interstitials can form a variety of stable aggregates. In many cases aggregation is mor likely than recombination, and sometimes irradiation at a given temperature can even produce more total damage than a comparable dose at a lower temperature.

Impurities, even in very low concentration, can have strong effects on the damage process. In KCl around liquid nitrogen temperature the primary intertials center, the V_1 or H_A, is actually an interstitial halogen atom trapped by a substitutional Na^+ impurity[46]. This center dominates until the radiation dose becomes great enough to saturate the impurity.

Another interesting case involves electron-trapping impurities present in low concentration. In KCl, at temperatures where interstitials are mobile but the V_k center is stable, the damage rate is greater in crystals doped with an impurity such as Pb than in the pure crystal[47]. This effect is reasonably attributed to the stabilization of interstitials trapped at the impurity. At higher temperatures, however, the effect becomes reversed and electron trapping impurities actually inhibit defect formation. Either of two factors can be responsible : Mobile F^+ centers, whose relative concentration is enhanced by the presence of electron traps, can annihilate interstitials trapped at impurities ; or mobile V_k centers may recombine at the impurities, as in the scintillation process, and thus become unavailable for participation in the nonradiative recombination process in the perfect lattice which produces Frenkel defects. The latter process has been shown to predominate in KCl and KI by means of comparisons with the measured thermal properties of V_k and F^+ centers[48].

IV. IONIZATION DAMAGE IN NONALKALI HALIDES

Crystals in this category have received only a small fraction of the attention given to alkali halides primarily because of problems with preparation, purity, and characterization. Several crystals have been shown to exhibit radiation damage properties similar to those of the alkali halides, and the operation of an excitonic mechanism is strongly implied. In fact, at this early stage the available data would lead one to guess that the excitonic damage mechanism is a general characteristic of those simple halide lattices for which the band gap is relatively large and the halide ions are located not too far apart. The occurrence of self-trapped holes in the X_2^- configuration is evidently a necessary but not sufficient condition for the operation of an excitonic mechanism ; the ammonium halides are evidently the exception, since they yield V_k centers quite readily but have so far shown no F centers. In this Section we shall briefly describe certain aspects of the damage process in MgF_2, $KMgF_3$ and SrF_2. It is fair to presume that the behavior illustrated by these examples will not prove to be confined to them alone.

At low temperatures and also around room temperature, x-ray irradiation of MgF_2 produces F centers with an efficiency of the order of 10^5 eV per F center[49,50]. However, in the range 80-200K the production rate is lower and a strongly saturating behavior is noted. These results can be interpreted in terms of a model in which only close pairs of F and H centers are produced. These are stable at low temperatures, but due to a small net attraction between vacancy and interstitial they tend to recombine at intermediate temperatures as the interstitial becomes mobile. At temperatures where the thermal energy is greater than the net attraction, the interstitial has appreciable chance of diffusing away from its vacancy and becoming stabilized at some other imperfection. A simple kinetic model has been shown to reproduce this behavior, that is, to be consistent with the observed temperature dependence[50].

The aspect of these results which we wish to emphasize is the extent to which they implicate a close-pair distribution of Frenkel defects. In MgF_2, which has the Cassiterite structure, there are no close-packed, symmetric directions for focused collisions comparable to the <110> directions in the NaCl structure. This should inhibit the separation of pairs, whether by a collision sequence or by a more nearly adiabatic process such as that sketched in fig.11.

Various color centers have been identified in irradiated $KMgF_3$[51]. For this material the intrinsic F-center production efficiency is also on the order of 10^5 eV per F center, and the temperature dependence of the damage process is generally similar to that of KCl. In $KMgF_3$ a weak, higher temperature suppression of F-center production is observed, similar to that mentioned for KCl and KI. Here, however, the data indicate that it is F^+ motion and not V_k motion which causes the effect. In the perovskite structure focused collision sequences can occur, but their length is expected to be less than in the alkali halides.

Recently a rather thorough preliminary investigation of radiation damage in SrF_2 has been carried out[52]. This is the best documented example of closely spaced vacancies and interstitials. At 5K x-ray irradiation creates F centers at an efficiency only slightly less than for KBr. Thus an excitonic mechanism is strongly implicated. The complementary hole center is the V_k and not the H as in alkali halides. H centers are observed, but their concentration is more than two orders of magnitude smaller than that of the V_k centers. Thus the F-I pair is evidently predominant, its net negative charge being compensated by a remote V_k center. Either the interstitial ion is the direct product of the excitonic mechanism, or the nascent interstitial atom immediately captures an electron, probably from its complementary F center. In contrast to KBr and KCl, the F^+ center can evidently readily capture an electron in spite of the close proximity of the negative interstitial.

The F band shows complex structure which is attributed to perturbation by the nearby interstitial ion ; partial linear polarization with <111> symmetry can be induced in the F band by optical bleaching. ENDOR measurements on the F center, while extremely complex, nevertheless confirm the predominance of a close F-I pair with a <111> axis. The most probable configuration involves the following sequence along a <111> direction: F center-Sr^{2+}-F^--I. The mechanism for placing an interstitial in such a position presents a dilemma. Although the replacement collision sequence would be expected to be quite short, it would involve <100> segments, at least three of which are required to reach the proposed interstitial position.

F centers anneal at the temperature where V_k centers become mobile, 110K. The V_k evidently recombines with the F-H pair, and the resulting F^+-I pair is stable only below about 40K. The temperature-dependence data indicate the presence of other smaller stages involving pairs of slightly greater separation, but above about 120K the rate of production of F centers under steady irradiation is negligibly small. Unlike in MgF_2, the interstitial appears unable to diffuse away from its companion vacancy.

It should be noted that SrF_2 gives no evidence of an inverse relationship between STE recombination luminescence and F-center production. This is not unexpected, since the F-center production efficiency is relatively high at low temperature. Only singlet STE luminescence has been observed in the alkaline earth halides[53], which raises the possibility that the triplet states always depopulate by a channel such as the indicated in fig.11.

Upon comparing these three fluorides with the alkali halides, one concludes that a focused collision sequence is by no means essential to the efficient production of vacancies and interstitials at low temperature. However, there does appear to be a correlation between the possibility of focused sequences and the production of remotely spaced pairs which are able to survive effectively at higher temperatures.

On behalf of a balanced perspective, we close this Section by mentioning two halides which are evidently entirely immune to vacancy production by an excitonic mechanism. They are the silver and thallous halides. Silver halides of course exhibit photochemical damage in the silver sublattice, the photographic process. They also show hole self-trapping[54], but again in the silver sublattice as Ag^{+2}. Thallous halides have been little investigated but show no strong damage effects.

REFERENCES

1. For general reviews of topics in this field, see Point Defects in Solids, ed.by Crawford J.H.Jr. and Slifkin L.M (Plenum Press, N.Y.,1972), Vol.I, particularly the chapter by Sonder and Sibley, and D.Pooley and W.A.Sibley.

2. Birks J.B., the theory and Practice of Scintillation Counting (MacMillan, N.Y., 1964)

3. The absorbed energy required to create an electron hole pair is approximately three times the band gap. See Goulding F.S and Stone Y, Science 170, 280 (1970).

4. Dietrich H.B. and Murray R.B., J.Luminescence 5, 155 (1972).

5. Dietrich H.B., Purdy A.E., Murray R.B. and Williams R.T., Phys.Rev. B, to be published.

6. For a review of hole centers, see Kabler M.N. in Point Defects in Solids, ed. by Crawford J.H., Jr. and Slifkin L.M. (Plenum Press, N.Y., 1972), Vol.I.

7. Jette A.N., Gilbert T.L. and Das T.P., Phys.Rev. 184, 884, (1969).

8. Keller F.J. and Murray R.B., Phys.Rev. 150, 670, (1966).

9. Hattori A., Tomura M. and Nishimura H., J.Phys.Soc. Japan 31, 611 (1971).

10. Balzer R., Peisl H. and Waidelich W., Physics Letters 27A, 31 (1968) and Phys.Stat.Sol. 28, 207 (1968).

11. For a comprehensive review of radiolysis in alkali halides, see Crawford J.H., Jr., Advances in Physics 17, 93 (1968).

12. Varley J.O.H., J.Phys.Chem.Solids 23, 985, (1962).

13. Ritz V.H., Phys.Rev. 133, A1452 (1964)

14. Klick C.C., Phys.Rev. 120, 760 (1960)

15. Williams F.E., Phys.Rev. 126, 70 (1962)

16. Hall T.P.P., Pooley D., Runciman W.A. and Wedepohl P.T., Proc.Phys.Soc. 84, 719 (1964)

17. Lushchik C.B., Liidya G.G. and Elango M.A., Sov. Phys.Sol.State 6, 1789 (1965) ; Goldstein F.T., Phys.Stat.Sol. 20, 379, (1967).

18. Kabler M.N., Phys.Rev. 136, A1296, (1964) ;
    Murray R.B. and Keller F.J., Phys.Rev.137, A942,
    (1965).

19. Pooley D., Proc.Phys.Soc. 87, 245, (1966) ; ibid
    257.

20. Keller F.J. and Patten F.W., Solid State Commun,
    7, 1603, (1969).

21. Pooley P. and Runciman W.A., J.Phys.C3, 1815
    (1970).

22. Ikezawa M., Shirahata K. and Kojuma T., Science
    Reports of Tohoku Univ. LII, 45, (1969)

23. Rabin H. and Klick C.C., Phys.Rev. 117, 1005,
    (1960).

24. Hughes A.E., Pooley D., Rahman H.U. and Runciman
    W.A., AERE Report R5604 (1967).

25. Fuller R.G., Williams R.T. and Kabler M.N., Phys.
    Rev. Lett. 25, 446 (1970) ; Williams R.T. and Ka-
    bler M.N., Phys.Rev., to be published.

26. Marrone M.J., Patten F.W. and Kabler M.N., Phys.
    Rev.Lett. 31, 467 (1973).

27. Kabler M.N. and Patterson D.A., Phys.Rev.Letters
    19, 652, (1967).

28. Blair J.M., Pooley D. and Smith D., J.Phys.C5,
    1537 (1972).

29. Marrone M.J. and Kabler M.N., Phys.Rev.Lett. 27,
    1283 (1971).

30. Wood R.F., Phys.Rev. 151, 629 (1966).

31. Schulman J.H. and Boag J.W., Phys.Stat.Sol. 3,
    516, (1963).

32. Ueta M., J.Phys.Soc.Japan 23, 1265 (1967).

33. Kondo Y., Hirai M. and Ueta M., J.Phys.Soc.Japan
    33, 151, (1972).

34. Hirai M., Kondo Y., Yoshinari T. and Ueta M., J.
    Phys.Soc. Japan 30, 440, (1971).

35. Dienes G.J., Hatcher R.D., and Smoluchowski R.,
    Phys.Rev. 157, 692, (1967).

36. Fowler W.B., Marrone M.J. and Kabler M.N., Phys.
    Rev., to be published.

37. Williams R.T., private communication.

38. Schulze P.D. and Hardy J.R., Phys.Rev. B6, 1580, (1972).

39. Smoluchowski R., Lazareth O.W., Hatcher R.D. and Dienes G.J., Phys.Rev.Lett. 27, 1288, (1971).

40. Torrens I.M. and Chadderton L.T., Phys.Rev. 159, 671, (1967).

41. Hersh H.N., Phys.Rev. 148, 928, 1966

42. Fischbach J.U., Fröhlich D. and Kabler M.N., J. Luminescence 6, 29, (1973).

43. Itoh N., Royce B.S.H. and Smoluchowski R., Phys. Rev. 137A, 1010, (1965).

44. Giuliani G., Perinati A., Reguzzoni E. and Chiarotti G., Solid State Commun. 3, 161, (1965).

45. Delbecq C.J., Hayes W. and Yuster P.H., Phys.Rev. 121, 1043, (1961).

46. Patten F.W. and Keller F.J., Phys.Rev. 187, 1120 (1969) ; Delbecq C.J., Hutchinson E., Schoemaker D., Yasaitis E.L. and Yuster P.H., Phys.Rev. 187, 1103, (1969).

47. Sonder E. and Sibley W.A., Phys.Rev. 140A, 539, (1965).

48. Dawson R.K. and Pooley D., Solid State Commun. 7, 1001, (1969).

49. Sibley W.A. and Facey O.E., Phys.Rev. 174, 1076, (1968).

50. Buckton M.R. and Pooley D., J.Phys.C5, 1553 (1972).

51. Riley C.R. and Sibley W.A., Phys.Rev. B1, 2789 (1970).

52. Haynes W. and Lambourn R.F., J.Phys. C6, 11 (1973).

53. Beaumont J.H., Haynes W., Kirk D.L. and Summers G.P., Proc.Roy.Soc. Lond. A 315, 69 (1970).

54. Marquardt C.L., Williams R.T. and Kabler M.N., Solid State Commun. 9, 2285 (1971)

COLOR CENTERS IN OXIDE GLASSES

D.L.GRISCOM

Naval Research Laboratory
Washington DC 20375
U.S.A.

INTRODUCTION

The problem of studying radiation damage processes in glasses is quite different from the problem in crystals. To begin with, all hope of exploiting crystal symmetry must be abandoned. By definition, glasses have no long range order. Glasses are isotropic. Thus, all spectroscopic information is automatically averaged over 4π steradians. Moreover, the quenched-in randomness of the glassy state tends to blur this information ever further. Nevertheless, these seemingly great difficulties are being surmounted. Not only are the fundamental radiation defects being identified, but they are also being employed as useful "probes" of glass structure.

There exists a substantial literature dealing with the general subject of radiation effects in oxide glasses. An entry to this is provided by several recent review articles[1,2]. The present notes are drawn from a somewhat more specialized review article[3] entitled "ESR Studies of Radiation Damage and Structure in Oxide Glasses Not Containing Transition Group Ions : A Contemporary Overview with Illustrations from the Alkali Borate System". The ESR studies are emphasized here because electron spin resonance has proved to be the most powerful single technique for _identifying_ the generic types of defects induced by radiation. Transition group ions have been excluded in order to concentrate fully on those defects which are intrinsic to the undoped base glass. (It is remarked, however, that the radiation co-

loration of oxide glasses is far less sensitive to doping with small concentrations of impurities than is
the case for alkali halides)[4]. Finally, it will be seen
that specialization to the alkali borate glasses entails
no loss of generality, since analogous defects are (or
in principle could be) observed in virtually all other
oxide glass systems.

DEFECT STRUCTURE IN RELATION TO GLASS STRUCTURE

Figure 1 presents a schematic view of an oxide
glass. Here, a three-dimensional random network, consisting of network forming metal atoms covalently bonded to oxygens, is portrayed as a stylized two-dimen-
sional structure in the usual manner[5]. It should be understood by this diagram that when four oxygens sur-
round the network former, R, R_A, or R_B, they are in a
tetrahedral arrangment with the network former at the
center. Silicon is always tetrahedrally coordinated,
except when there is an oxygen vacancy as in fig.1a.
Such oxygen vacancies result in positively charged
"point defects" which may trap electrons. When R = Si,
this trapped electron center is the E' center[6], and
ESR evidence has shown the wavefunction of the unpaired
spin to look something like the dashed "balloon" at the
bottom of fig.1b.

When glasses are prepared using B_2O_3, P_2O_5, GeO_2,
or Al_2O_3 as constituents, other situations can and do
occur. Most of these may still be discussed within the
framework of fig.1 if appropriate conceptual allowances
are made. For example, boron is frequently in tetrahedral coordination in complex oxide glasses, although
it also occurs in planar triangular units[5]. (The fraction of borons that are four coordinated can be ascertained by nuclear magnetic resonance[7]). Since atomic
boron has just 3 valence electrons, when it is found in
4-coordination it has gained another electron from another source (usually a network modifying cation, as
discussed below). The same is true of aluminum. Thus,
if R_B = B or Al in fig.1, the tetrahedral complex (R_B)
O_4 may be looked upon as a negative point defect whenever it should be isolated from a charge-compensating
interstitial cation such as the network modifier. Such
negative defects are hole traps. The approximate wavefunction of a hole trapped at such a site has been determined from ESR[8] to be something like the dashed
"dumbell" centered on the oxygen bridging between R_B
and an R to the right of fig.1b.

Explicit consideration is now given to the effects

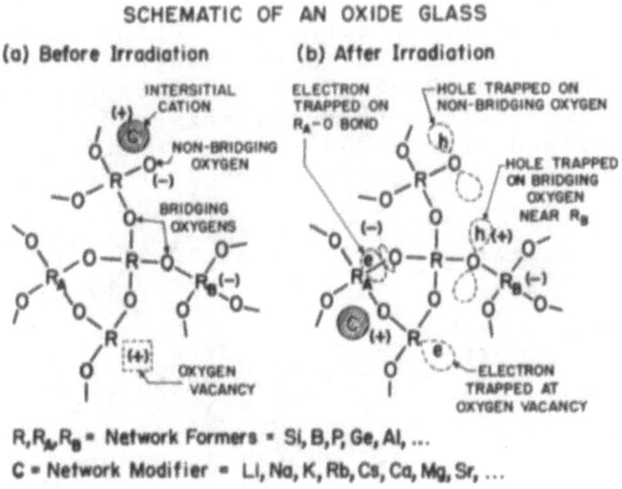

SCHEMATIC OF AN OXIDE GLASS

(a) Before Irradiation   (b) After Irradiation

R, R_A, R_B = Network Formers = Si, B, P, Ge, Al, ...

C = Network Modifier = Li, Na, K, Rb, Cs, Ca, Mg, Sr, ...

Fig.1 : Schematic diagram of an oxide glass showing several types of radiation-induced paramagnetic centers in their relation ships with pre-existing defects such as an oxygen vacancy, a non-bridging oxygen, and substitutional impurities. Dashed "balloons" enclose regions of high probability density for the trapped electrons (e) or trapped holes (h). This figure illustrates the locally charge-compensated nature of the trapped species. It should be noted, however, that the individual trapped holes and electrons are assumed to be much more distant from one another than shown here.

of additions of <u>network modifying</u> oxides such as alkali or alkaline-earth oxides. As seen above, such additions can lead to the 4-coordination of boron or aluminum. Thus one "molecule" of alkali oxide will provide an oxygen ion with sufficient electrons to form bonds with two borons or aluminums, leaving the two positively charged alkali ions in interstitial positions. If, however, a "molecule" of alkali oxide is added to an SiO_2 glass, the effect is to create two nonbridging (singly bonded) oxygens . In the absence of nearby charge-compensating cations, these nonbridging oxygens may be looked upon as another type of negatively charged point defect which can trap a hole. The wavefunction for this trapped hole center could not be other than that indicated at the top of fig.1b, since the oxygen π orbitals perpendicular to the R-O bond represent the lowest available energy states for a hole at this site.

Another type of electron trap can result when R_A has the same valency as R, but a greater electron afinity. An example is when $Ge(=R_A)$ is doped into quartz. For this case an electron can be trapped in a Ge-O antibonding orbital[9]. The stability of such a center would be increased by a nearby interstitial cation. In fact, cations often diffuse to different positions following low temperature irradiation and subsequent warming, thereby stabilizing an otherwise weakly trapped species[1,9]. This situation is illustrated in fig.1.

Finally, the insterstitial cations themselves may serve as electron traps. Recent ESR evidence for alkali-associated trapped electron centers in alkali borate glasses has been given[10]. Due to the general mobility of interstitial ions, "clusters" may form before, during, or after irradiation, resulting in trapping sites comprising a number of cations[10]. The antimorph to the cation electron trap is the anion hole trap. Although interstitial anions are a rare situation, there is evidence[11-15] that when alkali borate glasses are prepared with alkali halide additions, the halide ions take up interstitial positions and are often effective hole traps.

ESR SPECTRAL ANALYSIS

All of the radiation-induced paramagnetic states to be discussed have an effective spin of 1/2 and are adequately characterized by a resonance condition of the form

$$H_{res} = \frac{h\upsilon}{g\beta} - m_I \left(\frac{A}{g\beta}\right) \qquad (1)$$

where

$$g=[g_{zz}^2\cos^2\theta+(g_{xx}^2\cos^2\phi+g_{yy}^2\sin^2\phi)\sin^2\theta]^{1/2}$$

$$A=[g_{zz}^2A_{zz}^2\cos^2\theta+(g_{xx}^2A_{xx}^2\cos^2\phi+g_{yy}^2A_{yy}^2\sin^2\phi)$$

$$\sin^2\theta]^{1/2}/g$$

and m_I is the magnetic quantum number of a nucleus with which the spin is undergoing a hyperfine interaction. (In some cases, however, significant improvement has been obtained by appending 2nd order hyperfine terms to Eq.(1)).

Because the spectra observed in glasses are always averaged over all angles, Eq.(1) must be employed to compute the appropriate ESR powder pattern[16,17]. This is done by computing H_{res} for each of a large number of orientations corresponding to a uniform grid in cos θ - φ space (equal units of solid angle) and histogramming the results on a magnetic field scale[18]. These powder patterns are then convoluted with a "single crystal" broadening function (Gaussian or Lorentzian) of an appropriate width and the first derivative is calculated and displayed on a plotter. Accurate values for the spin Hamiltonian parameters are obtained by making trial-and-error adjustments until the computed spectrum is brought into good agreement with the experimental line shape.

THE OBSERVED SPECTRA AND THEIR INTERPRETATIONS

A. Oxygen-Associated Trapped-Hole Centers

The most commonly observed ESR spectra in irradiated complex oxide glasses arise from oxygen-associated trapped-hole centers. The fact that these centers are associated primarily with oxygens is inferred from the relatively small hyperfine interactions noted with the network formers (see table I)[19] and the negligible dipolar broadening due to the network modifiers (<4G in lithium borate glasses)[20]. That this class of paramagnetic defect centers is of the trapped-hole type was originally inferred from the fact that the principal values of the g tensor are greater than ·the free-electron value[21]. A more convincing proof of this conclusion was the observation that in halide-containing borate glasses hal$\frac{1}{2}$ hole-type centers are produced at

214

Table I Spin Hamiltonian parameters for hole-type centers in a variety of oxide glasses and the mineral phenacite.

Material	Nucleus	g_1	g_2	g_3 [a]	$\langle A \rangle$[b]$/A_s$	Ref.
Borate glass	^{11}B	2.002	2.010	2.035	0.013	20
Silicate glass	^{29}Si	2.003	2.009	2.019	0.010	22
Phenacite	^{29}Si	2.002	2.014	2.035	d	23
Germanate glass	^{73}Ge	2.002	2.008	2.051	d	24
Phosphate glass	^{31}P		2.010^c	a	0.009	25
Aluminate glass	^{27}Al		2.009^c	a	0.005	26
Niobate glass	^{93}Nb		2.01^c	a	0.002	27

a- Each of the glass spectra is evidently characterised by a distribution of g_3 values, accounting for the observed low-field shoulders. The "average" g_1 value for borate glasses was determined by computer simulation(ref.20).

b- $\langle A \rangle$ =(average measured hyperfine splitting)$= \frac{1}{3}\,\mathrm{Tr}|A|$. Values for A_s, the atomic Fermi contact coupling, were taken directly, or extrapolated, from the calculations of Hurd and Coodin(ref.28).

c- For phosphate, aluminate and niobate glasses, the only available g value is that for the centroid of the hyperfine multiplet structure; this should correspond approximately to the value of g_2 listed for the other centers.

d- Hyperfine structure due to ^{29}Si (5%) or ^{73}Ge (8% abundant), if present, may be obscured by superhyperfine structure or other effects.

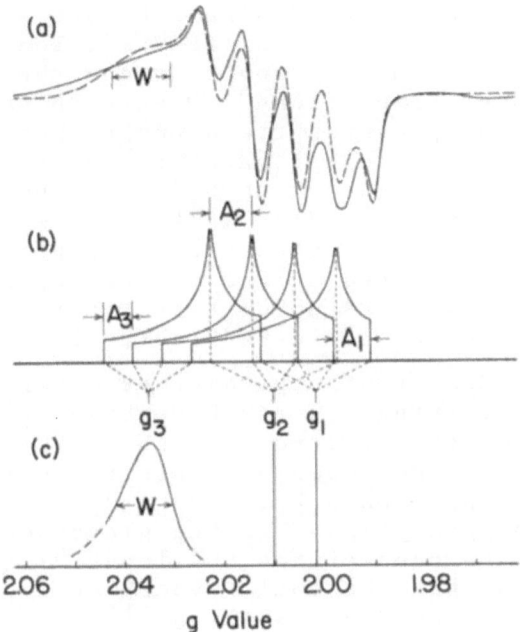

Fig.2 : An analysis of the X-band (9 GHz) ESR spectrum of the boron-oxygen hole center (BOHC) in irradiated alkali borate glasses. The unbroken curve in (a) is the experimental spectrum for a 20 % K_2O-80 % B_2O_3 glass x-irradiated at 77°K and bleached with IR light. The dashed curve in (a) is a computer simulation[20] based on the powder patterns shown in (b) and the distribution of g values shown in (c). Because the width W of the g_3 distribution, expressed in units of magnetic field, is greater than the hyperfine coupling constant A_3, no hyperfine structure is discernable on the low-field shoulder in (a).

the expense of the oxygen-associated centers normally
produced in the halide-free glasses[11]. Some properties
of the oxygen-associated trapped-hole centers will be
illustrated by an exposition of the borate-glass spe-
cie, the boron-oxygen hole center (BOHC).[20]

Figure 2a displays the X-band spectrum of the
BOHC typical of irradiated alkali borate glasses containing $\lesssim$ 25 mole % alkali oxide[20,21]. The unbroken
curve is the experimental spectrum and the dashed curve is a computer simulation[20]. Figure 2b indicates the
four overlapping powder patterns corresponding to the
four values of the nuclear magnetic quantum number for
[11]B (I=3/2, natural abundance = 81 %). The model was
also corroborated by studies of samples enriched to
96 % in [10]B (I=3)[20,21]. Good agreement between the experimental and computed curves in the low field region
was obtained only by assuming a statistical distribution of g_3 values as indicated in fig.2c.

The spin Hamiltonian parameters determined in
this computer fitting process have been used to formu-
late models for the structure of the BOHC[19,20,29,30].
At present it has not been possible to decide whether
the hole is trapped on a bridging oxygen (fig.3) or a
nonbridging oxygen (fig.4) in the glass network. By
analogy with the case of the BO_3^{2-} ion[29,31] it seems
fairly certain that the small hyperfine interaction
with a <u>single</u> boron nucleus (A_{iso} ([11]B) $\lesssim$ 10G) is via
the mechanism of <u>core polarization</u>. If fig.3 is the
correct picture, there must be a much smaller interaction with the second boron. Such a situation is conceivable, since both three-and four-coordinated borons are
present[7]. In either case, the observed g values can be
accounted for by relations of the type.

$$g_1 \ (=g_z) \ \approx \ g_{\text{free electron}} \qquad (2a)$$

$$g_2 \ \approx \ g_{\text{free electron}}[1+ \frac{c_2 \lambda}{\Delta}] \qquad (2b)$$

and $$g_3 \ \approx \ g_{\text{free electron}}[1+ \frac{c_3 \lambda}{\Delta}] \qquad (2c)$$

where c_2 and c_3 are constants of the order 0.1 - 1
calculable from theory, λ is the spin orbit coupling
constant for the O^- ion ($\sim$ 0.01 eV), and Δ is the e-
nergy splitting indicated in fig.3. (Similar relations
involving Δ_1 and Δ_2 in place of Δ apply to fig.4). Statistical variations in B-O-B bond angles (fig.3) rela-

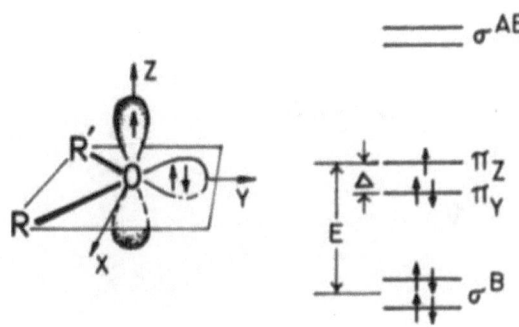

Fig.3 : Schematic of a bridging-oxygen trapped hole center. Left : steric picture : electronic energy levels. As indicated, the unpaired spin would be in the non-bonding π_z orbital. If the assumed sp hybridization of oxygen orbitals is correct, it is estimated[20] that $\Delta \sim 0.5$ eV, and E is believed to be ~ 2.5 eV on the basis of optical data.

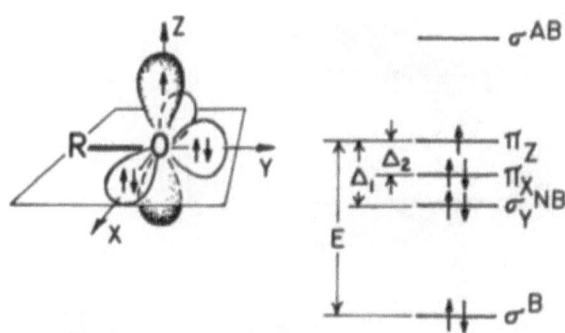

Fig.4 : Schematic of a non-bridging-oxygen trapped hole center. Left : steric picture. Right : electronic energy levels. The mechanism responsible for the splitting Δ_2 is not known.

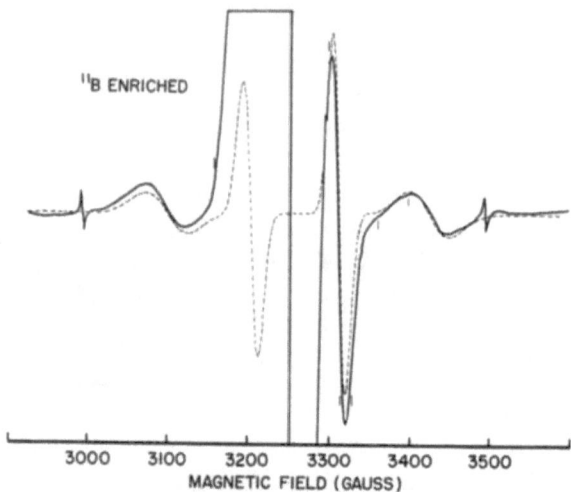

Fig.5 : X-band ESR spectrum of a 17.7 % K_2O-82.3 % B_2O_3 glass, enriched to 98.4 % ^{11}B, x-irradiated and observed at 77°K. Dashed curve is a computer simulation of the BEC contribution. The BOHC is responsible for the intense absorption centered near 3250 G. (Weak lines split by $\sim$ 500 G are due to atomic hydrogen).

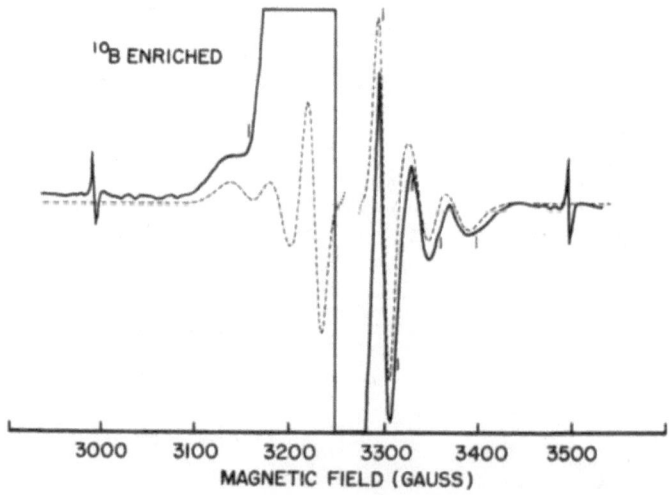

Fig.6 : X-band ESR spectrum at a 17.7 % K_2O-82.3 % B_2O_3 glass, enriched to 96.4 % ^{10}B, irradiated and observed at 77°K. Dashed curve is a computer simulation of the BEC contribution, based on parameters wholly consistent with those used to compute the spectrum of fig.5. For simplicity in computation, the central, $m_I = 0$ component was suppressed.

ted to vitreous disorder could give rise to the apparent distribution of g_3 values.

In view of the fact that alkali silicate glasses
have no three-coordinated silicons[5] and alkali borate
glasses of $\lesssim$ 25 % alkali content have negligible numbers of non-bridging oxygens,[7] it seems possible that
fig.3 may apply to the BOHC while fig.4 may apply to
the oxygen-associated hole-type centers in the silicate
glasses. (A center with similar spin-Hamiltonian parameters in the mineral phenacite[23] seems fairly certain
to be due to a hole on a non-bridging oxygen on a four-
coordinated silicon). This would mean that table I may
include both types of centers and that they are indistinguishable on the basis of g values alone.

B. Dangling-Orbital-Type Trapped-Electron Centers

Perhaps the best known example of a trapped-elec-
tron center in oxide glasses is the E' center[6], which
is observed in pure SiO_2 glass,[32] as well as crystalline quartz. Essentially, it consists of an electron trapped in a dangling sp^3 silicon orbital projecting into
an oxygen vacancy. The borate analog of the E' center,
the boron electron center (BEC), was a recent discovery[33]. The BEC was not observed in pure B_2O_3 glass but
was found to be stable in alkali borate glasses below
$\sim$ 77°K in darkness.

The ESR spectral analysis of the BEC illustrated
in figs 5 and 6 is of interest for several reasons.
First, it shows how the techniques of isotopic substi-
tution and computer simulation may be used to confirm
the identities of defects in glasses even when their
ESR spectra are partially obscured by other resonances.
A more subtile point relates to the question of why the
outer hyperfine lines are broader and "weaker" than the
inner lines. As confirmed by the computer simulations,
this effect is due to a fairly broad gaussian distribution of A_{iso} values characterizing the ensemble of defects in the glass - another manifestation of vitreous
disorder.

The computer simulations, based on Eq.(1) augmen-
ted by second order hfs terms, provided reasonably accurate measurements of the spin Hamiltonian parameters.
The average g shift was found to be small and negative
($\Delta g_{iso} \sim -0.0005$), as for the E' center. For the glasses of figs. 5 and 6 it was further determined that[33]

$$|<\psi|2s>|^2 = (A\| + 2A\bot)/3A_s = 0.13 \qquad (3a)$$

and

$$|<\psi|2p>|^2 = (A\| - A\bot)/3A_p \lesssim 0.51 \qquad (3b)$$

where ψ is the BEC wavefunction and A_s and A_p are the s-state and p-state coupling constants for atomic boron.[34] The considerable uncertainty in $|<\psi|2p>|^2$ results from the obscuring effects of the distribution in A_{iso} values. Nevertheless, Eqs.(3) have been useful in formulating models for the BEC. There is a strong basis[33] for believing the BEC is a dangling orbital defect, rather than an electron in a boron-oxygen antibonding orbital. Clearly the unpaired spin is highly localized on a boron nucleus. However, the possibility of its participating in a weak 3-electron bond with an alkali ion is supported by some observations and cannot be ruled out[33]. A_{iso} is sensitive to both the type and quantity of alkali ions present. For the potassium borate glasses, $|<\psi|2s>|^2$ drops from 0.14 to 0.11 as the molar % K_2O is raised from 15 to 25. The latter effect can be related to changes in the boron-oxygen structural groupings postulated on the basis of other evidence.[35]

C. Alkali-Associated Trapped-Electron Centers

When the BEC population is compared to the BOHC population, it is found[33] that the BEC can account for only 15 ± 10 % of the trapped electrons for T $\lesssim$ 77°K and none of those electrons still trapped at room temperature. In the low temperature regime it now appears that the majority of the remaining defect electrons are trapped on alkalis or alkali "clusters" in the sodium and potassium borate glasses. Evidence for this is somewhat diverse,[10] but is simply summarized : Broad, relatively shapeless resonances underlying both the BOHC and the BEC have been found which display average widths nearly identical with the published values[36] for the atomic s-state hyperfine splittings of sodium for the sodium borate glasses and of potassium for the potassium borate glasses. Computer simulations of the spectra have suggested that the alkali-associated centers typically involve more than one alkali and possibly as many as a dozen. Thus, one visualizes a defect analogous to the F center in alkali halides with a wavefunction spread out over an array of alkali ions, probably centered on a negative ion vacancy. Warming to room temperature or exposure to visible light following x-irradiation at 77°K destroys the ESR spectrum of these alkali-associated centers, while 30-50 % of

the original hole-center resonances remain. This effect
has been attributed[10] to the retrapping of electrons
in singlet-state pairs on still larger alkali-ion clusters. Possibly, this occurs in <u>lithium</u> borate glasses
even without warming or bleaching.

D. <u>Interstitial-Proton Trapped Electron Centers</u>

Hyperfine doublets (A $\sim$ 500G) due to atomic hydrogen are generally observed in most oxide glasses which
are not "water free" and have been irradiated at cryo-
genic temperatures (see figs.5 and 6). A discussion
of the H$^{\bullet}$ defect in fused silica has been given[37]. The
point to be emphasized here is that hydrogen generally
plays a <u>minor</u> role in the overall trapping scheme ;
in figs.5 and 6 there are $\sim$ 10^4 times fewer H$^{\bullet}$ centers
than alkali-associated electron centers.

E. <u>Interstitial-Anion Trapped-Hole Centers</u>

Except in glasses of high PbO contents[38], true
interstitial O$^-$ defects in oxide glasses are' a rarity.
However, it now seems quite clear that doping alkali
borate glasses with alkali halides results in the presence of hal$^-$ ions, which upon trapping a hole become
halogen atoms[14,15]. After x-irradiation at 22°K, new
ESR spectra were observed in chloride-and bromide-con-
taining glasses which could be attributed to Cl$^{\circ}$ and
Br$^{\circ}$, respectively (figs. 7 and 8).

The computer simulations of figs.7 and 8 were carried out under the following constraints : (i) the
trial values for A$\parallel$ and A$\perp$ were not to vary substantially from those calculated on the basis of published
values[36] of A_p for the respective halogen atoms. (ii)
The computations should take into account both abundant
isotopes of the respective elements . (iii) The g values should be given by relations appropriate to the
isoelectronic O$^-$ ions[39]:

$$g \parallel \,\, \tilde{\sim} \,\, g_{\text{free electron}} \qquad \qquad (4a)$$

$$g \perp \,\, \tilde{\sim} \,\, g \parallel \, [1+ \frac{\ell\lambda}{\Delta}] \qquad \qquad (4b)$$

(iv) A gaussian distribution of Δ values must be presumed, consistent with expectation for vitreous disorder (see fig.9). (v) Published values of λ should be
employed and ℓ sould be a constant of the order unity[39].

Some interesting results were that a reasonably
broad distribution of Δ values was <u>necessary</u> to achie-

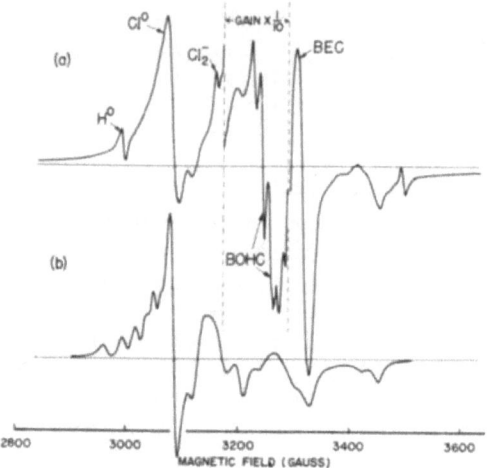

Fig.7 : (a) X-band ESR spectrum of an 8.3 % KCl-12.7 % K$_2$O-79 % B$_2$O$_3$ glass immediately following x-irradiation at 22°K
(b) A computer simulation of the Cl° contribution to this spectrum.

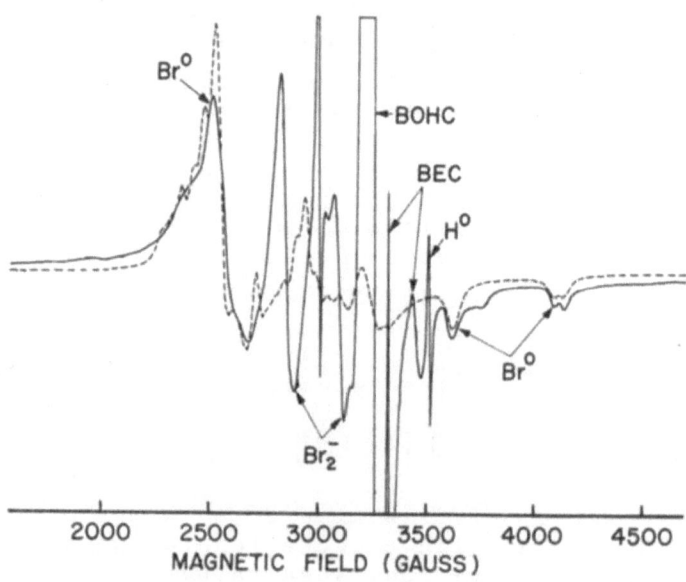

Fig.8 : X-band ESR spectrum of an 8.3 % KBr-12.7 % K$_2$O-79 % B$_2$O$_3$ glass immediately following x-irradiation at 22°K. Dashed curve is a computer simulation of the Br° contribution.

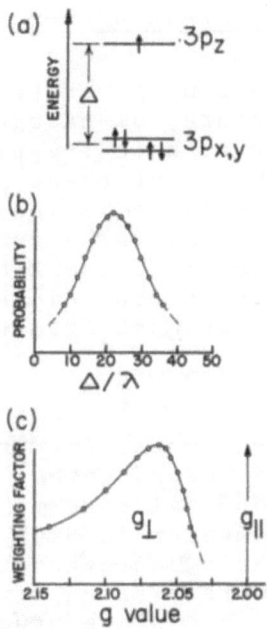

Fig.9 : (a) Schematic diagram defining the splitting Δ of the p_x^2 p_y^2 p_z^1 ground state of atomic chlorine isolated in a rigid matrix. An environment having nearly axial symmetry about the z axis is assumed.

(b) Distribution of Δ values estimated to chacaracterize the ensemble of chlorine atoms contributing to the spectrum of fig. 7a.

(c) g values and weighting factors (g only) derived from (b) and used to compute the spectrum of fig.7b.

ve satisfactory simulations and that the average values of Δ were the same for Br° as for Cl° ($\sim$ 1.2 eV). This tends to confirm the inference that these splittings are determined by an interaction with the glassy matrix, possibly by the formation of a weak three-electron bond with a network oxygen.

F. Interstitial Di-Halogen-Molecular-Ion Trapped-Hole Centers

When halide-containing borate glasses are irradiated at room temperature, di-halogen molecular ions are formed as the result of hole trapping. I_2^-, Br_2^-, and Cl_2^- have been observed ; of these, Cl_2^- has been found to be stable for indefinite periods of time. The primary means of identifying these defects once again was ESR. Figs.10 and 11 exhibit the X-band and Ka-band spectra, respectively, of the Cl_2^- species. Successful computer simulations at both frequencies effectively confirm the model[11].

OPTICAL SPECTRA

While the present notes are primarily intended to cover ESR studies of color centers in oxide glasses, a few remarks on the optical spectra of the centers are appropriate. The E_1' center in SiO_2 is well known to have an absorption band centered at 5.7 eV[6]. Its borate-glass analog is more difficult to assign, but appears on the basis of incomplete studies[40] to absorb in a very broad band peaking near 4.5 eV. The BOHC apparently has a comparatively low oscillator strength but is tentatively thought to absorb near 2.5 eV. The alkali-associated centers are probably[10] responsible for the familiar[41,42] radiation-induced absorption peaking in the range 1.5 - 2.4 eV. With the latter exception, the optical bands in the halide-free alkali borate glasses are extremely difficult to resolve or correlate with the ESR spectra. By contrast, the chloride-containing glasses have displayed Cl_2^- absorption bands[13] with peak positions and resolutions comparable to those observed in KCl. In addition, a strong Raman band corresponding to the expected position for the Cl-Cl stretching mode was also obtained for such a glass[12].

PROCESSES

A detailed understanding of radiation damage mechanisms in oxide glasses is only now becoming accessable with the recent ESR identification of most of the fundamental defect centers (See above). With this objective, a number of alkali borate and silicate glasses

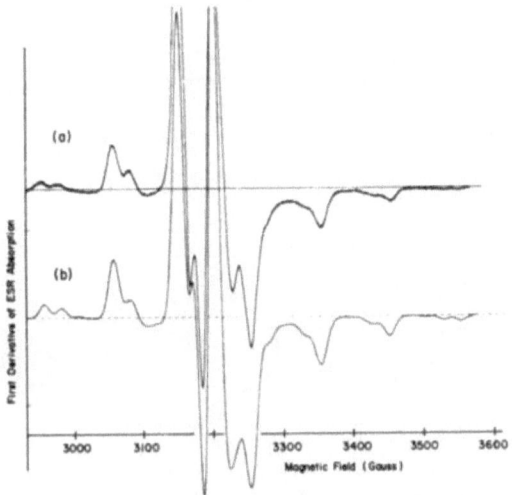

Fig.10 : X-band ESR spectrum of an irradiated glass prepared from NaCl and B_2O_3. (a) experimental spectrum (b) computed spectrum for Cl_2^- centers[11].

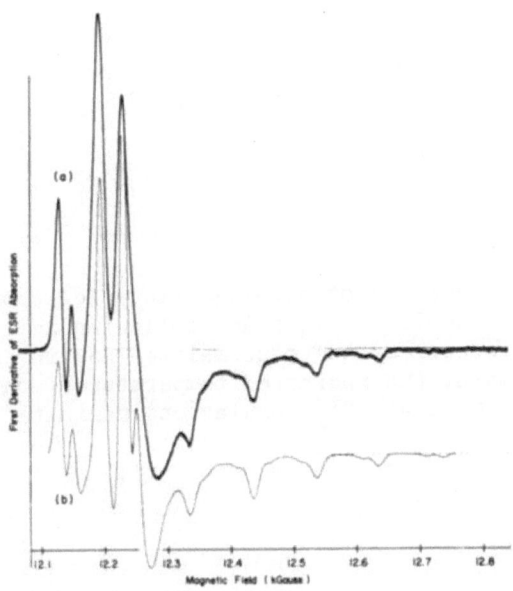

Fig.11 : Ka-band ESR spectrum of an irradiated glass prepared from KCl and B_2O_3 (a) experimental spectrum (b) computed spectrum for Cl_2^- centers[11].

226

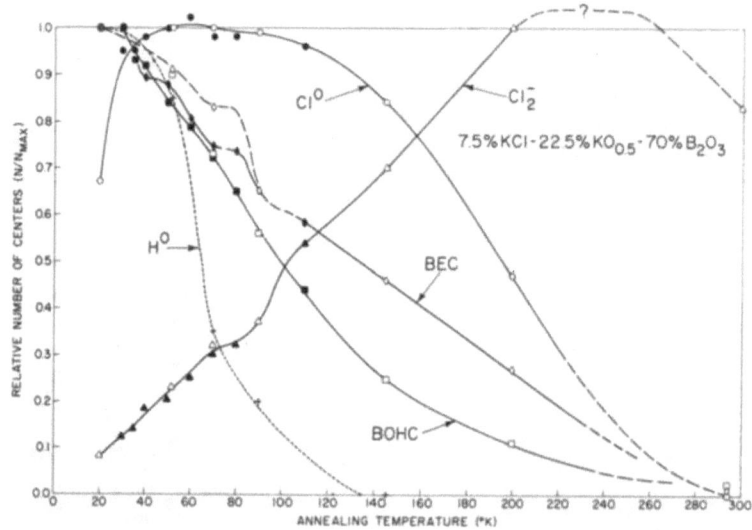

Fig.12 : Thermal behavior of paramagnetic defect populations in a
8.3 % KCl-12.7 % K_2O-79 % B_2O_3 glass following x-irradiation near
25°K. Data points correspond to normalized ESR intensities after
10-minute anneals at the indicated temperatures. On an absolute
scale, the BOHC, Cl°, and Cl_2^- populations dominate.

have been x-irradiated near 25°K in the dark and the responses of the ESR spectra to isochronal anneals have been observed[15]. Typical data are shown in fig.12. While a thorough kinetic analysis of these results is still in progress, several qualitative conclusions are immediately possible. Free holes generated by ionizing radiation near 25°K are initially trapped at the following sites with the indicated relative trapping cross-sections :

$$I^- > BOHC^- \gtrsim Br^- > Cl^- > F^- \approx zero$$

About 15 % of the electrons are trapped at BEC^+ sites ; the remainder are presumed to go to alkali-related sites. Upon warming, holes leave the (apparently shallower) BOHC sites, and the following reactions ensue :

$$BOHC + (phonons) \rightarrow BOHC^- + h^+$$
$$h^+ + hal^- \rightarrow hal^o$$
$$hal^o + hal^- \rightarrow hal_2^-$$

These reactions go to completion at room temperature and the BEC traps are also emptied. Annealing behavior above room temperature seems to be governed by the presence of relatively deep alkali-associated electron traps which do not exhibit ESR spectra. Either the BOHC or the Cl_2^- in potassium borate glasses are destroyed by heating to $\sim$ 420°K ; the same centers are stable up to $\sim$ 440-460°K in lithium borate glasses. It is inferred that the lithium-associated electron trap is deeper, as might be expected. One final generalization which may be drawn from fig.12 is that the glasses are generally characterized by extremely broad ranges of trapping depths. This is just one more example of the effects of vitreous disorder.

Some preliminary optical studies of transient ionization damage in SiO_2 glass have been undertaken.[43] It appears that high concentrations of E' centers are formed by 3-nanosec pulses of 500 keV electrons. Virtually all of these are annihilated in times $\sim$ 10 μsec with an attendent blue luminescence. These phenomena have been tentatively attributed to transient breakage and rehealing of Si - O bonds.

REFERENCES

1. LELL E., KREIDL N.J. and HENSLER J.R., in : Progress in Ceramic Science, vol.4, Ed.J.Burke Pergamon Oxford, 1966

2. BISHAY A., J.Non Crystalline Solids 3, 54, 1970

3. GRISCOM D.L. J.Non Crystalline Solids (in press)

4. SCHULMAN J.H. and COMPTON W.D., Color Centers in Solids, Mac Millian, New-York, 1962

5. RAWSON H., Inorganic Glass Forming Systems, Academic Press, New-York, 1967

6. WEEKS R.A., J.Appl.Phys. 27, 1376, 1956 ; NELSON C.M. and WEEKS R.A., J.Am.Ceram.Soc., 43, 396, 1960 ; ibid, 43, 399, 1960 ; SILSBEE R.H., J.Appl. Phys. 32, 1459, 1961 ; FEIGL F.J., J.Phys.Chem. Solids 31, 575, 1970

7. BRAY P.J. and O'KEEFE J.G., Phys.Chem.Glasses 4, 37, 1963 ; BRAY P.J. in Interaction of Radiation with Solids, Ed.A.Bishay, Plenum Press, New-York, 1967

8. O'BRIEN M.C.M., Proc.Roy.Soc. (London) A 231, 404, 1955

9. MACKEY J.H., Jr., J.Chem.Phys. 39, 74, 1963

10. GRISCOM D.L., J.Non Crystalline Solids 6, 275, 1971

11. GRISCOM D.L., TAYLOR P.C. and BRAY P.J., J.Chem. Phys. 50, 977, 1969

12. HASS M. and GRISCOM D.L., J.Chem.Phys. 51, 5185, 1969

13. GRISCOM D.L., J.Chem.Phys. 51, 5186, 1969

14. GRISCOM D.L., Solid State Comm.11, 899, 1972

15. GRISCOM D.L. and PATTEN F.W., Am.Ceram.Soc.Bull. 51, 367, 1972 ; to be published

16. SANDS R.H., Phys.Rev. 99, 1222, 1955

17. TAYLOR P.C., KRIZ H.M. and BAUGHER J.H., Chem.Reviews, manuscript in preparation

18. TAYLOR P.C. and BRAY P.J., J.Mag.Res.2, 305, 1970

19. GRISCOM D.L., TAYLOR P.C. and BRAY P.J., J.Chem. Phys. 53, 469, 1970

20. GRISCOM D.L., TAYLOR P.C., WARE D.A. and BRAY P.J. J.Chem.Phys. 48, 5158, 1968

21. LEE S. and BRAY P.J., J.Chem.Phys. 39, 2863, 1963

22. SCHREURS J.W.H., J.Chem.Phys. 47, 818, 1967

23. LOZYKOWSKI H., WILSON R.G. and HOLUJ F., J.Chem. Phys. 51, 2309, 1969

24. PURCELL T. and WEEKS R.A., Phys.Chem.Glasses, 10, 198, 1969 ; Nalamolu Gopalarao, Ph.D.Thesis, Brown University, June 1969

25. KARAPETYAN G.O. and YUDIN D.M., Fiz.Tver. Tela 3, 2827, 1961 (Eng.Transl. : Sov.Phys. -S.S.3, 2063, 1962) ; Y.Nakai, Bull.Soc.Japan 37, 1089, 1964

26. LEE S. and BRAY J.P., Phys.Chem.Glasses, 3, 37, 1962

27. KIM Y.M., REARDON D.E. and BRAY P.J., J.Chem.Phys. 48, 3396, 1968

28. HURD C.M. and COODIN P., J.Phys.Chem.Solids, 28, 523, 1967

29. SYMONS M.C.R., J.Chem.Phys. 53, 468, 1970

30. TAYLOR P.C. and GRISCOM D.L., J.Chem.Phys. 55, 3610 1971

31. TAYLOR P.C., GRISCOM D.L. and BRAY P.J., J.Chem. Phys. 54, 748, 1971

32. WEEKS R.A. in Interaction of Radiation with Solids, Ed. A.Bishay, Plenum Press, New-York, 1967

33. GRISCOM D.L., J.Chem.Phys. 55, 1113, 1971

34. REINBERG A.R., J.Chem.Phys. 41, 850, 1964

35. KROGH-MOE J.,Phys.Chem.Glasses 3, 101, 1962 ; 6, 46, 1965

36. AYSCOUGH P.B., Electron Spin Resonance in Chemistry Methuen, London, 1967 p. 438

37. WEEKS R.A. and ABRAHAM M.M., J.Chem.Phys. 42, 68, 1965

38. KIM Y.M.and BRAY P.J., J.Chem.Phys. 49, 1298, 1968

39. BRAILSFORD J.R., MORTON J.R. and VANNOTTI L.E., J. Chem.Phys. 49, 2237, 1968

40. GRISCOM D.L.,unpublished

41. BEEKENKAMP P., Thesis, technische Hochschule, Eindhoven, Netherlands, 1956 (published in Philips

Res.Rept.Suppl. n°4, 1966)

42. ARAFA S. and BISHAY A., Phys.Chem.Glasses 10, 192, 1969

43. GRISCOM D.L. and SIGEL G.H. Jr., Bull.Am.Phys.Soc. 13, 1474, 1968 ; SIGEL G.H.Jr. and GRISCOM D.L., Bull.Am.Ceram.Soc. 48, 447, 1969, SIGEL G.H. Jr. submitted for publication.

RADIATION EFFECTS IN ORGANIC MATERIALS

A.CHARLESBY

Royal Military College of Science
Shrivenham - Swindon, Wiltshire

SUMMARY

Basic distinctions between the irradiation of organic and aqueous systems arise not only from the nature of the chemical bonds involved, but also from the dielectric constant and polarisability of the molecule. In organic molecules, absorbed energy can break homopolar bonds giving rise to radicals, or may give ionic intermediates. A number of mechanisms can intervene before or after these processes and serve to modify the final reaction products.

In certain organics, notably monomers, a chain reaction may occur, leading to large chemical changes at relatively low doses. Such polymerisation reactions may also take place in the solid state depending on morphology and defects. In other, high molecular weight materials, radiation can promote chemical changes which though small, have a large effect on the subsequent physical properties. These reactions may be modified by additives in low concentration, capable of enhancing or reducing the radiation-induced charges. In several instances these high molecular weight molecules of simple chemical structure may serve as models for radiobiological research.

At a much earlier stage than the chemical reaction, the electrons resulting from ionisation give rise to physical changes such as radiation-induced conductivity and thermoluminescence. A number of complementary techniques are available to study the behavior of electrons in organic liquids and solids. These are briefly outlined, and attention is drawn to aspects not yet adequately clarified.

INTRODUCTION

Before discussing in detail some aspects of radiation effects in organic and plastic materials, it may be helpful to situate the main concern of my lectures within the general context of this meeting. The subject termed Radiation Chemistry deals primarily with the chemical changes produced by exposure to high energy radiation, but at the same time there are many important physical changes both temporary and permanent, and indeed these are becoming of increasing interest, as judged by recent published papers and conferences. Many of these varied effects also occur in radiobiology, so that the term radiation chemistry has perhaps become too restrictive.

The subject itself has passed through a number of major phases. The earlier history of radiation chemistry was largely concerned with the irradiation of gases, due primarily to the weak sources available, mainly alphas of low penetration.The major effect produced was the production of ions which could be readily measured by electrical means. As more powerful sources of penetrating radiation became available (especially fast electrons, gammas) attention was focussed on liquids which were more readily studied by available chemical techniques than were solids. At first attention was mainly paid to the products formed and here there was a sharp division between irradiated aqueous systems and organic molecules, ascribable to the different type of bonds involved, but also to the very different polarisability and dielectric constants of these materials. Later more interest was devoted to radiation mechanisms and especially to reaction constants. In organic liquids the observed changes were at first accounted for on the basis of radical intermediates resulting from excitation and there was no evidence of the ions which figured so prominently in irradiated gases. Later experiments such as the polymerisation of isobutene at low temperature show that ionic reactions do in fact occur but with much lower efficiencies than do the radical reactions ; typical values for radical production are $G \sim 3$ but only $G \sim 0.1$ for ions, although G values ~ 3 are observed for ions in gases. The reason for this great decrease in ion production in the condensed phase is that the number of collisions suffered by the electrons per unit path is very much greater

than in gases ; electrons lose their energy much more
rapidly and may therefore reach thermal energies while
within the effective coulombic field of their parent
ion. Such geminate electrons may then be recaptured to
produce indirectly further excited species at the expense of ions. The critical range r_c at which thermal
energies (kT) and coulombic energy are equal is given
by the simple equation

$$\frac{e^2}{\varepsilon r_c} = kT$$

where e is electrical charge and ε the dielectric constant. In typical organic materials such as n alkanes
and benzene $\varepsilon \sim 2$, while in aqueous solutions $\varepsilon \sim 80$,
so that in the latter r_c is far smaller and a far hig-
her percentage of the electrons can escape. High dielectric constant is associated with high polarisability and electrons in an aqueous environment can form polarised groups, hardly present in simple organics. Other
organic molecules such as acids and alcohols with a
higher dielectric constant can be expected to occupy
an intermediate position between the reactions of simple paraffins and aqueous solutions.

    In irradiated organic solids many of the primary
reactions such as radical production are not in essen-
ce different from those for the corresponding liquid,
but due to the relative immobilisation of the molecules
their subsequent fate can be greatly changed. In such
materials radicals can live for a long time and have in
fact been studied intensively by such techniques as
electron spin resonance. For a final stable product to
be formed two radicals must meet, and since they are
widely spaced one either has to assume a certain degree
of molecular motion or alternatively small groups (notably H) must be invoked to account for this apparent
mobility. A remarkable illustration is the solid state
polymerisation of certain monomers irradiated from the
solid state. In fact the behaviour of irradiated organic solids can provide a valuable means of studying
molecular mobility and defects in crystalline and amorphous materials.

    An important feature of radiation induced reactions
in organics is that in many cases the radical production is proportional to dose (until of course there is
a very significant build up radiolytic products and
reduction of primary materials) with G values for change of the order of unity.In other systems where a chain

reaction is involved, radiation serving primarily as initiator, G values for loss of primary compounds may amount to many hundreds. In the latter systems even relatively small doses can have a profound effect. On the other hand, many molecular compounds, notably aromatics, can be very resistant to the effects of radiation ; this radiation protection may even be imparted to other organic materials with which they are mixed. This phenomenon is usually ascribed to energy transfer but this term may be misleading as other explanations can be advanced to account for the observations.

TYPE OF RADIATION AND DOSE RATE

At present the most convenient sources of high energy radiation are accelerators providing fast electrons, and radioactive isotopes such as Co^{60} giving gammas. These differ primarily in their depth of penetration and in the dose rate but are otherwise comparable in their effect. Energetic heavy particles such as fast protons and alphas are less frequently used and they often result in different reactions and products, or similar products with very different yields per unit absorbed energy. These differences are however explicable in terms of the lower velocity and hence greater interaction with the medium of the heavy charged particles. A 1 MeV electron with a penetration of 0.5cm in unit density material produces one ionisation event per distance of approximately 10^3 angstrom, so that each such event is surrounded by unmodified material. With alphas of the same energy ionisation takes place about 10^3 times as effectively so that the secondary electrons, excited molecules, ions and radicals occur within a dense column with a very high concentration of products, which can react with each other rather than with the environment. Scavenger effectiveness is therefore greatly reduced at the concentration usually available. However, even in the case of electrons and gammas each primary ionisation can result in several secondaries in close proximity ; within each spur the average concentration of reactive groups is higher than the overall average even though it is far lower than in the material as a whole.

Fast neutrons by colliding with and ejecting light atoms, primarily H, effectively behave as sources of heavy particle irradiation with corresponding highly localised, high concentrations of reactive species.

In theory it is possible to obtain very high concentrations of reactive species using electron radiation of sufficiently high intensity. In practice the electron dose rate required would not be achievable but some modification of reactivity might nevertheless be expected at the highest available electron currents ; the extent to which this effect will differ from the products obtained at low intensities depends on the lifetime of these reactive species.

LOW MOLECULAR WEIGHT ORGANIC COMPOUNDS

A major feature in the radiation induced changes in simple organic compounds is that in spite of the high energy available from the incident beam (far greater than the strength of any homopolar bond), a considerable degree of selectivity is observed as denoted by the final product. This can be taken to mean either that the energy is not deposited at random or that once energy is absorbed it can migrate within the molecule or even between molecules before giving rise to ionisation or excitation. According to the uncertainty principle an energy loss of say 35 ev corresponds to an uncertainty in time of 10^{-16} secs and for a high velocity particle an uncertainty in position of up to 300 Å, far greater than the dimensions of a small organic molecule. The time taken for a fast electron or photon to pass a small molecule is less than 10^{-17} secs, whereas the time per molecular vibration is about 10^{-13} sec. This means that the energy transfer to the molecule may take place at any one of a number of molecular configurations and this may be one cause of the range of potential radiolytic products, in addition to the different energy transferred. There is adequate time for absorbed energy to be dispersed over a molecule and to accumulate selectively in one or other bond particularly sensitive to disruption before the molecule changes its configuration. A notable example is in branched alkanes :

$$nC_5H_{12} ; \quad G(H_2) = 4.2 \quad G(CH_4) = 0.4 \quad G(C_3H_8) = 0.42$$

$$(CH_3)_3 - C - CH_2 - CH_3 ; \quad G(H_2) = 2.0 \quad G(CH_4) = 1.2$$

where the side branches $-CH_3$ are removed selectively by C-C scission as compared with the destruction of the main chain likewise involving C-C bond fracture.

A widely quoted example of this transfer of energy within the molecule is referred to as the sponge effect. Benzene and other aromatic molecules have a large number of potential energy levels and can absorb and disperse energy with little or no decomposition.

Many aromatic molecules show this response to radiation
and triphenyl for example is used in organically mode-
rated reactors. The range over which protection by an
aromatic group extends may be estimated in several ways.
An early result shows that dodecane exposed to reactor
radiation is rendered insoluble by a dose of some 27
pile units (about 1350 Mrads). With a 2 naphthyl group
attached in the 1, 4 and 6 positions in C_{12} the doses
needed to achieve the same effect are 43, 62 and 66 pile units. This implies that protection is offered over
a range of some 3 to 4 carbon atoms within the same
molecule.

Excitation energy may also be transferred from one
molecule A to another B, providing of course that the
appropriate energy level of excited B is equal to or
lower than that of A. This means that B would show a
higher degree of chemical change than would be warran-
ted from the energy it absorbs directly from the incident radiation. This energy may not only be used in
causing chemical changes but may also be emitted in the
form of fluorescence from an excited singlet state or
phosphorescence from an excited triplet. These early
stages in energy absorption are of course fundamental
in many problems of chemical physics in other reactions
than those resulting from exposure to high energy radiation.

If the excited state is maintained for a sufficient
time, bond fracture may occur resulting in the formation of 2 molecular fragments, usually radical in nature. However, this type of fracture is not essential
and in simple paraffins for example unsaturation and
hydrogen molecules may be produced in a single stage
process.

$$- \overset{\displaystyle H}{\underset{\displaystyle H}{C}} - \overset{\displaystyle H}{\underset{\displaystyle H}{C}} - \longrightarrow -\overset{\displaystyle }{\underset{\displaystyle H}{C}} = \overset{\displaystyle }{\underset{\displaystyle H}{C}} - \quad + H_2$$

In view of the high energies available one might perhaps expect the formation of unusual radicals, but in
fact this appears to be rarely the case.

Once these radicals have been formed they may react with each other either to reconstitute the original
molecule or to form new products. Here we must consider
the so-called cage effect. In a condensed medium where
individual molecules are constrained within a small volume, the chance of recombination is obviously much higher than in a gas and different products are obtained.

In a solid matrix such as a crystalline lattice where
motion is even more restricted, the probability of interaction to reconstitute the original molecule is much
higher than in a liquid. In amorphous solids changes in
reaction rate may be expected as one passes through
transition points such as the glass temperature, allowing radicals on separate molecules to come together.
These considerations may account in part for differences in the yields of radiolytic products from identical
compounds irradiated below and above the melting point.

We may next consider the oxygen effect. It is
known that many radical reactions are far more marked
when an organic molecule is irradiated in the presence
of oxygen. Several explanations can be given of this
role of oxygen. Oxygen may react by capturing electrons
and therefore modifying ionic reactions. It is also
possible to envisage oxygen as reacting with a radical
fragment and stabilising it, whereas in the absence of
oxygen two radical fragments from the same molecule
might recombine to restore the original status. Chain
reactions may also be involved as in the formation of
peroxides from irradiated cyclohexenes. Oxygen can be
considered as a diradical $\cdot O-O\cdot$ which gives rise to
auto-oxidation without loss of radical population.

$$R\cdot + O-O\cdot \rightarrow R-O-O\cdot$$
$$RO_2^{\cdot} + RH \rightarrow RO_2H + R\cdot$$

A further factor likely to cause changes in product yield is the temperature of irradiation. This can
arise in several ways. As the temperature rises there
is a change in excitation, since there is a smaller
probability of electrons escaping the coulomb field of
the parent ion. The molecules themselves have higher
vibrational energies and can therefore more readily de-
compose to give radical products. Activation energies
for subsequent reactions will also influence the yield.
This temperature effect of radiation response has been
known for some time in radiobiology where the lethal
dose (LD 50) decreases with rise in temperature. Since
a similar dependence is observed in simple organic compounds, it is not necessary to introduce any other explanation based on a specific biological change.

Ion molecule reactions are known to play an important part in the radiolysis of many organic compounds
and although they were overshadowed for some time by
the more obvious radical reactions, they could be distinguished from them by their occurrence at low tempera-

ture. The capture of an electron by a positive ion gives a highly excited molecule which can release its excess energy by a structural change or by the emission of fluorescence or phosphorescence radiation.

Collisions between polyatomic ions and neutral molecules produce long-lived complexes which can react very rapidly. Often the transfer of a hydrogen atom or a larger group is involved.

$$CH_4^+ + CH_4 \rightarrow CH_5^+ + .CH_3$$
$$CH_3^+ + CH_4 \rightarrow C_2H_5^+ + H_2$$

An electron which escapes from its parent ion can be captured by a neutral molecule to give an anionic group which then reacts further. Compounds with a high electron affinity such as alcohols or containing halogen are particularly sensitive to this type of behaviour. A valuable reaction is that which occurs with N_2O which, after accepting an electron, decomposes to give a stable molecule N_2 whose concentration can then be used to measure electron concentration.

THE ROLE OF H

In many reactions the role of H has not received the same amount of attention as has the possibility of energy transfer. Many reactions, especially in the solid state, require some degree of radical mobility, as distinct from molecular mobility, and there are several key experiments showing that this occurs by a series of H addition and abstraction reactions.

$$R_1H \rightarrow R_1^\cdot + H$$
$$H + R_2H \rightarrow H_2 + R_2^\cdot$$
$$H_2 + R_2^\cdot \rightarrow R_2H + H$$
$$H + R_3H \rightarrow H_2 + R_3^\cdot$$
$$R_1^\cdot + R_3^\cdot \rightarrow R_1R_3 \qquad etc.$$

This sequence of reactions may allow the eventual formation of two radicals $R_1^\cdot$ and $R_3^\cdot$ in close proximity, when they may react together forming a single molecule R_1R_3. This process may take some time and evidence for it can be found from the slow reduction of total radical concentration as measured by electron spin resonance. The rate depends not only on the concentration and rate of production of R_1 directly by radiation but also on the H_2 concentration, not necessarily produced by radiation. Further evidence for this reaction can be obtained by adding D_2 gas, whereupon RD compounds and

HD are produced.

In a mixture of alkane and aromatic molecules there is a high likelihood of hydrogen being added to the ring rather than abstracting the hydrogen from the alkane

$$Ar+H \rightarrow ArH$$

A classical experiment involves the irradiation of cyclohexane in the presence of various concentration of benzene. Even when the benzene is present at a low concentration corresponding to an electron fraction of only 0.07, the G value for hydrogen evolution is halved. It is generally assumed that primary energy deposition is proportional to electron density so that it is clear that some form of transfer must be involved. This is often referred to as energy transfer but may equally well be accounted for by the intervention of H atoms. Thus the reaction constant for H abstraction from hexane is 0.5×10^7 1/mole sec. while for H addition to benzene it is 1.8×10^8 1/mole sec. Thus 1 molecule of benzene is as effective in reacting with hydrogen as 36 molecules of hexane. It may well be that the intervention of hydrogen in a number of these reactions accounts for a high degree of selection in the final radiolytic reaction.

At low doses, yield is usually proportional to dose (as is indeed necessary if the quoted G values are to have any real meaning). At higher doses this linearity often fails, due to secondary reactions between the products. An example is in the irradiation of simple alkanes, which not only gives H atoms by CH bond fracture, but H_2 and unsaturation by a molecular process. Unsaturation can be used as a measure of dose, but tends to a maximum (low) value as the dose is increased. This may be explained as the capture of H atoms by the increasing concentration of $-C = C-$, giving

$$-\overset{\displaystyle H}{\underset{\displaystyle H}{C}} - \overset{\displaystyle \cdot}{\underset{\displaystyle H}{C}} - \quad \text{which would otherwise be produced by H abstrac-}$$

tion from the alkane chain. As expected, this secondary reaction is diminished with alpha radiation, which because of the high local concentrations in its track, yields more H_2 and less H.

In a somewhat similar way the radical concentration as measured by e.s.r. tends to a limiting value, of the order of 10^{17} per cm^3. This corresponds to an average distance between radicals of the order of a hundred Angstrom. It has not yet been decided whether this is the range of secondary reactive products, of H atoms, or of electrons which would otherwise be recaptured by their primary ions.

IRRADIATION OF HIGH MOLECULAR WEIGHT ORGANIC STRUCTURES

In low molecular weight organic materials when no chain reaction is involved only very low yields can be expected excepting at extremely high doses. For example a dose of r Mrads provides 0.625×10^{18} Gr changes/g, i.e. 1.04×10^{-6} Gr M per molecule of molecular weight M. If $M \sim 100$, $G \sim 3$ and $r = 1$ Mrad, only 3 molecules in 10,000 will be affected. On the other hand with a chain reaction as in polymerisation or a high molecular weight compound ($M \sim 10^5$) marked changes in chemical and physical properties can be observed for doses of a few Mrads or even less. Some of these reactions can therefore be used on an industrial scale and at an economic cost.

In discussing this type of work we may distinguish between :

(1) production of polymer by irradiation of monomer

(2) modification of polymer where no chain reaction is involved

(3) modification of polymer by a chain reaction (when even a relatively low initial molecular weight may be acceptable).

RADIATION INDUCED POLYMERISATION IN LIQUID MONOMER

In conventional polymerisation three separate stages are involved, initiation, propagation and termination, which may be modified by chain transfer. Irradiation of the monomer in the liquid state produces polymerisation by a chain reaction but radiation only intervenes in the initial step (and possibly the termination). The propagation reaction occurs with the same parameters as in polymerisation initiated by purely chemical means. A wide range of polymers has been produced by irradiation of the monomer, under a variety of conditions.

Radiation acts as a very convenient initiator. It can be used over a wide range of temperatures without significantly affecting propagation, so that temperature can be used to control the propagation step, thereby permitting a degree of control not possible by purely chemical initiation. It also leaves no residues as occur when chemical initiators are present. The molecular weight of the polymer is controlled by the constants for propagation and termination ; the termination step may result from the encounter of the growing chain with another compound (possibly initiating another chain) or when two growing chains meet and terminate each other, either by addition or disproportionation. In this respect radiation (by affecting the concentration of growing chains) may also intervene in the termination step. The time available for growth depends on the concentration of growing chains with radical ends hence on the dose rate. Analysis shows that in this case the number of polymer molecules is proportional to the total dose, but that the average molecular weight varies inversely as the half power of the radiation intensity. This analysis assumes steady state conditions which would not prevail if the polymerisation is allowed to proceed until a significant fraction of monomer is used up or if the solution becomes viscous and the polymeric chains lose their mobility. This simple relation between number of polymer molecules and radiation dose ceases when chain transfer agents are present and each chain initiated by radiation may be terminated and a new chain initiated without the intervention of radiation.

Very different effects may be observed in other irradiated systems. Among these one may mention the case where the polymer is not soluble in the monomer or in a solution of the monomer. The growing chain then tends to fold in on itself and the radical end becomes buried within the polymer folds, and is not accessible for the addition of further monomer. The polymerisation reaction can then cease while reactive groups are still available to be resuscitated at some later stage, as when the solvent is changed.

When irradiation of monomer takes place in solution the polymer chain may be initiated from monomer radicals, from radicals produced in the solvent, or energy absorbed in the solvent may be transferred directly or via intermediates to the monomer. Here again a strict proportionality between number of polymer molecules and energy absorbed in the monomer itself no longer holds.

Although most polymerisation reactions studied with radiation occur via radical reactions, ionic reactions can also be observed under suitable conditions, e.g. at low temperature. After initiation the growing chain is stabilised by a gegen ion and propagation ceases when a

gegen ion is captured by charge transfer, proton transfer, or other charged species. In this case there is
no interaction between two separate growing chains and
the yield within obvious limits is proportional to the
dose and to the radiation intensity. In the case of
isobutene for example, very low G values are observed
for initiation of about 0.1, less than 1/10 of those
found in typical radical polymerisations. These measurements were one of the first indications that ionic
reactions do occur but with low efficiency. G values
of ionic species derived by entirely different means
(as will be described later) indicate the same low efficiency of production of ions in simple organics in
the condensed phase.However, when ionic polymerisation
occurs in the presence of a very fine powder, considerable increases in the G values are found and the dependence of radiation intensity no longer varies as I^1
but rather as $I^{0.5}$. This might be taken to indicate
that the fine powder stabilises the electrons by capture and therefore acts as a massive gegen ion. Ionic
polymerisations are very sensitive to the presence of
water and there is alternative evidence to show that
these powders may be effective by removing traces of
water from the polymerising system.

SOLID STATE POLYMERISATION

The drastic morphological changes involved in the
transformation from a large number of monomer molecules
into a single polymer chain might be expected to mili-
tate against any noteworthy polymerisation in the solid
state, but this is not the case. A number of monomers
irradiated in the solid phase and even as single crystals, can be readily induced to polymerise, often far
more rapidly than when the monomer is present as liquid.
indeed some monomers polymerise very readily as crystals
but hardly if at all as liquids. Numerous monomers have
been irradiated at low temperature or as solids, either
crystalline or glassy, to determine the process involved and it would appear that several alternative mechanisms are relevant depending on the monomer, the lattice constants of the crystal, on the mobility within the
solid and on the presence of defects. In trioxane for
example an irradiated single crystal produces oriented
polymer, presumably because the molecules are appropri-
ately aligned. In acrylamide on the other hand, crystal
defects dominate the reaction although even with this
monomer the reaction proceeds far more rapidly in the
crystalline than in the liquid state. Polymerised regions within the crystal can be readily seen by phase mi-

croscopy. The regions tend to lie along slip planes although the individual polymer molecules show no such orientation. This optical technique is therefore an excellent method for determining and analysing location of defects in organic crystals.

Other factors which can dominate the polymerisation reaction in solids include the effect of pressure, where in certain monomers propagation can be stopped only to be resumed when the external pressure is released. In other monomers however external pressure enhances the reaction rate.

Temperature, especially near the melting point, also has a considerable effect with a very high rate of polymerisation just below and negligible polymerisation just above the melting point. In post-radiation polymerisation the polymer chains initiated by radiation carry a reactive end group which can continue to grow after external radiation has ceased. The size of the crystals is likewise of importance ; for trioxane for example polymerisation occurs throughout the body of the crystal and is faster in large crystals. This would imply that defects are not essential to the reaction since the surface of a crystal contains a large defect area. Phase changes may also result in marked changes in polymerisation rate.

MODIFICATION OF POLYMERS BY IRRADIATION

Although the primary chemical effects of radiation on long chain polymers are not inherently very different from those occuring in low molecular weight compounds this subject has received a considerable amount of attention for several reasons :

(a) Small chemical changes can produce physical changes readily assessable by a number of techniques developed for polymer research.

(b) The range of conditions under which the material is irradiated can be varied and the effect of conditions on the radiation process thereby deduced.

(c) Many of these reactions take place in the solid state so that the effect to irradiation of organic materials, crystalline, amorphous and oriented, can be determined.

(d) Long chain polymers can serve as a simple model for biological materials and indeed many of the radiation effects observed in radiobiology are closely parallel to those found in simple long chain polymers.

(e) Several of the processes involved have found large scale industrial use.

In discussing these radiation effects it is necessary to distinguish between polymers or monomers which react by a chain mechanism in which radiation is the initiating step and those where no chain reaction is involved and the marked physical changes are largely due to the high molecular weight involved.

Many long chain polymers when subjected to relatively small doses of radiation, of the order of a few Mrads, show great changes in such properties as solubility, swelling and mechanical properties. This is primarily due to the fact that they have molecular weights often of the order of 10^5, a thousand times greater than those in low molecular weight organic materials. To modify one molecule therefore requires a dose 1000 times less. The two major changes are those resulting from cross linking (due to a side chain fracture) and degradation (due to main chain fracture). The two simplest examples of these are polyethylene and polymethyl-methacrylate.

CROSSLINKING REACTIONS

When polyethylene is irradiated hydrogen atoms are evolved leaving radicals

```
 H H H            H . H
-C-C-C-          -C-C-C-      +H
 H H H            H H H
```

Radicals on two adjacent chains can then combine to give a crosslink, reducing the number of separate molecules by one.

```
    H H H
   -C-C-C-
    H | H

    H | H
   -C-C-C-
    H H H
```

As this process continues the material is converted at first to higher average molecular weight and eventually into a three-dimensional network which can be completely insoluble. The process is essentially similar to the vulcanisation of rubber but involves no additives. The bridges are direct carbon-carbon links as compared with the sulphur bridges observed in rubber.

Apart from its scientific interest the process has considerable industrial importance and is widely used, with an annual production in the neighbourhood of 10^8 per year. It is also noteworthy that this represents a reversal of the usual production sequence in that product is first formed into its final physical shape, and then treated by irradiation to promote the chemical modifications. Usually the chemical process precedes the fabrication step.

The density of crosslinking can be assessed in a number of ways, such as the solubility and swelling of the irradiated material. The intermediate radical can also be determined by electron spin resonance techniques. The gradual decay in radical concentration after irradiation is a measure of the rate at which radicals combine with each other to promote crosslinking.

The study of irradiated polyethylene allows us to investigate a number of problems of more general interest. To produce a crosslink two radicals must be present in close proximity. If these are produced by the same radiation event (ionisation or excitation), the reaction would be extremely rapid and there would be little chance of intervening by the use of additives in low concentration. On the other hand, if the two radicals eventually forming a crosslinking are produced separately some means must be found to allow them to move together. Evidence indicates that this radical motion is due to the interaction of H and H_2 with the polymer and the radicals. By a series of abstraction and addition reactions these effectively serve to move radicals within the polymer until they match up and can form crosslinks. For a crosslink to form the polymer chains must also be in the right orientation relative to one another, at least in the immediate vicinity of the radicals. The location of these crosslinks may therefore depend on the morphology of the polymer and there is evidence that the crosslinks occur preferentially in the amorphous regions or in folds at the edge of crystallites. The reaction is also modified when irradiation takes place in the presence of oxygen or of additives in low concentration. This enables us to study reactions in the solid state comparable to those which have received considerable attention in the irradiation of mixed liquids. There exists a further pos-

sibility of irradiating oriented polymer chains. The
reinforcement of materials such as rubber by the addition of fine particles may also be studied quantitatively to determine whether the mechanism of reinforcement is due primarily to chemical bonding between polymer and particle or to the physical changes resulting
from the diversion of mechanical stresses by very fine
powder within the lattice.   ·

At the same time as crosslinking polyethylene
shows reactions, notably the formation of unsaturation
by a molecular process and main chain scission as in
degrading polymers. These effects are smaller than those resulting from crosslinking. At very high doses
crystallinity in polyethylene is destroyed and polyethylene may be converted into a brittle non-crystalline material. Thus starting from the same partially
crystalline polymer one can transform it successively
and quantitatively into a rubber-like, into a cheesy,
and eventually into a glassy structure. Irradiation
is therefore a means of comparing these three forms of
solid state.

Many other polymers show crosslinking ; a notable
example is silicone (dimethyl siloxane polymer) which
from a viscous liquid can be transformed into a very
weak rubber-like transparent material. However, in the
presence of very fine powders a much tougher material.
is obtained.

CROSSLINKING IN SOLUTION

Water-soluble polymers, irradiated in an aqueous
environment, can also be crosslinked by radiation. They
then become swollen gels, the degree of swelling depending on the initial concentration and on the dose.
Three interesting features are (a) the doses needed to
form ·a crosslinked network are lower than when the polymer is irradiated in the solid state ; (b) below a
certain concentration, of the order of 1 % or less, no
gel can be formed even at very high doses ; (c) above
this critical concentration the dose needed to form a
gel (network) structure is lower, the lower the polymer concentration, i..e. the further the molecules are
apart.

The explanation for this behaviour appears to be
as follows. During irradiation radicals are formed not
only in the polymer molecules, but also in the water,
yielding H and OH radicals. These react with the polymer molecules, forming radicals thereon, which allow
additional crosslinking. The lower the concentration of

molecules, the greater the share each will have from these aqueous fragments. Each crosslink requires two radicals to interact, and there will be competition between radicals on two separate polymer chains, and two radicals on the same chain. The reaction of the latter does not provide a crosslink effective in network formation, but instead forms an intramolecular link, which reduces the molecular dimensions, which become looped and form microgel on a molecular scale. Not only does this produce a different type of molecular configuration, but it also allows an estimate to be made of molecular dimensions in solution. This will of course depend on such factors as pH. The reaction can also be modifed by the introduction of H or OH scavengers in the solvent.

DEGRADING POLYMERS

These are long chain polymers such as polymethyl-methacrylate, cellulose, and polyisobutylene, which do not show a significant degree of crosslinking but instead suffer main chain fracture with a consequent reduction in average molecular weight and mechanical strength. By control of the dose one can therefore reduce molecules to any desired average molecular weight. Here again the process may be inhibited to a large extent by the use of protecting additives. The reduction in molecular weight can indeed be used as a method of determining radiation dose.

The distinction between crosslinking and degrading polymers is not absolute but those in which degradation predominates appear to be primarily polymers with bulky side groups whose presence inhibits the reformation of main chain fractures induced by irradiation. Alternatively the steric hindrance can weaken the main chain and energy may be transferred primarily to these more sensitive bonds.

Usually doses of at least several Mrads are needed to produce a very marked weakness in mechanical strength of degrading polymers, but tetrafluorethylene polymer (Teflon) is remarkable in that it shows a considerable reduction in strength at very much lower doses. This polymer is not readily studied by conventional polymer techniques due to its insolubility, but tentative suggestions are that the strength may depend on a relatively few linking segments between crystalline regions which are particularly sensitive to radiation or oxidation.

MODIFICATION OF POLYMERS INVOLVING A CHAIN REACTION

Graft copolymers can be readily produced using radiation. The simplest process involves exposing polymer A in the presence of monomer B to high energy radiation which induces radical formation in the polymer on to which monomer B can graft and polymerise by chain reaction to give long side chains. The properties of the copolymer depend on the nature of A and B as well as on the size, number and distribution of side chains B. However, with this technique homopolymer of B is also produced and must be eliminated. Alternative methods are the irradiation of A in the presence of oxygen producing unstable peroxides which may subsequently be decomposed thermally in the presence of monomer B to provide the side chains. Irradiation of polymer and monomer in suitable solvents is another technique. Considerable scientific effort has been devoted to this topic but industrial uses are still rather few, largely because of the problem of finding appropriate uses for such copolymers. One possibility is the formation of thin membranes, e.g. for water desalination or other forms of purification. Styrene and other monomers have also been introduced into wood or concrete to enhance their mechanical properties although in these cases copolymerisation is not involved. However, these applications are limited by costs of monomer and technical difficulties of irradiation on site.

A very different type of chain reaction involves crosslinking by a chain reaction as in polyester-styrene mixtures. Although the molecular weight of the polymer is relatively low, curing can still be achieved at modest doses due to the chain nature of the reaction. In saturated polymers such as polyethylene each link requires two radicals at least one of which must be produced directly by radiation. It is therefore easy to compute the necessary dose knowing the G value for radical production. In these other systems a chain reaction is possible since unsaturation is present both in the polyester chain and in the monomer. Thus a radical present on one polyester chain can induce a growing graft chain which then reacts with a second polyester molecule (giving a crosslink) and at the same time gives a free radical on the second polyester to continue the reaction. Hence a polymerisation chain through styrene, requiring only one initiating radical, can traverse a number of polyester chains linking them together. The statistics of such a chain crosslinking reaction are quite different from those of random crosslinking but requires the availability of an adequate supply of double bonds both on the polyester and in the styrene. A chain reaction may also be found with allylic compounds in which there are two double bonds per monomer unit. Curing of thin films of paints and lacquers of relatively low molecular weight may also be undertaken at relatively low doses due to the presence of an appropriate distribution of unsaturated groups. Recently evidence has been found of a chain crosslinking reaction with polyethylene and acrylic acid and this can best be explained as due to some form of dimerisation of the acid giving more than one double bond per molecule. Other sensitisers have also been reported, e.g. sulphur monochloride with polyethylene.

PROTECTION VIA A CHAIN REACTION

In all these reactions radiation protection can be obtained by the inclusion of suitable additives such as radical scavengers. Each additive molecule, e.g. I, reacts with a polymer radical preventing further reaction with adjacent molecules. The degree of protection can be readily assessed by measuring the increased dose needed to achieve the same physical or chemical change. For example, in the case of the curing of polyester systems each protecting additive molecule enhances the dose by an amount equal to that needed to produce one extra radical. In other cases the degree of protection is much greater than can be explained on this basis and each additive molecule appears to react with and stabilise a number of radiation induced radicals. This high protective efficiency occurs particularly in the case of molecules with sulphydryl groups and has been explained in terms of a succession of H capture and release reactions which would otherwise react with polymer molecules to produce further reactions. Without the additive radiation produces one radical and a H atom, the latter producing a second radical elsewhere by H abstraction.

$$RH \xrightarrow{\quad\quad} R\cdot + H$$

$$H + RH \rightarrow R\cdot + H_2$$

In the presence of the protecting additive S the H released in the first reaction is captured and held loosely so that it can be used subsequently to replace the H lost elsewhere.

$$RH \xrightarrow{\sim\sim} R\cdot + H$$

$$S + H \rightarrow SH$$

$$R\cdot + SH \rightarrow RH + S$$

The additive therefore stores available H and replaces it elsewhere without being used up. This type of reaction may explain very efficient protective reactions also occurring in biological systems.

ELECTRICAL PROPERTIES OF IRRADIATED ORGANICS

It has been known for many years that plastic materials show considerable changes in their electrical properties when irradiated. Investigations into these changes can produce information on the earlier stages of radiation effects which eventually lead to the observed chemical changes ; it also provides information on the behaviour of charges in insulating materials, a subject of considerable interest in solid state physics and electrical engineering. Here we consider irradiation as a very effective method of producing charge carriers throughout the body of insulating solids.

The early observations showed that insulating plastic materials have a greatly increased conductivity during irradiation but in addition a slowly decreasing current following irradiation. This post-radiation current, which may last for hours and even days, has been ascribed to electrons initially ejected by radiation becoming trapped and only slowly released by thermal means. As the subject has developed two alternative theories have been propounded to describe the motion of these charges :

(1) the band model in which the electron is delocalised in an energy band extending over many atoms. This is an extension of the theory developed and used so successfully in semiconductor theory.

(2) the hopping model in which a number of discrete traps are available. After ionisation the electron can move from one trap to another by a hopping motion during which it will drift in one direction under the influence of an external field. The nature, number and depth of these traps is of primary importance.

Subsequent work on radiation induced conductivity has emphasised the distinction between geminate and non-geminate electrons. The former are trapped within

the effective coulomb field of the parent ion and when
released can migrate back to it. On the other hand,
non-geminate electrons have left the field of their
parent ions and are assumed to be trapped at random in
the matrix. The low G value of ionisation (typically
0.1) in the condensed phase indicates that following
ionisation most electrons in organic materials are geminate. A number of techniques have been developed to
determine the fate of these electrons but unfortunately
these have been largely treated as separate investigations, dealing with different materials, irradiated
under different conditions. At this stage correlation
between these developed techniques would appear to be
highly desirable.

The major techniques available include the following :

(1) Conductivity measurements during and after
    irradiation.
(2) Ion yields and electron mobility.
(3) Thermoluminescence.
(4) Optical absorption.
(5) Optical bleaching.
(6) Electron spin resonance.
(7) Electrically stimulated luminescence.
(8) The use of scavengers, e.g. for electrons.

In this paper it will only be possible to outline
a few of these.

CONDUCTIVITY

During irradiation the current is approximately.
proportional to the applied field and varies with the
radiation intensity I, usually following an $I^{0.8}$ power
dependence. This is difficult to explain in simple
terms. If each electron released by radiation or after
untrapping drifts in the direction of the field and is
permanently trapped by an impurity or reaches the electrode, the current should vary as $I^{1.0}$. If however an
electron drifts until it combines with a positive charge produced by radiation the current should vary as
$I^{0.5}$. A combination of both can produce an approximately 0.8 power dependence but only over a limited range
of intensities. The situation is rendered more complex
by the possibility of space charge, while for steady
state conditions the negative electrode must be capable
of injecting the appropriate number of electrons into
the insulator.

The post radiation current decreases approximately as $(t+a)^{-1}$ (t is time after cessation of radiation). This time dependence is likewise difficult to account for on any simple theory. In one set of experiments due to Yahagi, a $I^{0.5}$ dependence for current during radiation has been reported and this, together with the decay relation, can be derived satisfactorily from the theory. It involves two constants k (recombination constant between electron and positive charge) and μ (electron mobility) and their values deduced from his results are in reasonable agreement with theory. However, the considerable amount of published work cannot be reconciled with these ideas, nor can the dependence on temperature. This basic problem still remains to be solved.

ELECTRON MOBILITY

A free charged particle moving in an insulator will drift in the direction of an applied electric field and the drift velocity per unit field is defined as the mobility μ. This may be derived by a radiation pulse technique to produce the carriers, whose motion is subsequently studied as a function of time, field, etc. The values determined are far lower than would be expected from simple drift kinetics and this may be ascribed to (a) the charge particle being a massive ion rather than an electron, and (b) a series of trapping/ untrapping processes such that the average time available for motion is greatly reduced. By very fast pulsing techniques the mobility of the electron itself can be derived before it is collected by a molecule to form an ion. The electron mobility is still far below the expected value, presumably due to trapping. Alternatively the motion of an electron is greatly affected by the polarisability of the medium through which it moves. This brings us directly into a basic problem of radiation chemistry, i.e. the interaction of electrons with a medium when the electrons no longer have enough energy to cause excitation or ionisation. Measurements of electron mobility have been primarily carried out with very pure low molecular weight hydrocarbons, notably n hexane, where any traps would be expected to have a short life and therefore not comparable with those in solid organic materials. The important parameter here is the average distance an ejected electron travels before thermalisation and trapping. Various methods have been adopted, including strong electric fields (which may assist the electon to escape) or electron scavengers which at a sufficiently high concentration can react with the electron before it is recaptured. It is interesting to note that as compared with n alkanes, the distance travelled is greater in cyclic molecules such as neopentane, and smaller in unsaturated molecules, presumably due to their polarisability. The G values for electron production follow the same pattern. Temperature also influences G(e). Since not all electrons have the same range, further work is designed to derive the number/range relationship, and ultimately to determine the rate of energy loss by electrons travelling at low velocities.

THERMOLUMINESCENCE

If a solid is irradiated at a sufficiently low temperature and then allowed to warm up it can emit light which for the organic materials discussed here occurs as a series of broad peak grouped at different temperatures. The overall light intensity/time or temperature curve is known as the glow curve, and the peaks are associated to a large extent with temperatures at which there are morphological changes in structure of the solid. However, the spectrum of light emitted (e.g. in the case of polyethylene) does not correspond to the material but rather to the fluorescence and phosphorescence spectra of an impurity of an aromatic character. If this is removed, e.g. by washing, and replaced by another aromatic molecule, the same glow peaks are observed but the emission spectrum is changed. By measuring the thermoluminescence one can therefore obtain information both on the basic material irradiated as well as on these extraneous molecules which need only be present in low concentration

The mechanism used to explain these observations is that electrons ejected by radiation are trapped within the matrix. At a high temperature electrons can escape from their traps and migrate to the impurities producing excited species which lose their excess by fluorescence and phosphorescence. Since these spectra correspond to the uncharged impurity one either has to assume that a positive charge has previously migrated to the impurity (although the electron itself has been trapped elsewhere), that the impurity is ionised directly, or that the electron and positive charges recombine elsewhere in the material and the energy, possibly in the form of an exciton, is transferred to the impurity giving off light.

It is also necessary to explain the width of each peak within the glow curve. This could either be due to the long time needed for the reaction to occur, during which time the temperature has risen, or alternatively to the existence of a series of traps of slightly different depth giving a range of temperatures for thermal release of the trapped electron. These two explanations can be distinguished by a partial preheating technique. If the temperature is raised to within one of the peak maxima, and the specimen is then cooled before a second and full temperature rise, a single trap depth would be partially emptied by the preheating and the same intensity distribution would be observed, though at a lower total intensity. On the other hand if each peak corresponds to electron traps over a small range of depths, those traps corresponding to the lower part of the peak would be completely emptied and there would be little or no luminescence up to the temperature previously reached. Above this temperature the remainder of the peak would be completely reproduced. Experiments show that the latter is approximately the case. Very little luminescence is observed until a temperature within the peak is reached close to that attained on the preheating cycle. It must therefore be concluded that within each temperature peak there is in fact a range of different electron depths.

Even if the specimen is allowed to stand at a low temperature, e.g. liquid nitrogen, light is still emitted at a slowly decreasing intensity. This preglow fades approximately as $(t+a)^{-1}$. Here again the kinetics are difficult to explain on any conventional reaction mechanism and the suggestion has been made that preglow is due to tunnelling. Most electrons are trapped within the coulombic field of the parent ion and for any combination of trap depth and distance there will be a probability of tunneling followed by recapture. Calculations show that this will in fact produce the observed dependence on time. For electrons trapped at 0.5 eV the corresponding distance is approximately 75 Å. Electrons trapped much closer will be recaptured within microseconds of the exposure while those trapped much further out will have lifetimes considerably greater than are used in the experiment.

Table I

Trap Depth (E_0) and Mean Lifetime τ for geminate
recapture by tunnelling

E_0 (ev)	$\tau =$ 10^2	10^4 Distance (A)	10^6 secs
0.1	184	202	219
0.3	92	106	115
0.5	72	78	85

In fact the preglow seen during an experiment would only be due to electrons trapped within a narrow range of
distances from their parent. These calculations give
the relationship between range and trap depth needed to
explain the preglow. Research on ion formation in the
presence of scavengers gives an estimate of the former,
from which the mean trap depth can theoretically be derived if the basic models are accepted.

An alternative explanation of the preglow is based
on thermal detrapping in the neighbcrhood of the parent ion. To determine whether this explanation is valid one heats the specimen up to a much higher temperature, sufficient to remove one or more of the glow
peaks, and lets it cool again. Although all the trapped electrons corresponding to these higher temperature peaks (and presumably deeper traps) have been eliminated the preglow still continues, albeit at a reduced intensity.

RELATIONSHIP BETWEEN CONDUCTIVITY AND THERMOLUMINES-CENCE

On the electron trapping concept, both post-radiation conductivity and thermoluminescence are due to the
thermal untrapping of electrons. As the irradiated specimen is warmed one might therefore expect to see an
increase in conductivity at temperatures corresponding
to the glow peaks. In fact the conductivity rises with
temperature, with only very minor increases at these
temperatures. The most likely explanation is that any
combination of an electron with a positive charge is
equally likely to give rise to fluorescence and phosphorescence whatever the distance involved, so that
thermoluminescence will arise mainly from geminate recombination. Conductivity on the other hand depends on
the distance an electron travels before capture. Gemi-

nate recombination will on the average play no part. Thus thermoluminescence is essentially an indication of geminate electrons, conductivity of non-geminate electrons.

OPTICAL ABSORPTION

Changes in the optical absorption spectrum of materials are observed following irradiation ; these have been obtained largely from frozen aqueous systems, or hydrocarbon glasses. By using non-polar hydrocarbons such as 3 methyl pentane, incorporating suitable electron scavengers such as naphthalene, the kinetics of the system may be investigated, and the spectra of the ions deduced. These spectra can be further modified by exposure to light, which causes a transfer of the electron from scavenger to the matrix.

Changes in optical spectra can also be seen in polyethylene for example irradiated at liquid nitrogen temperature. The difference from the unirradiated specimen shows a broad absorption spectrum with little in the way of characteristic peaks. If this spectrum consists of a superposition of the spectra of electrons trapped at different depths, corresponding to the different glow peaks, it should be possible to distinguish between them by partial heating up to successive peaks, the specimen then being recooled to a standard temperature for remeasurement. The loss in optical density would now correspond to the electrons untrapped up to that temperature. The experiment shows a small but significant change in the spectrum corresponding to each thermal cycle, especially at the ultraviolet end of the spectrum, but the changes show only a broad spectrum change.

The electrons may also be removed from their traps, and the optical absorption spectrum largely removed, by optical bleaching, but no detailed dependence on wavelength has been derived.

ELECTRON SPIN RESONANCE

The presence of trapped electrons in an irradiated organic solid may be detected by E.S.R. techniques, operating at a low microwave power to avoid power saturation. The signal decays on standing even at a low temperature and this could be correlated with the preglow observed in thermoluminescence.

A polyethylene specimen irradiated and examined at liquid nitrogen temperature shows an E.S.R. spectrum containing an intense singlet which is ascribed to an electron trapped physically, e.g. not involved in a chemical reaction with surrounding molecules. In high density (high crystallinity) polyethylene the initial G value is 0.46 and only 0.12 for low density polyethylene, an indication that the electron is trapped primarily in the crystalline regions. The trapped electron concentration increases with dose to a maximum value of about 4.2×10^{16}/g at a dose of about 0.5 Mrads and then decreases. This decrease has been ascribed to a reaction of the physically trapped electrons with radiation produced radicals R· rather than with more permanent radiation-induced chemical changes since the effect of a prior dose can be eliminated by warming. The range over which electrons can react with radicals can be estimated from this data and with $G(R·) \sim 3$ amounts to about 60 Å.

The E.S.R. absorption of the electron trapped at 77°K decays on standing towards an asymptotic value, but a rise in temperature causes a further drop towards another asymptotic value. The decay appears to be first order at each temperature drop, unlike the preglow thermoluminescence characteristic ; this might be expected from geminate electrons trapped at various distances from the positive ions.

ELECTRICALLY STIMULATED LUMINESCENCE

When the glow curve described above takes place in the presence of an electric field it can be significantly enhanced. This enhancement only lasts for a short time and the intensity then sinks to approximately the same level as it would have in the absence of an external electric field. Successive applications and removals of this external field show that this is a one-time only process; the electrically-stimulated contribution to luminescence dies off very rapidly. If however the field is reversed there is a second large increase in luminescent intensity for a time. This must involve some directional effect in electron trapping and resultant luminescence and is best explained as due to electron untrapping being assisted or hindered by the external field. For randomly distributed electrons this would have little effect but for geminate electrons where there is already present a directional influence electrons situated to one side of their parent ion will be removed from their trap by one direction of the field, while those on the other side will require the opposite polarity of electric field.

NATURE OF THE CHARGE CARRIER AND OF THE TRAP

The possibility exists that some of the observations on charge motion are due to mobile holes or more massive ions rather than to electrons, and to a large extent this can be ascertained by the use of the appropriate scavengers. In liquids the electron rapidly becomes attached to a molecule so that two different regimes appear depending on the lifetime of the electron and of the ion. In solids some ionic motion may be possible in glasses but in long chain molecules this is no longer feasible except possibly for the motion of H, but this would probably be present as an atom rather than as an ion.

In initial ionisation electrons are the charge carriers and in the condensed phase these rapidly lose energy and become thermalised at various distances from their parent ion, depending on the medium and on the statistics for interaction between electron and the molecule which it encounters. Once an electron has been thermalised at a distance r from its parent, the probability of escape is given by the Onsager relation $p = \exp(-r_c/r)$. The relation between the rate of energy loss for these sub excitation electrons is not known as it is for electrons with higher energy and it is therefore necessary to determine experimentally the range for thermalisation by the use of electric field or scavengers ; a number of empirical number/range distributions are then chosen to fit the observed data. These show the expected variations with temperature and with the shape and polarisability of the molecule and also indicate that a simple relation such as a Maxwellian distribution in range is not valid.

When an electron scavenger is present the subsequent recombination of an electron with an ion to give radicals is diminished and fewer products are formed from such radical reactions. Table 2 shows that in cyclohexane for example N_2O, which has a high capture cross section for electrons, reduces the formation of H_2.

<u>Table 2</u>
N_2O as scavenger for electrons $(N_2O+e \rightarrow N_2+O^-)$

Effect on $G(H_2)$ yield from c. hexane

N_2O conc (M)	Ratio N_2O c $C_6 H_{12}$	$G(N_2)$	$G(H_2)$
0	0	0	5.6
4×10^{-3}	0.43×10^{-3}	1.14	4.95
2×10^{-2}	2.1×10^{-3}	2.2	3.80
5×10^{-2}	5.4×10^{-3}	3.0	3.57
1×10^{-1}	10.8×10^{-3}	3.75	2.78
2.1×10^{-1}	22×10^{-3}	4.26	2.78

One can readily accept that the recombination of electron and positive charge occurs when these are separated by distances of the order of 60-100 Å, either by tunnelling or by thermal release of electron. However, some of the scavengers for electrons are effective at relatively low concentrations and it is somewhat difficult to see why the electron should be affected by an uncharged molecule at more than a few atomic spacings. It may of course be that when the electron undergoes ionisation, or is released from its trap, it suffers a large number of collisions until it enters the immediate vicinity of the uncharged scavenger, but in this case the average capture distance should be correspondingly larger when dealing with a positive charge.

The nature of the trap calls for comment. Electrons may be stabilised in the neighbourhood of a polarisable molecule, but this is far less likely in such materials as cyclic or spherical alkanes of very low polarisability. One possibility in solids is that the electron is trapped in local defects, especially voids. Due to the difference in dielectric constant of a vacuum and a hydrocarbon (even with a dielectric constant ~ 2), such traps may be expected to occur. In this case the freeing of a trapped electron may be due not so much to thermal or tunnelling release as to the disappearance of the trap by molecular motion. This would explain the association of thermoluminescent peak with morphological changes in the polymer.

REFERENCES

More detailed surveys and references may be found in the following texts :

Haissinsky M., Actions Chimiques et Biologiques des Radiations (Masson, Paris) (Series) 1956 on.

Swallow A.J., Radiation Chemistry of Organic Compounds Pergamon 1960

Charlesby A., Atomic Radiation and Polymers Pergamon 1960

Chapiro A., Radiation Chemistry of Polymeric Systems Wiley 1962

Spinks J.W.T., Woods R.J., Introduction to Radiation Chemistry Wiley 1964

Gould R.F., Irradiation of Polymers Conference, American Chem.Soc. 1967

Gaumann T., Hoigne J., Aspects of Hydrocarbon Radiolysis Academic Press 1968

Burton M., Magee J.L., Advances in Radiation Chemistry Wiley Interscience 1969 on (Series)

Dole M., Radiation Chemistry of Macromolecules Academic Press 1972

Swallow A.J., Radiation Chemistry Longman 1973

Kinell P.O., Roanby B., Runnströn-Reio V., E.S.R. Applications to Polymer Research Nobel Symposium 22, Wiley 1973

RADIATION DAMAGE IN METALS AND SEMICONDUCTORS

R.S. NELSON

AERE Harwell
Didcot, Berkshire, U.K.

INTRODUCTION

In these lectures we will outline the mechanisms responsible for irradiation damage to both metals and semiconductors. We will primarily be concerned with high damage levels where defect agglomeration occurs and for this reason we will rely heavily on observations based on electron microscopy. For a fuller understanding of the field the reader is referred to the numerous review articles and books which are listed within the references.

ATOMIC DISPLACEMENT

The elastic collisions which occur between incident irradiation, be it fast neutrons or heavy ions, and the atoms of a solid result in energy transfers ranging from fractions of an eV to many tens of keV. The exact energy spectrum of those atoms recoiling from such primary encounters depends intimately on the scattering cross-section which in turn depends on the interatomic potential between the colliding particles. However, in this instance we will not do well on this point as we will be mainly concerned with subsequent events.

Provided a recoiling lattice atom receives an energy in excess of a minimum value, about 25 eV, called the displacement energy, E_d, it can leave its lattice site to become permanently displaced within the solid. The exact magnitude of this energy not only de-

pends on the solid in question but also on the recoil
direction within the lattice. In other words, it is
somewhat easier to displace an atom in certain well
defined directions than in others. This problem has
been considered both theoretically and experimentally
for a variety of materials (Banbury et al[1] ; Erginsoy
et al.[9] ; Lomer and Pepper[15]). For instance fig.1 shows
a displacement threshold map for b.c.c. iron as calculated in a computer simulation study. It is readily
apparent that in this case the easiest displacement
direction is the <100> at around 17 eV whilst atoms
recoiling in the region near the centre of the unit
triangle require many times this energy. In general
however, and especially in the context of ion implantation, it will suffice to take 25 eV as a good average value for virtually all materials.

In most cases of interest lattice atoms recoil
from primary collisions with kinetic energies far in
excess of the displacement energy, and as such are capable of penetrating many atomic distances into the
surrounding lattice. To a good approximation, as these recoil atoms start from lattice sites, their penetration into the solid is expected to result from a
succession of uncorrelated collisions and is therefore amenable to calculation. As the mean energy falls,
the scattering cross-section will become so large that
the mean free path between successive collisions will
be essentially equal to the distance between adjacent
atoms. In such a situation it is quite clear that a
cascade of secondary collisions will be initiated very close to the original primary event, see fig.2. It
is the spreading of such collision cascades which is
therefore important to radiation damage, as those regions remaining in the wake of the cascades will contain many displaced atoms.

From the foregoing it is readily apparent that in
order to understand radiation damage we must first focus our attention on the details of the collision cascade and the initial configuration of displaced atoms.
The details of the first of these has been covered extensively by Nelson[25] and it will suffice to present
an outline of the more important features. The majority of our knowledge has been gained from both theoretical and experimental studies on metals ; although recently, due to the interest in the ion implantation
of semiconductors, a substantial effort has been directed towards Si, Ge and GaAs.

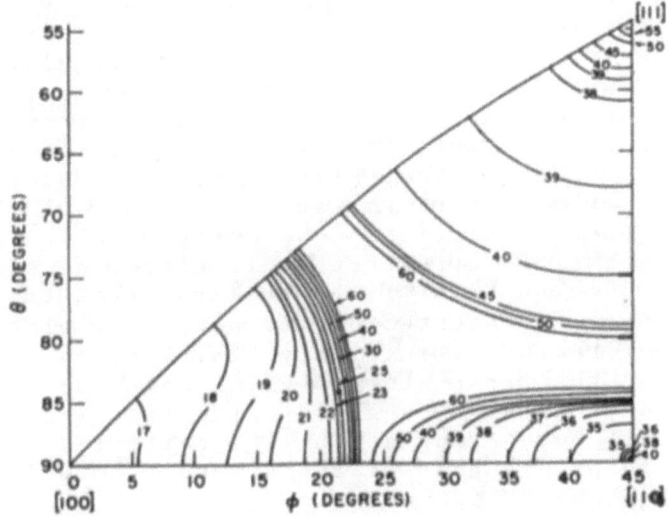

<u>Fig.1</u> : Contours of displacement energy threshold within the unit triangle

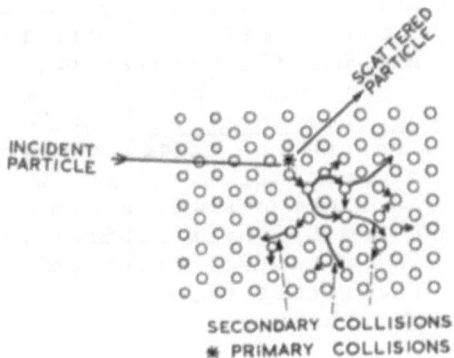

<u>Fig.2</u> : The interaction of an incident particle with a solid showing the primary collision and subsequent secondary collisions.

Originally, theories of the collision cascade assumed the solid to have no structure and the multiplication of collisions was assumed to be completely random. However, it is now known that the regular nature of the crystal lattice plays a vital part in the spreading of the collision cascade. It was Silsbee[31] who first pointed out that the ordered atomic array would impose a directional correlation between successive collisions and that energy and momentum would be focused into those directions consisting of close-packed rows of atoms. To illustrate this effect consider the simple collision illustrated in fig.3 between two identical atoms, represented by two perfectly elastic hard spheres of radius R, initially separated by a distance D. Suppose the first atom is given momentum in the direction AP, making an angle θ_1 with the line of centres AB. It will move along AP until a collision occurs when its centre has reached P. The second atom will then move off along PB at an angle θ_2 to AB. For the case of small angles a simple geometrical argument shows that

$$\text{if } D > 4R, \text{ then } \theta_2 > \theta_1$$
or
$$\text{if } D < 4R, \text{ then } \theta_2 < \theta_1 \tag{1}$$

Thus, in a row of such atoms, provided the atomic radius is greater than one quarter the interatomic spacing, successive angles will converge towards zero until all collisions are effectively "head-on". Under these conditions, therefore, momentum has been focused into the line, and for this reason such sequences of correlated collision events have been called <u>focused collision sequences</u> or <u>focusons</u>. In a real crystal it is necessary to relate the effective radius of an atom to the interatomic potential. At the atomic separations of interest the nuclear charges of the colliding atoms are heavily screened by their orbital electrons and a purely exponential potential provides a reasonable approximation to the interaction, for example a Born-Mayer potential of the form

$$V(r) = A \exp(-r/a) \tag{2}$$

where A and a are constants for a particular atom. Under these conditions, to a first approximation, near head-on collisions can be treated as if the atoms were perfectly elastic hard-spheres. Such an approximation clearly over-simplifies the problem but has proved ve-

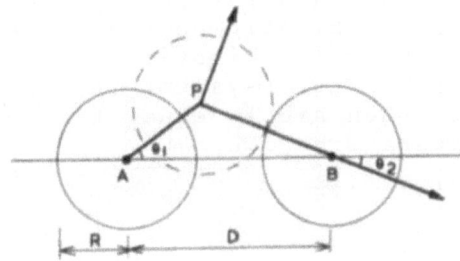

Fig.3 : Simple collision between two atoms of radius R initially separated by D.

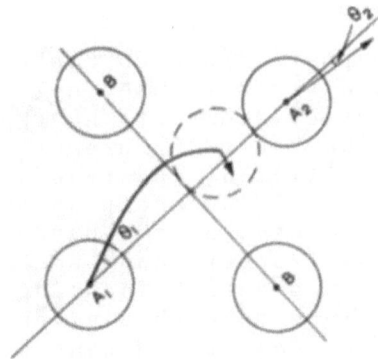

Fig.4 : Plan view showing the atomic trajectory of a <100> collision sequence.

ry successful in obtaining order of magnitude estimates.
The interaction is likened to that between two billiard
balls of radius R such that in any collision their centres are separated by a distance ZR. Then transforming
the collision co-ordinates into the centre of mass
system, the total kinetic energy of the system is equal
to 1/2 E, where E is the kinetic energy of the moving
atom in the laboratory system. In a head-on collision
this enegy is then simply equal to the potential energy of the interaction i.e.

$$V(2R) = \frac{1}{2} E \qquad (3)$$

whence using (2)

$$R = \frac{1}{2}a \; \ell n(ZA/E)$$

Using this expression we can relate the geometric focusing condition contained in (1) to a practical case.
The first point to notice is that the hard-sphere radius is inversely proportional to the logarithm of e-
nergy R < 1/4 D and focusing cannot occur. Then as the
energy decreases the radius becomes slowly larger until
at a critical energy corresponding to R = 1/4 D the correlation between successive collisions permits a reduction in angle. This energy is known as the focusing
energy and is given simply by

$$E_f = 2A \; exp \; (-D/2a) \qquad (4)$$

This hard-sphere approach has provided us with a
simple basis to discuss focused collision sequences.
However, in practice atoms are no rigid spheres and we
must allow for the "soft" nature of the interaction.
In a realistic collision, the true scattering angle
corresponding to a given impact parameter is somewhat
less than that predicted by the hard-sphere model. This
means that the hard-sphere approach overestimates the
focusing and so in practice focusing will occur at an
energy somewhat less than that given by (4). On the o-
ther hand, with a realistic potential, atoms ahead of
the main energy packet will start moving before its arrival ; this leads to a shortening of D and a consequent
increase in focusing. Lehmann and Leibfried[13] have considered these effects in some detail, but their treatment is too sophisticated for our present purposes.
However, a simpler version due to Sigmund[32] gives the
following reasonably simple analytical expression :

$$E_f = 2A \exp[- \frac{D}{4a} \{1+(1+ \frac{8a}{D})^{1/2}\}] \qquad (5)$$

In a real crystal it is not only sufficient to satisfy the above conditions for the propagation of focused collision sequences but also that the surrounding atoms do not interfere to such an extent as to destroy the sequence. A simple consideration of both f.c.c. and b.c.c. lattices shows that in each case their close-packed rows, i.e. the <110> and <111> directions respectively, are situated in an environment which allows the unhindered propagation of collision sequences. But on the other hand, in the diamond lattice, such isolated rows do not exist and the propagation of such sequences is unlikely.

In order to put the foregoing discussion on a quantitative basis let us consider a row of copper atoms equi-spaced by a distance equal to the close-packed <110> direction of the f.c.c. copper lattice, 2.55 Å. Using the Born-Mayer constants used by Vineyard et al[35] of $A = 2.2 \times 10^4$ eV and $a = 1/13 \; D^{110}$, we obtain a focusing energy using (5) of $E_f^{110} = 35$ eV. The next step is to consider how far such focused collision sequences are likely to travel. As already pointed out atoms are not just hard-spheres, with the consequences that as sequences propagate through the lattice, both potential and kinetic energy will be dissipated. Further, as in reality the atoms of a crystal are in a continual state of thermal vibration, perfect focusing can never be attained and energy is inevitably lost to neighbouring atoms as a consequence of lateral displacement. A full discussion may be found in Nelson[25]; the net result being that for copper at room temperature the mean energy lost per collision is roughly constant at 3 eV, and so a sequence starting off at the focusing energy has a maximum range of some 10 collisions, in other words about 25 Å.

So far we have only considered simple focused collision sequences which propagate along the close-packed rows of the crystal lattice. If we also consider the possibility of such sequences propagating along the next close-packed rows of atoms, i.e. the <100> directions in both the f.c.c. and b.c.c. structures, their focusing energies as calculated by (5) turn out to be too low for practical importance. However, in both cases, atoms which recoil close to these <100> directions have their trajectories essentially focused by the surrounding atoms, see fig.4. Once again, it is possible to focus energy into the line by a succession of such collisions. This effect is called <u>assisted focusing</u> and has been treated theoretically by Nelson and Thompson[24] and by Leibfried and Dederichs[14]. The range of such assisted focused collision sequences is generally less than expected for simple sequences.

In the context of radiation damage we must assess the influence of focused collision sequences both on the numbers of atoms displaced and on their initial distribution. From the foregoing we know that whenever the energy of the cascade falls below the focusing energy, the random multiplication process ceases and successive collisions are strongly correlated so that pulses of energy are focused into certain low index directions. We can therefore assume that once this situation is reached the number of further displacements is strictly limited. In the random collision model we must first of all estimate the total fraction of energy within the collision cascade that is dissipated via nuclear rather than electronic losses. Using the Lindhard expressions for energy loss, we can simply estimate the relative fractional energy loss of the primary recoil atom itself as a function of energy. Then by integration we can obtain the total fractional energy loss to nuclear collisions for the primary from its starting energy E_p until it comes to rest. However, we must remember that the secondary and subsequent recoils also lose a fraction of their energy to electronic excitation and so the total energy dissipated in nuclear collisions within the cascade is still further reduced from that fraction lost by the primary alone. Fig. 5 show an estimate of the net total energy lost to such nuclear collisions $(\Delta E)_n$ within a collision cascade in Cu initiated by primaries of given energy plotted as a function of primary energy. Then in the hard-sphere approximation without focusing, the final number of displaced atoms is simply given by $(\Delta E)_n/2E_d$. On the other hand, a more realistic treatment which accounts for the inadequacy of the hard sphere approximation would give $0.8(\Delta E)_n/2E_d$.

In fact recently the IAEA has adopted a proceedure based on these ideas developed by Norgett, Robinson and Torrens[26] for the case of medium mass metals such as iron etc. The formulae is summarised as follows :

The number of displaced atoms is given as

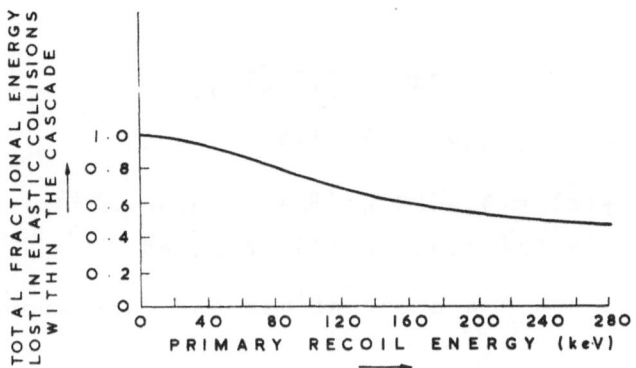

<u>Fig.5</u> : The net total energy lost to elastic collisions within a
collision cascade in copper plotted as a function of primary recoil energy.

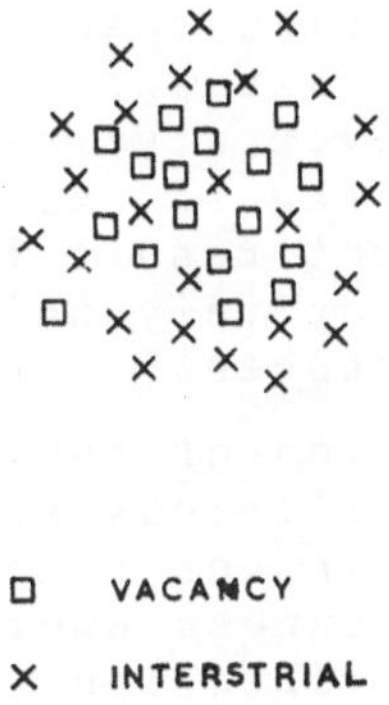

☐   VACANCY

✗   INTERSTRIAL

<u>Fig.6</u> : The initial distribution of vacancies and interstitials
left in the wake of a collision cascade.

$$n_d = \beta E_{damage} \quad \text{displacements/primary} \qquad (6)$$

where $\beta = 10 \text{ keV}^{-1}$ and E_{damage} is an estimate of the energy deposited into atomic processes given by

$$E_{Damage} = \frac{E}{[1+kg(\varepsilon)]} \qquad (7)$$

where $k = 0.1337 \ Z^{2/3} \ A^{1/2}$

$$g(\varepsilon) = 3.4008 \ \varepsilon^{1/6} + 0.40244 \ \varepsilon^{3/4} + \varepsilon$$

$$\varepsilon = E/[86.931 \ Z^{1/3}] \quad E \text{ in eV}$$

where Z and A are the atomic and mass numbers respectively.

With regards to the spatial distribution of displaced atoms it is well known that in some instances focused collision sequences propagate via a series of replacement collisions such that interstitial-vacancy pairs can be separated by many atomic distances. Further, at the extremity of the cascade the majority of collision sequences will be directed radially outwards from the cascade centre with the inevitable result that the inevitable result that the damaged region will in the first instance have a vacancy rich core surrounded by a region rich in interstitials, see fig.6. This will be especially true for the heavier metals where, due to the large scattering cross-section, the central core region of the collision cascade is rather small. However, in lighter metals such as aluminium the collision cascade is spread over a very much larger volume and also, because focusing is expected to be rather weak in aluminium, the net result is that the collision cascade contains a relatively uniform distribution of interstitials and vacancies.

From the creation of the primary recoil the high energy secondary collisions are completed in about 10^{-13} sec. However, this leaves a highly damaged region with an accumulation of excess energy which can only be dissipated by thermal vibration. This results in the creation of a local hotspot, called a thermal spike. The lifetime of such a spike depends on the efficiency of heat conduction to the surrounding lattice. In this case, however, the energy is initially given to the atomic system, and, as the transfer of energy between the atomic and electronic systems is relatively slow, we must consider the cooling of the spike to proceed by atomic processes rather than by the usual equilibrium electronic conduction processes. After about 10 atomic vibration periods, the velocity distribution of the atoms in the heated region approximated to that expected from Maxwell-Boltzmann statistics and as such becomes amenable to calculation. Experiments by Thompson and Nelson[33] and Nelson[23] have specifically studied the sizes, temperature and life-times of thermal spikes in a variety of solids ; a few of these are listed in fig.7. In general, as the temperatures of such spikes are only a few hundred degrees above that of the surrounding lattice and their life-times are only a few times 10^{-12} sec, it is unlikely that significant defect rearrangement will occur as a result of thermal spikes, except perhaps for the recombination of some close interstitial-vacancy pairs.

ATOMIC MOTION AND DYNAMIC DEFECT CONCENTRATION

It is well known that simple point defects such as vacancies and interstitials, once they are created, can migrate randomly throughout the lattice under thermal activation. The details -such as activation energies for migration and formation has been the subject of much research and discussion over the past 25 years. The reader is referred to the International Conference on "Vacancies and Interstitials in Metals" held at Jülich in 1968 (see Seeger et al[28]) for a comprehensive review and discussion of the subject. However, in the context of the present discussion let us assume that interstitials migrate very much more rapidly than do vacancies- typical activation energies being ~ 0.1 eV and ~ 1.0 eV respectively.

At very high temperatures where vacancy emission from grain-boundaries and surfaces is sufficiently large that thermodynamical equilibrium can readily be maintained, the equilibrium vacancy concentration is given by

$$C_v^o = A \exp (-E_f^v/kT) \qquad (8)$$

where A is an entropy factor ~ 1 and E_f^v is the formation energy of the vacancy. It should be noted that as the formation energy of the interstitial is so large (~ 10 eV) the concentration which can be maintained in thermal equilibrium is negligible. At lower temperatures it is quite probable that the radiation induced vacancy concentration will exceed the thermodynamal value, and it is in this regime that we are in fact more interested.

272

Metal	Ion	Ts (°K)	τ_s (10^{-12} sec)
Au	Xe	910	3
	A	600	6
Ag	Xe	530	6
	A	523	4
	Ne	518	7
Cu	Xe	490	5
Zn	Xe	150	1
Bi	Xe	600	9
Ge	Xe	1060	3

Fig.7 : The temperature, size and lifetime of thermal spikes produced by 45 keV Xe^+, A^+ and Ne^+ ions.

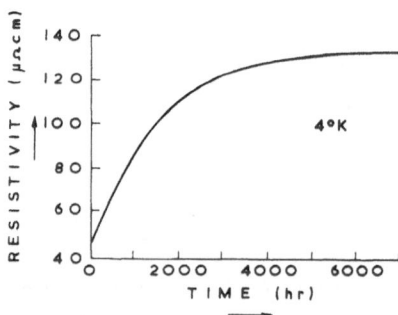

Fig.8 : The saturation of resistivity of α-Pu plotted as a function of irradiation time held in liquid helium.

Let us therefore examine the defect concentrations which build up during irradiation in the temperature range where the irradiation induced concentration is expected to dominate. As we have already pointed out the irradiated sample is usually maintained at some temperature above 0° K with the result that the individual defects created within the collision cascade can migrate under the influence of thermal activation. On the other hand, even at 0° K some defect rearrangement is possible due to the mutual elastic interaction between defects. For instance, computer calculations have shown that there is a critical separation between an interstitial and a vacancy which must be attained before the defects do not spontaneously recombine. This sets an ultimate limit on the density of damage that can be produced even at temperatures too low for thermally activated migration. Experimentally this is manifest in the saturation in the change of some physical property, such as electrical resistivity on irradiation; see fig.8 for instance. From an analysis of such results it is possible to make estimates of the average volume around a point defect which is unstable against spontaneous recombination. Typically this is of the order of 50-100 atomic volumes, which therefore sets an absolute limit on the number of individual point defects that can exist at 10^{-2}.

However, in most circumstances of practical importance we are concerned with temperatures where both the individual interstitials and vacancies can migrate. Let us make the assumption that there is no recombination within the collision cascades such that the numbers of displaced atoms as calculated above all contribute to give a uniform average damage rate throughout the material of K displacements/atom/sec. Clearly if both defects are mobile, they can in fact be lost from the system. It is usual to consider two loss mechanisms, namely the mutual recombination of interstitials and vacancies and loss to fixed sinks such as dislocation lines, etc. Then following the work of Lomer (1954) and Dienes and Damask[8] and Sharp[29] we can write down the following differential equations :

$$\frac{dv}{dt} = K - K_v C_v - \nu_i Z (C_v + C_v^\circ) C_i \tag{9}$$

$$\frac{dC_i}{dt} = K - K_i C_i - \nu_i Z (C_v + C_v^\circ) C_i \tag{10}$$

where C_v and C_i represent the vacancy and interstitial concentrations, $K_v C_v$ and $K_i C_i$ represent the loss of vacancies and interstitials to fixed sinks, ν_i is the interstitial jump frequency, Z is the recombination coordination number ~ 1 and C_v^o is the thermal vacancy concentration. $K_v = a_v \nu_v \lambda^2$ and $K_i = \alpha_i \nu_i \lambda^2$ are the proportionality constants for loss to sinks where α_v and α_i are the fixed sink densities for vacancies and interstitials and λ is the defect jump distance ($\lambda^2 \sim 10^{-15}$ cm^2).

These two equations can be solved for any time interval after the irradiation has commences, however, it is of interest to consider the conditions which prevail once equilibrium has been reached. In practice it is readily seen that the loss by recombination dominates at low temperatures, whereas the loss to fixed sinks is the dominant mechanism at higher temperatures. Under these circumstances we can essentially solve the above equations for these separate regimes and furthermore if we presume that $\alpha_v = \alpha_i = \alpha$ we arrive at the following expressions

$$c_v = (K/Z\nu_v)^{1/2} \quad \text{for loss by recombination} \quad (11)$$

$$c_v = K/(\alpha \nu_v \lambda^2) \quad \text{for loss to fixed sinks} \quad (12)$$

In reality α_v is not always equal to α_i and as we will see later, this fact is responsible for one of the most important technological manifestations of radiation damage at the present time -namely void formation.

If we insert reasonable values into the above expressions we find that the equilibrium instantaneous defect concentrations are in fact quite small, e.g. about $10^{-7} - 10^{-6}$ for steel irradiated in a fast reactor. Another way of expressing defect concentrations is through the diffusion coefficient. In most circumstances thermally activated diffusion relies on the motion of vacancies through the material, the activation energy for the process being the sum of the formation energy and the migration energy. However, during irradiation vacancies are created athermally and thus diffusion will occur provided there is sufficient thermal energy to satisfy their motion. The diffusion coefficient can be written as

$$D_v = \nu_v C_v \lambda^2$$

then substituting in for C_v we find

$$D_v = \lambda^2 (K\nu_v/Z)^{1/2} \quad \text{for recombination dominant} \tag{13}$$

$$D_v = K/\alpha \quad \text{for ion to fixe sinks} \tag{14}$$

A typical set of enhanced diffusion coefficients for different damage rates and dislocation densities are shown in fig.9 for Ni together with the thermal diffusion coefficient ; the curves for $K = 10^{-6}$ dpa/µc corresponds to a typical fast reactor dose rate whereas the curves for $K = 10^{-2}$ dpa/µc correspond to that expected during irradiation in an ion accelerator. It is interesting to note that for high dislocation densities where the major loss of defects is to fixed sinks, the enhanced diffusion coefficient depends only on the damage rate and is independent of the activation energy for migration. Under these circumstances vacancy controlled enhanced diffusion of impurities will also be independent of the activation energy for migration of the impurity.

THE CLUSTERING OF POINT DEFECTS IN METALS

As the irradiation induced interstitials and vacancies migrate throughout the solid, not only does recombination occur, but also both interstitials and vacancies can cluster. If the binding energy between vacancies or between interstitials is sufficiently large then such clusters remain stable and grow. Let us first consider the situation at very low damage doses where individual collision cascades are well separated. Then due to the dynamic segragation of defects within the cascades, which occur predominantly in heavy metals, the vacancy rich core can re-arrange to form a compact defect cluster, so leaving the interstitials and residual vacancies free to migrate away from the damaged region. The precise morphology of these clusters, i.e. whether they form three dimensional voids, faulted dislocation loops, unfaulted dislocation loops or stacking fault tetrahedra, depends on a variety of parameters such as stacking fault energy and surface energy. For instance, Sigler and Kuhlmann-Wilsdorf[30] have made a number of predictions for a variety of pure metals depending on the number of vacancies within the agglomerates, see fig.10. However, due to the inadequacy of our knowledge of the parameters involved, we must treat the quantitative results of such computations with some reservation. In light materials, especially those associated with poor focusing such as aluminium or graphite, where the segregation of vacancy and intersti-

276

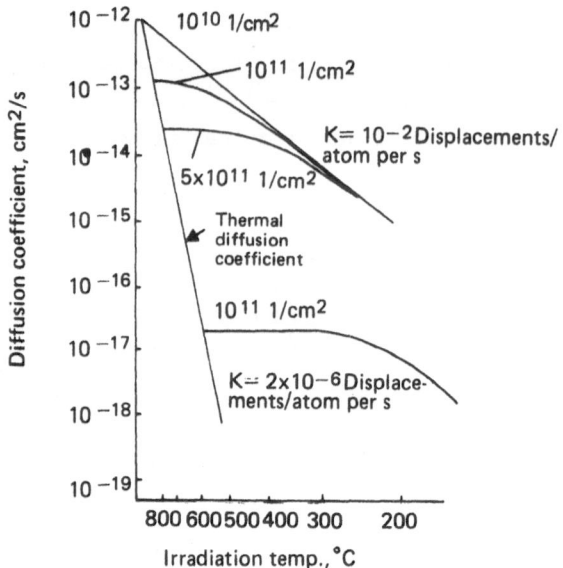

Fig.9 : Radiation enhanced diffusion coefficient for different dislocation densities at damage rates of 10^{-2} and 2.10^{-6} displacement/atom/s.

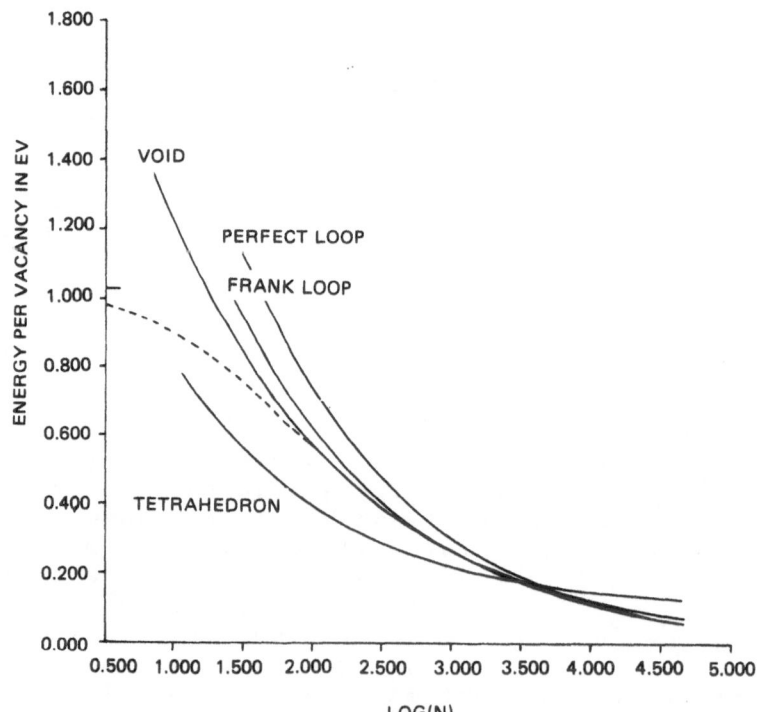

Fig.10 : Specific energy per vacancy plotted as a function of number of vacancies for different defect agglomerates in Cu.

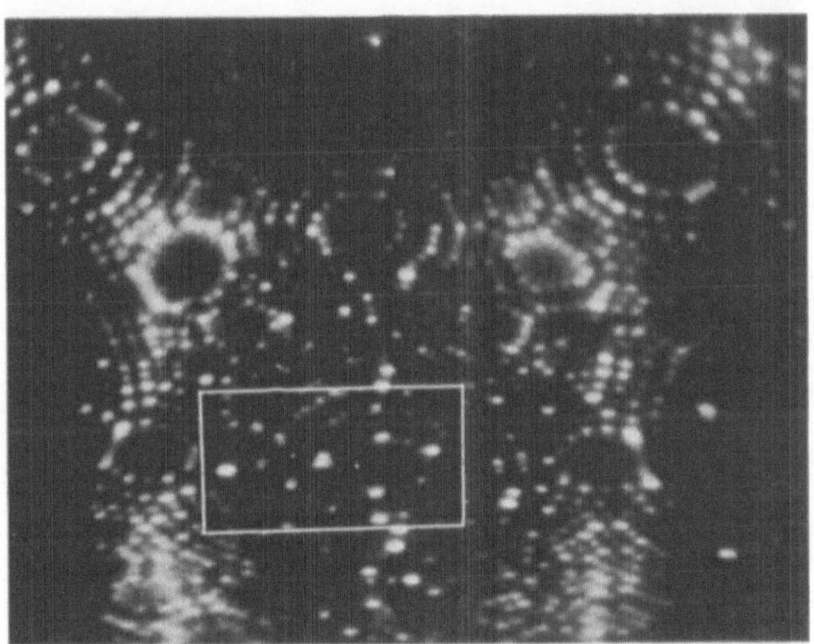

Fig.11 : A field ion microscope photograph of a dislocation loop in irradiated Ir, the two white dots define the ends of the dis-location.

278

Fig.12 : Black spot defects in Cu irradiated with 30 keV Cu^+ ions, each spot corresponds approximately to one incident ion. Analysis of the diffraction contrast suggests that such defects are vacancy type dislocation loops.

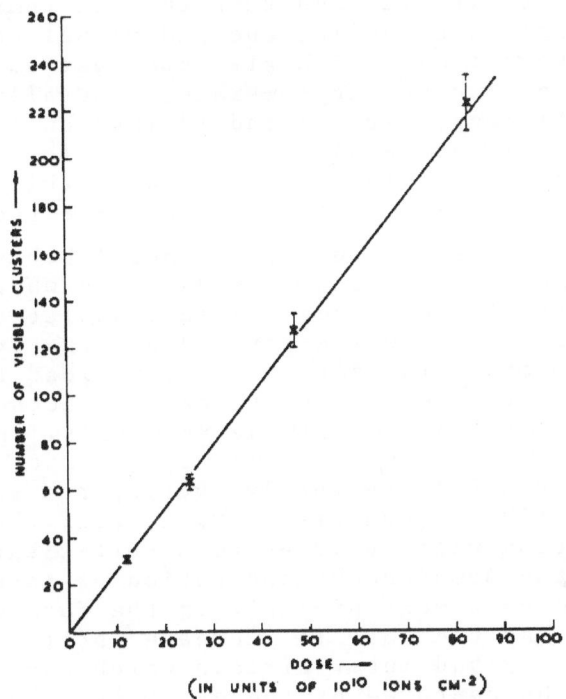

<u>Fig.13</u> : The number of clusters in 30 keV Cu$^+$ ion bombardment of Cu with diameters greater than 25 A plotted as a function of dose showing the linear relationship.

tial defects is somewhat limited, the heterogenous
nucleation of defect clusters is less likely, with the
result that perhaps every vacancy and interstitial is
free to migrate through the lattice in equal numbers.
Some experimental evidence for these ideas has been
provided by field-ion and electron microscopy. Both
techniques suggest that in heavy metals such as copper,
gold, silver, iridium and tungsten, at temperatures
where defects are mobile, the individual collision cascades leave regions which give contrast consistent with
some sort of vacancy agglomerate, generally a small
dislocation loop, figs.11 and 12 (Hudson[11] ; Wilson[36]).
However, in light metals bombarded to low doses, no
such evidence for cluster formation within the collision cascade has been forthcoming as expected.

Next we must discuss the general behaviour as the
bombardment dose steadily builds up to where the individual collision cascades overlap. Whilst the individual cascades are well separated a large fraction of
the freely migrating defects will be lost to the surface, which in most cases of interest acts as the most
dominant sink. However, in heavy metals, as the dose
builds up the isolated vacancy agglomerates will build
up linearly until eventually overlap and saturation pre-
prevails, fig.13. Concurrent with this, the interstitial
concentration will build up to a sufficiently high level that the homogeneous nucleation of interstitial
clusters occurs most probably in the form of loops as
in fig.14. As the dose is increased still further those
new vacancies and interstitials which escape recombination will be captured by the existing defect clusters
rather than sustain further nucleation. The clusters
will therefore steadily grow until they interact and e-
ventually form a complicated dislocation entanglement
as illustrated in fig.15. This situation represents an
ultimate saturation in configuration of damage as further irradiation will simply result in the rearrangement
of this entanglement by the processes of slip and climb.

On the other hand, in light metals such as aluminium, where the heterogeneous nucleation of vacancy
clusters within cascades is considered unlikely, substantial self-annihilation occurs in the early stages.
However, as the damage level increases it is thought
that homogeneous nucleation of both interstitial and
vacancy clusters occurs simultaneously (Beevers and
Nelson[2]).

Let us next consider the situation at slightly hi-

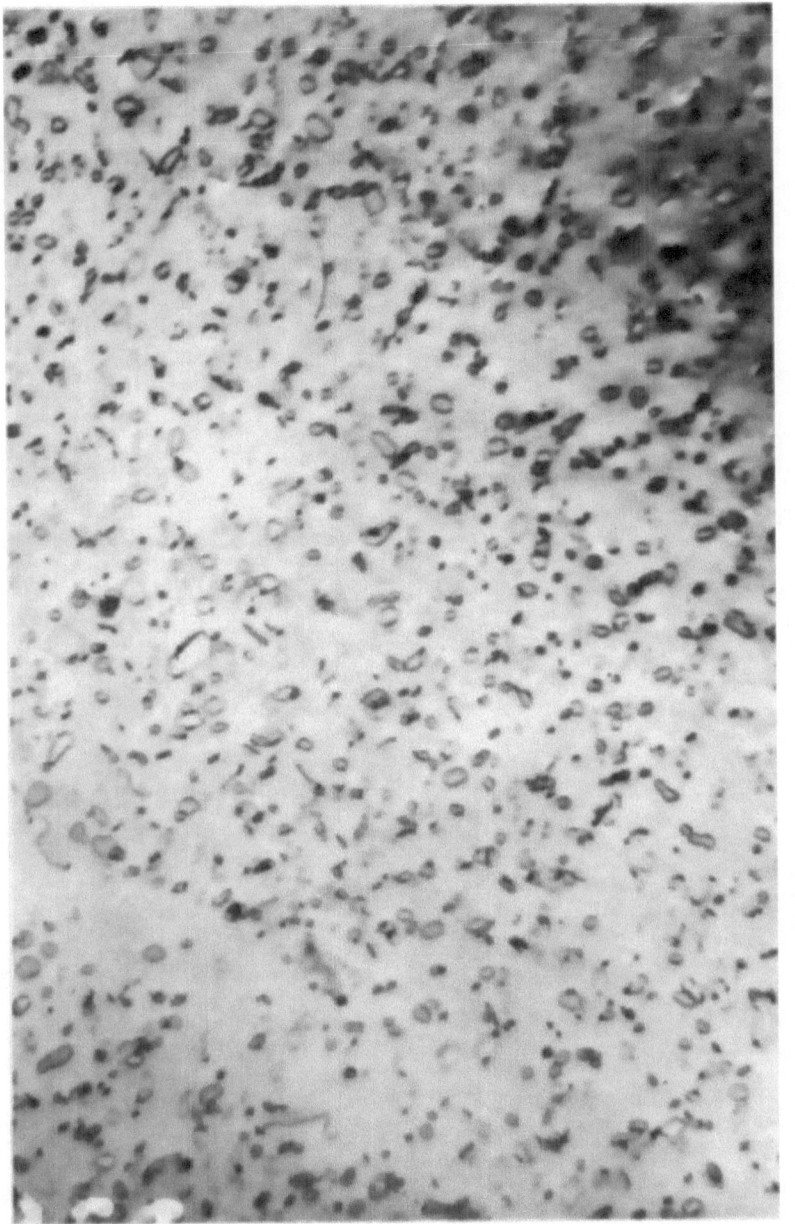

Fig.14 : Interstitial dislocation loops in α-particle bombarded Cu.

282

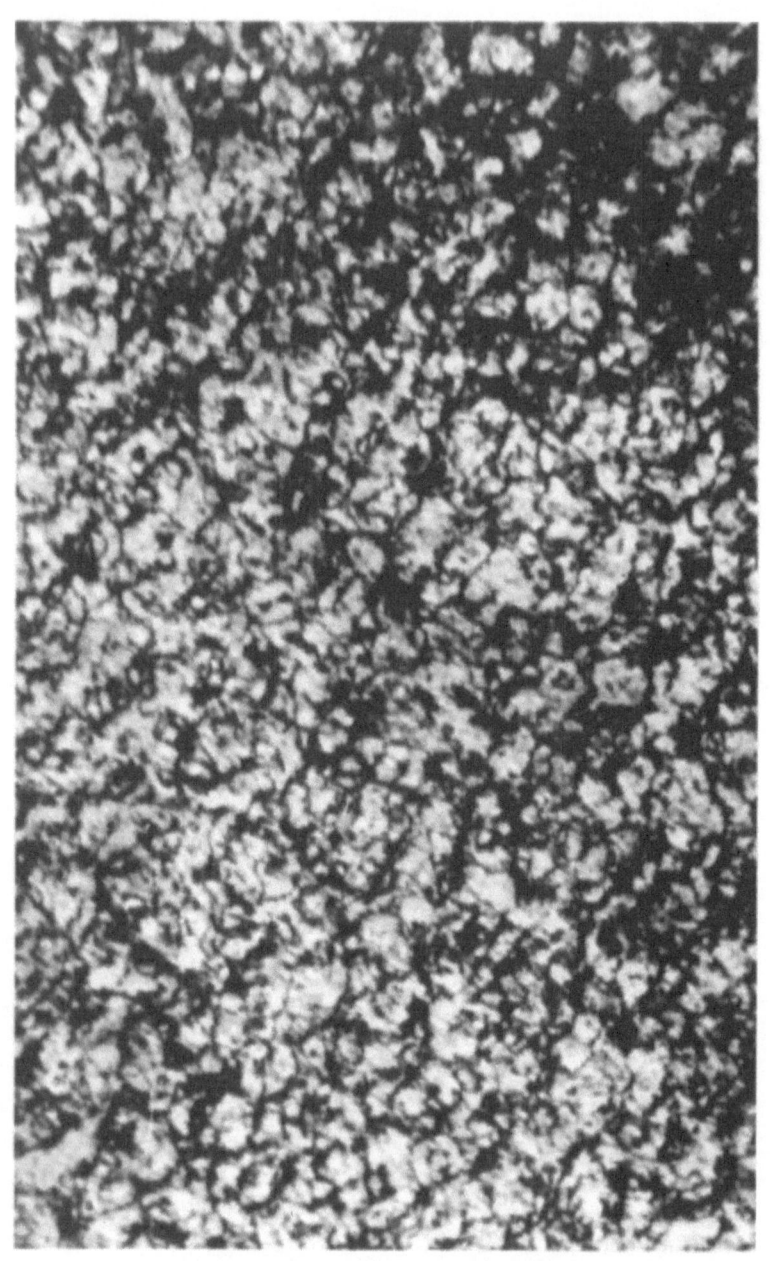

<u>Fig.15</u> : High dose ion bombarded Cu showing the growth of interstitial loops into a complex dis-
location entanglement.

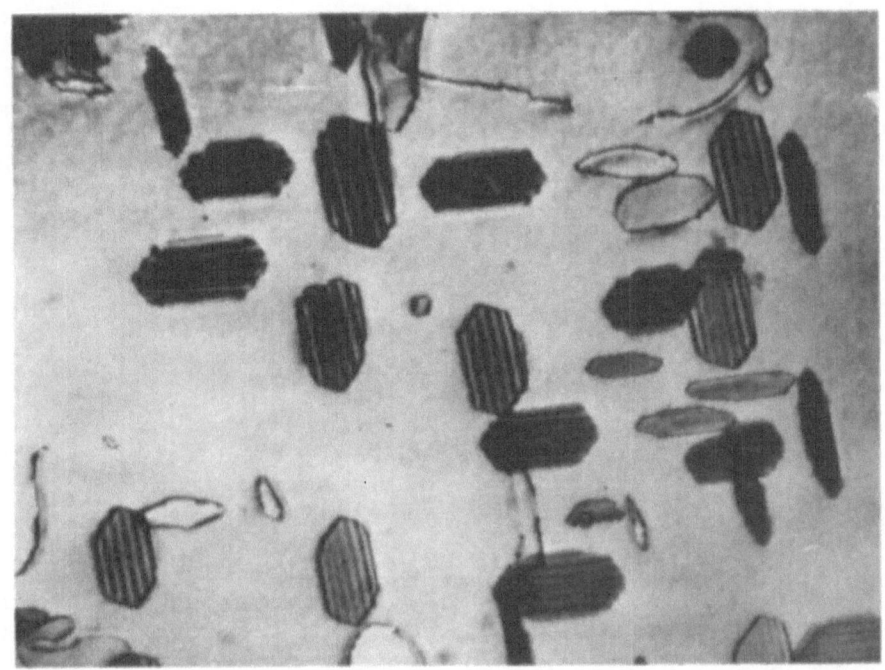

<u>Fig.16</u> : Faulted interstitial loops in proton irradiated Cu at
250°C

gher temperatures too low for the thermal activation of vacancies from extended sources, but where the discrete vacancy rich regions created within displacement cascades spontaneously dissociate, e.g. 200°C in copper. Under these conditions vacancy cluster formation is unlikely, however provided some vacancies can escape to nearby sinks such as an interface or become trapped at for instance dissolved gases, some interstitials clusters will in fact form and survive. For instance, the large intrusive fauted loops which occur during proton irradiation of copper at $\sim$ 250°C, fig.16 (Mazey and Barnes[20] ; Tunstall et al.[34]). It is now of interest to consider what happens when for instance, copper is irradiated at about 250°C, to very much higher doses. As the irradiation proceeds both interstitials and vacancies recombine and are both lost to fixed sinks. If all sinks have equal strength for both defects, e.g. $\alpha_v = \alpha_i$ then the situation would remain essentially stable, except for random fluctuations - e.g. saturation. However, in practice α_v is not equal to α_i for although both interstitials and vacancies are attracted to dislocations via their elastic interaction, the larger distorsion field associated with an interstitial results in a stronger interaction (particularly to the dilated region) and thus a preferential drift. The dislocations thus act as sinks for both interstitials and vacancies but are slightly more effective for interstitials ; the relative strength of this bias for interstitials can be estimated from our basic knowledge of point defect interactions and yields a value of about 1 % (e.g. Bullough[3]). Suitable dislocations are always present during irradiation, some because they were present before the irradiation and others as a result of interstitial clustering during irradiation as described above. The growth of interstial clusters into loops will then be facilitated by the preferential drift to them of subsequently produced interstitials. This slight preferential loss of interstitials to the dislocations means that the net point defect flux to any other neutral sink will be vacancy in character. For example, suppose we have a small three dimensional vacancy agglomerate -e.g. a small gas bubble nucleus- there would then exist a net flux of vacancies entering through its surface as a direct consequence of the dislocation line preference for interstitials. Such agglomeration of vacancies is called a "void".

The phenomenon of void formation has recently received substantial study for its technological significance to fast reactor design. However, in this instance it will simply be sufficient to outline some of the more important features. We have already evaluated the dynamic effect concentrations that are likely to be produced under various conditions. The condensation rate of interstitials and vacancies into a neutral sink of density a is simply given as

$$\lambda^2 \nu_i C_i \alpha \quad \text{(interstitials)}$$

$$\lambda^2 \nu_v C_v \alpha \quad \text{(vacancies)}$$

Then the net flow of interstitials to dislocations as a consequence of the interaction preference is :

$$\lambda^2 (\nu_i C_i \alpha_i - \nu_v C_v \alpha_v)$$

which in steady state conditions where $\nu_i C_i = \nu_v C_v$ gives :

$$\lambda^2 \nu_v C_v (\alpha_i - \alpha_v)$$

The net condensation of vacancies onto neutral sinks such as voids is then also given by this expression which is in fact also the "swelling rate". Substituting for C_v as derived above we then have two expressions for swelling corresponding to whether or not recombination plays a dominant role ;

e.g.
$$\frac{\Delta V}{V} = \frac{\lambda^2}{Z} \nu_v^{1/2} K^{1/2} t (\alpha_i - \alpha_v) \quad \text{recombination dominant} \quad (15)$$

$$\frac{\Delta V}{V} = \frac{Kt(\alpha_i - \alpha_v)}{\alpha_v} \quad \text{fixed sinks dominant} \quad (16)$$

where t is the time of irradiation at dose rate K.

This simple picture provides us with the fundamental concepts of void formation. However, it is extremely simplified and many parameters have been excluded. The interested reader is referred to the numerous detailed papers, for example Bullough and Perrin[4], Bullough and Nelson[5].

STABILITY OF PRECIPITATES DURING IRRADIATION

It has been known for some time that irradiation can both order and disorder alloys, for instance Cu_3Au is readily disordered during irradiation at room temperature. It is therefore very likely that an ordered precipitate such as Ni_3Al in a Ni specimen will suffer disordering during irradiation at room temperature. During ion bombardment we can expect that the disordering effect of the displacement cascades essentially destroy the ordered precipitate lattice so that localised regions of high solute concentration are created. In the absence of diffusion such a state will persist, for example, irradiation of Ni-Al alloys containing precipitates. However, when diffusion occurs, the small disordered regions created within in the precipitate will re-order whilst those on the perimeter will result in the loss of solute, but diffusion, to the surrounding matrix. Figure 17 illustrates the effect of precipitate dissolution in a Ni-Al alloy, both bright and dark field micrographs are shown, and the regions still containing precipitates were shielded during irradiation.

An approximate expression for the dissolution rate of such a precipitate has been derived elsewhere (Nelson et al[21]).

i.e.
$$\frac{dV}{dt} = 4\pi r^2 \psi K \qquad (17)$$

where V is the precipitate volume, r its radius, ψ a dissolution parameter $\sim 10^{-6}$ cm, and K is the damage rate.

Equilibrium conditions will prevail when the rate of dissolution equals the growth rate due to solute atoms to the precipitates. For simplicity, let us assume that every precipitate has a radius r and that there are n precipitates distributed uniformly throughout the irradiated volume. If the total concentration of solute atoms at any one instant is C, we can define a boundary condition such that the total solute contained in precipitates and in solution equals C,

i.e.
$$C = \frac{4}{3}\pi r^3 pn + c \qquad (18)$$

where p is the atomic fraction of solute atoms constituting the precipitate phase and c is the concentration held dynamically in solution during the irradiation. From the work of Ham[10] the growth rate of precipitates from an atomic concentration c in solution is given by

$$\frac{dV}{dt} = 3(D+D')cr/p \qquad (19)$$

where (D+D') is the sum of the thermal and irradiation

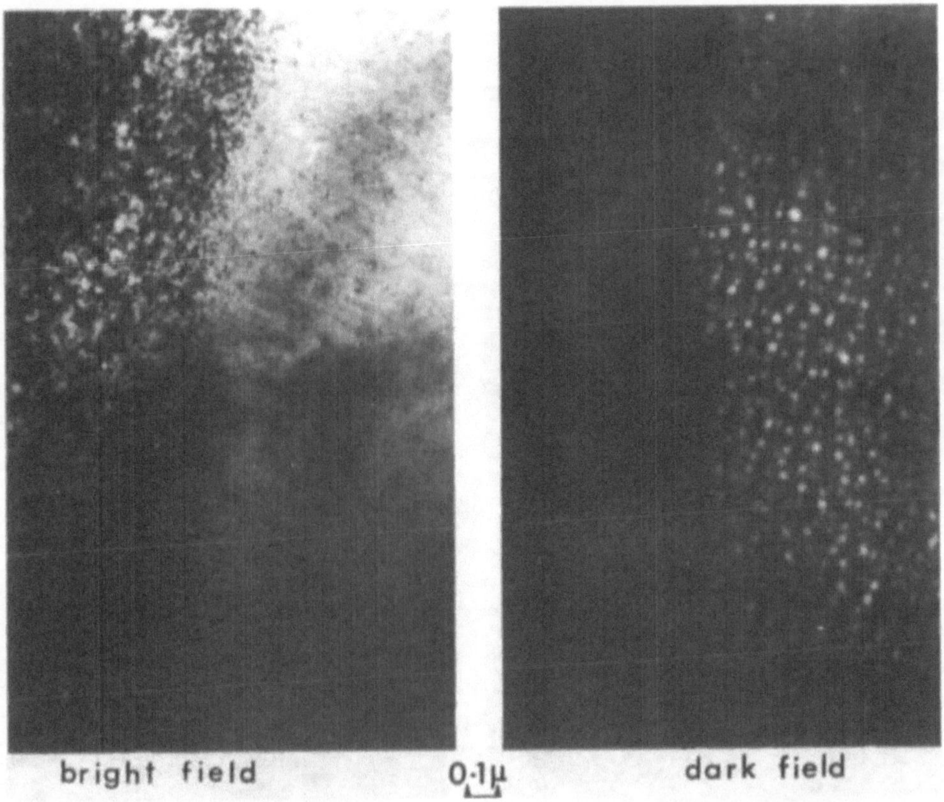

bright field   0.1μ   dark field

Fig.17 : Bright and dark field pictures showing the dissolution of γ' precipitates in Ni-Al alloy. The region still showing precipitates was shielded during irradiation.

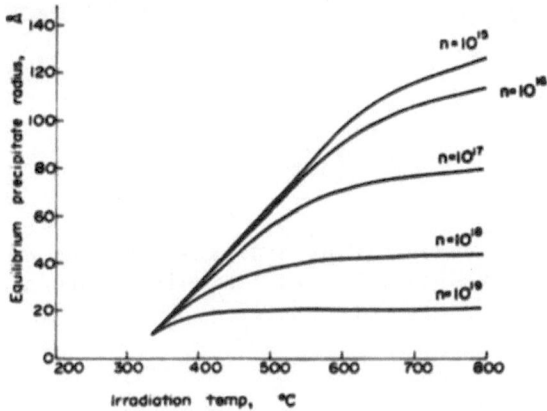

Fig.18 : The equilibrium radius of precipitates in Ni as a function of irradiation temperature, with C = 0.135 and $\psi = 10^{-6}$ cm.

288

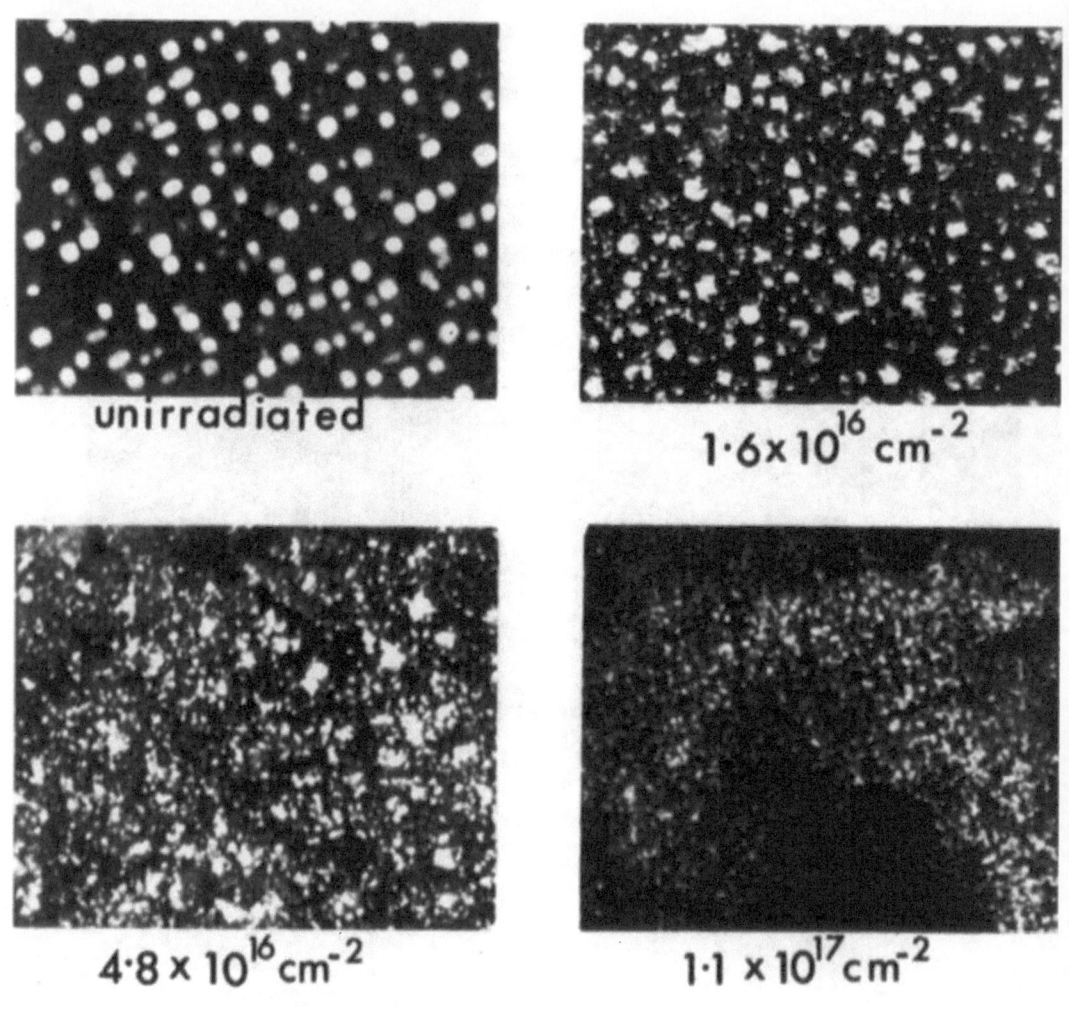

unirradiated

1.6×10^{16} cm^{-2}

4.8×10^{16} cm^{-2}

1.1×10^{17} cm^{-2}

0·1μ

Fig.19 : The change in precipitate size and density in Ni-Al as a function of size ; K = 10^{-2} displacements/atom sec, T=550°C.

enhanced diffusion coefficients. If we assume that to a first approximation the dissolved atoms simply contribute uniformly to the concentration in solution, equating equations (18) and (19) then yields the following expression for the net rate of change of precipitate radius.

$$\frac{dr}{dt} = -\nu K + \frac{3(D+D')C}{4\pi pr} - (D+D')r^2 n \qquad (20)$$

This cubic equation cannot be solved simply but from a computational solution we can derive the curves shown in fig.18 which shows the equilibrium radius of Ni$_3$Al precipitates in Ni as a function of irradiation temperature for different precipitate densities, with C = 0.135.

Experimental verification of these ideas have been aquired using ion bombardment. A series of Ni-Al alloys containing precipitates were irradiated as a function of increasing temperature to doses about 10-100 times that which was necessary to disolve the precipitates at room temperature. At all temperatures up to some critical value T_c the precipitates disorder ; above this temperature they persist, but often in modified form. In the case of Ni-Al alloys irradiated at a damage rate of $\sim 10^{-2}$ displacements/atom.sec, the critical temperature was 300-325°C. In other words, at this temperature the radiation-enhanced diffusion coefficient was just sufficient to allow the diffusion of solute atoms to the precipitate to balance the loss by dissolution. The predicted irradiation-enhanced diffusion coefficient at 300°C at this dose rate is about 5×10^{15} cm^2/sec, and a glance at eq.(20) therefore suggests that, in the case of Ni$_3$Al in Ni-Al alloys, as the precipitate shrinks to atomic dimensions an appropriate value for ψ is about 10^{-6} cm.

Fig.19 shows a series of micrographs of Ni-Al initially containing 250 Å γ' particles, taken as a function of increasing damage dose at a temperature somewhat above T_c. It is readily apparent that new precipitates are dissolved, so that ultimately all precipitates have attained approximately the same size. It is interesting to note that even in the case of alloys which have been solution-treated and quenched, the final state of precipitation is very similar to that shown in fig.19. In other words, it appears that whatever the initial starting configuration, the final state is one which is set solely by the irradiation condi-

290

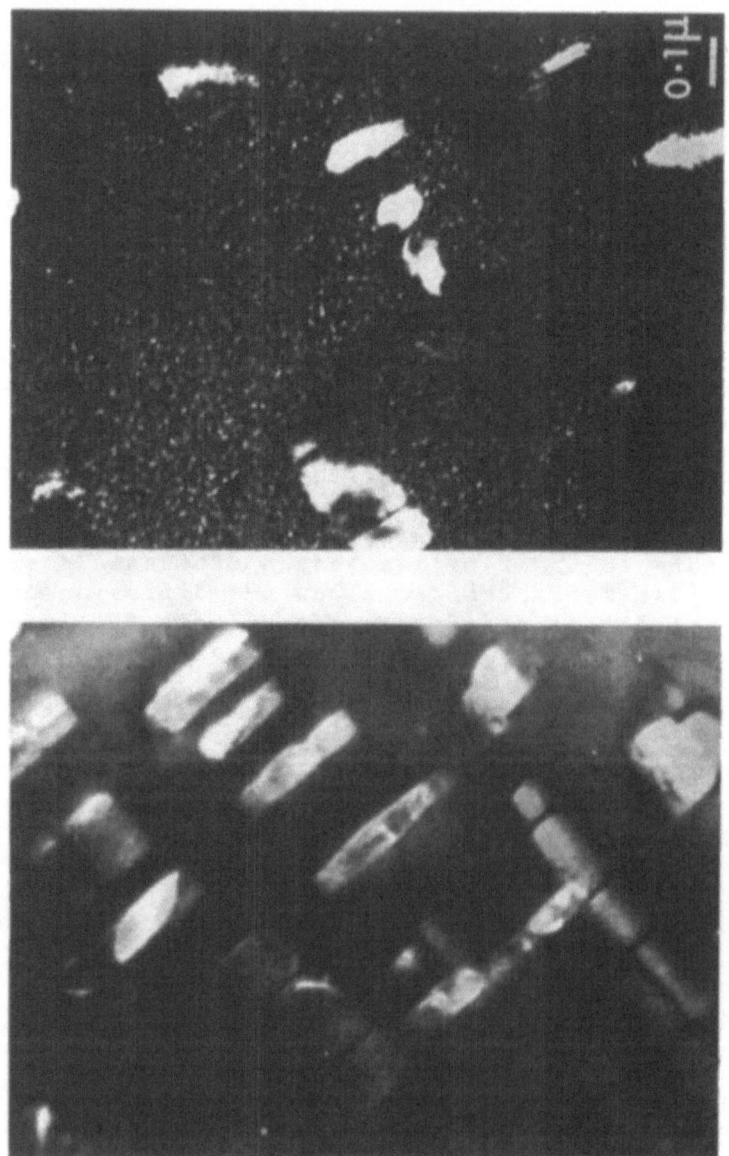

0·1µ

Fig.20 : Ni-Al alloys showing irradiation induced precipitation from solution after only ~ 5 dpa/sec.

tions.

For instance, in fig.20 we show an example of an
extreme case where a Ni-Al alloy was heat-treated to
give large 1200 Å precipitates prior to irradiation.
After irradiation at 550°C to only 5 displacements/atom
a very high density of small precipitates have formed
in the background from the solute previously in solution.

RADIATION DAMAGE IN SEMICONDUCTORS

Irradiation effects in semiconductors has provided
a fruitful field of research for many years, both from
the point of view of understanding the physical nature
of semiconductors and for the technological applications. To date, the majority of work has been concerned
with the individual point defects, created by relatively low dose electron irradiation, such as vacancies,
di-vacancies and vacancy-impurity complexes. However,
recent studies have been concerned with higher dose
such that defect agglomerates form a major part of the
irradiation damage structure. In particular, recent
world-wide interest in the technique of ion implantation to dope semiconductors has provided the major driving force behind the quest for an understanding of the
damage structures which result from ion bombardment.
More recently, the high voltage electron microscope
(HVEM) has been used to create electron irradiation
damage in semiconductors to very high doses, where once again point defect agglomeration in the form of
clusters becomes the dominant damage defect.

In this review we will outline the experimental
observations on the clustering of irradiation induced
defects in semiconductors as revealed by transmission
electron microscopy. The interpretation of the results
will be discussed in the light of recent theories of
defect clustering. Furthermore, as the most commonly
studied semiconductors (Si, Ge and GaAs) behave qualitatively the same, in order not to make the review too
cumbersome we will, in general, confine ourselves to a
discussion of damage in Si.(See Nelson[22]).

Heavy Ion Damage

It is now well established that crystalline Si,
Ge and GaAs are rendered amorphous* during high dose

---

* In this context amorphous material is thought to be composed
of a random arrangement of crystalline agglomerates, each agglomerate maintaining the basic 5 atom diamond type structure.

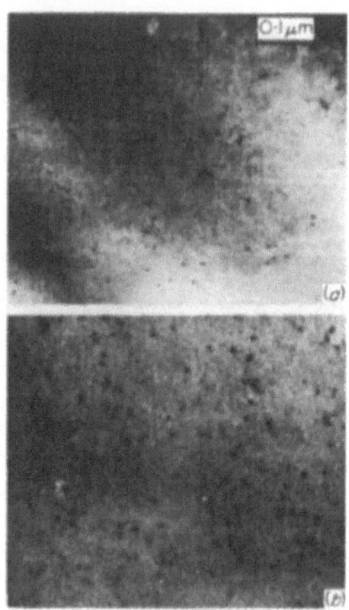

<u>Fig.21</u> : Damage regions in crystalline Ge irradiation (a) room temperature (b) < 30°K. (from Parsons 1965).

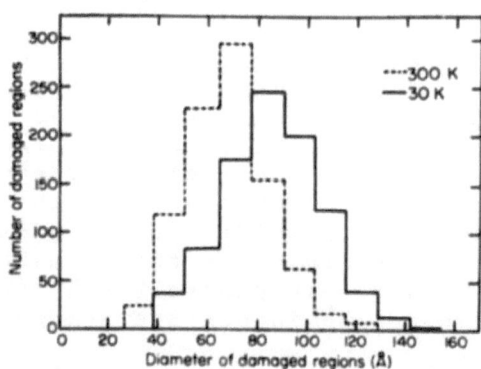

<u>Fig.22a</u> : Histogram of damaged regions in Ge at 300°K and 30°K
(from Parsons 1965)

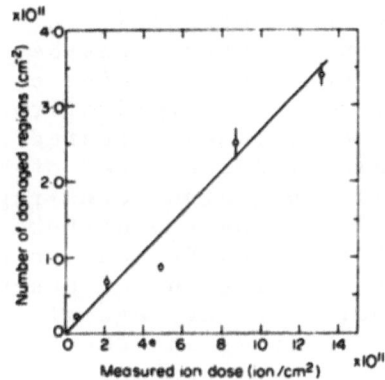

<u>Fig.22b</u> : Linear relationship between the number of damaged regions in Ge with ion dose (from Parsons 1965)

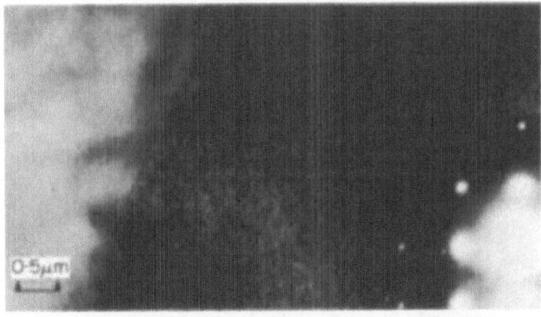

<u>Fig.23</u> : Transmission electron micrograph of the dense dislocation entanglement in Si produced during Ne bombardment at 300°C Note there is no evidence of amorphous material.

heavy ion irradiation. However, the detailed mechanism by which this phase is built-up has been a point of discussion.

The early work of Parsons[27] on O⁻ ion irradiated Ge in the electron microscope has been the inspiration for much of the subsequent studies on Si and GaAs. Using both bright and dark field microscopy he was also to show that the bombardment created small regions of contrast ($\sim$ 70 Å) which consisted of highly disordered material (fig.21). Such disordered regions could be imaged in dark field using the diffuse ring pattern of amorphous Ge and this led Parsons to suggest that they could be described as regions of essentially amorphous material. He studied the average size of these regions at different temperatures and the increase in their number with dose. Figure 22a shows size histograms for bombardment at 300°K and 30°K, whereas fig.22b shows the linear relationship between number and dose. Ultimately, the disordered zones touched and overlapped until the whole bombarded region was rendered amorphous.

A similar behaviour has been reported for ion irradiated (Mazey et al[19,18]). However, because of the lighter mass of Si, the collision cascades are less dense with the result that the disordered regions are smaller and more difficult to identify as amorphous. In the work of Mazey et al[19] the formation of the amorphous phase in Si was found to be strongly dependent on the irradiation temperature. For instance, at 200°C it required an ion dose in excess of two orders of magnitude greater than that required to form a complete amorphous phase at room temperature. Furthermore, an electron microscope study of Si samples irradiated at elevated temperatures during heavy ion bombardment revealed complex dislocation entanglements rather than disordered zones, see for example fig.23.

We can now state with some confidence that small disordered (essentially amorphous) regions are created heterogeneously throughout the material at the sites of energetic collision cascades, or whether they will appear as a consequence of homogeneous nucleation if the damage rate is sufficiently high relative to the defect migration rates, or whether they are indeed the only defect clusters to be created, we will try and elucidate in the course of the review.

Light Ion Damage

Boron Implantation

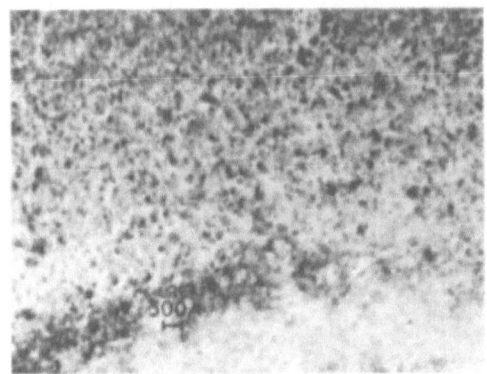

Fig.24 : Transmission electron micrograph of small regions of contrast in Si after B$^+$ ion bombardment at room temperature.

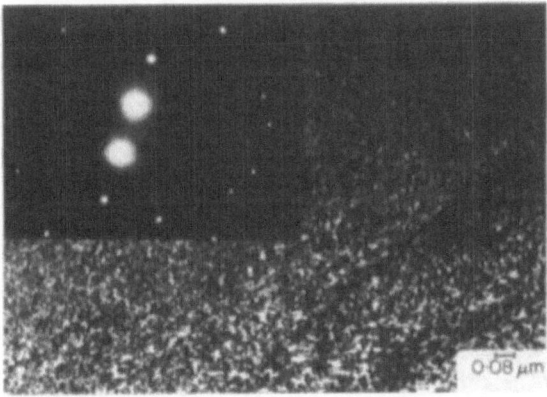

Fig.25 : Transmission micrograph of a proton irradiated Si sample at room temperature.

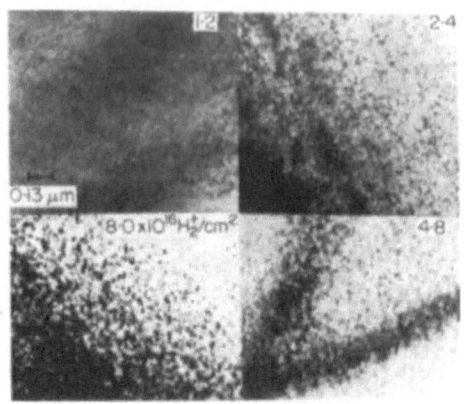

Fig.26 : Transmission micrograph of defect clusters in proton irradiated Si at room temperature as a function of dose.

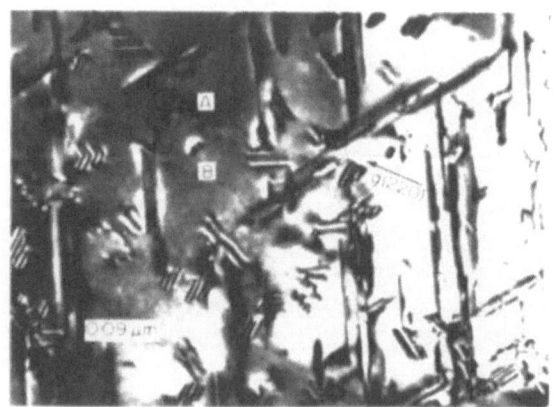

Fig.27 : Proton damage in Si at 450°C showing discrete loops and rods etc.

As previously described, in the case of light io
such as B a substantial fraction of defects are created as a consequence of low energy transfers, the number of large collision cascades being rather small.
In such a situation we might expect some difference to
that already described for heavy ions. Chadderton and
Eisen[6] have studied the nature of defect clusters in
Si during B ion implantation at room and liquid nitrogen temperatures. In both cases a very fine microstructure developed, however, this was so fine that it was
very difficult to ascertain its detailed nature, see
figure 24. It was postulated -primarily because of the
amorphous rings in the diffraction pattern- that both
disordered zones and small defect clusters existed together. Qualitatively similar results have been obtained by Matthews[16] . However, at liquid nitrogen temperatures he found that a dose in excess of about 10^{15}
40 keV B^+ ions cm^{-2} was sufficient to render the Si
completely amorphous in the bombarded area. On the o-
ther hand, a dose of more than an order of magnitude
greater was necessary to cause an amorphous layer during a room temperature bombardment.

Perhaps an even more extreme case of light damage
is that produced during proton bombardment. In this situation the primary encounters occur via essentially
unscreened Coulomb collisions and the number of large
energy transfers is very much reduced, even compared
with B^+ irradiation. The formation of amorphous material within energetic collision cascades is unlikely
and the majority of defects are created in ones and
twos. Figure 25 illustrates a typical electron micrograph of a room temperature bombardment to a very high
dose -about 10 displacements/atom- which knows a qualitatively similar situation to that for B^+ irradiation at the same temperature, but perhaps the defect
clusters are somewhat clearer.

Proton damage in Si has, in fact, been studied in
some detail as a function of dose and irradiation temperature in the accelerator-microscope link as described in section (Matthews[17]). Figure 26 shows a series
of micrographs taken at room temperature which indicates the evolution of the damage structure as the dose
is steadily increased. It should be noted that to the
doses reached in these experiments no visible sign of
amorphous rings were detected in the diffraction pattern. As the temperature is increased, the dislocation
network becomes steadily coarser until at about 400°C
the damage is manifest at discrete dislocation loops,

rods and complex faulted structures ; as can be seen in fig.27. We shall leave a discussion as to the nature of such clusters until later.

Electron Damage

Recent studies by Matthew[17] using the Harwell 1 MeV electron microscope have provided some interesting data on the clustering of defects during high dose electron irradiation. Due to the fact that the electron beam can be focussed to a spot $\sim$ 1 μm, it is possible to create a defect production rate as high as 10^{-2} displacements/atom/sec in an HVEM specimen. This compares with the damage rate produced at the peak of a 100 keV bombardment with B^+ ions at $\sim$ 10 μA cm^{-2}. The HVEM is equipped with both a cold stage and a hot stage so irradiations can be perform over a range of temperature. A further advantage of irradiation in the HVEM is that the evolution of damage structure can be studied dynamically as in the case of the accelerator-microscope link.

A first priority in this work was to ascertain whether or not it was possible to produce amorphous Si during electron irradiation. In other words, are collision cascades a pre-requisite for the formation of amorphous material ? A Si sample was cooled to near liquid nitrogen temperature and irradiated with 1 MeV electrons at a damage rate of 3 x 10^{-3} displacements/atom/sec to about 10 displacements/atom. Such a dose and dose rate are far in excess of those necessary to create an amorphous layer during B^+ ion irradiation at the same temperature. Repeated attempts to form amorphous Si have failed, the diffraction pattern remaining perfectly crystalline. However, a fine dislocation entanglement forms as shown in fig.28. A series of experiments were then carried out at different irradiation temperatures to ascertain the changes which occur as a function of electron dose. Figures 29, 30 and 31 show a selection of micrographs which illustrate the results. We can see the general features are very similar to the case of proton irradiation outlined previously. Figures 32 and 33 show the change in defect density and size as a function of dose during a room temper ture irradiation ; whilst figure 34 shows the increase in rod density as a function of dose at 450°C.

Defect Structures in Annealed Si

As we have already shown the structures which are produced after room temperature bombardment of Si reveal little detailed information on the nature of the

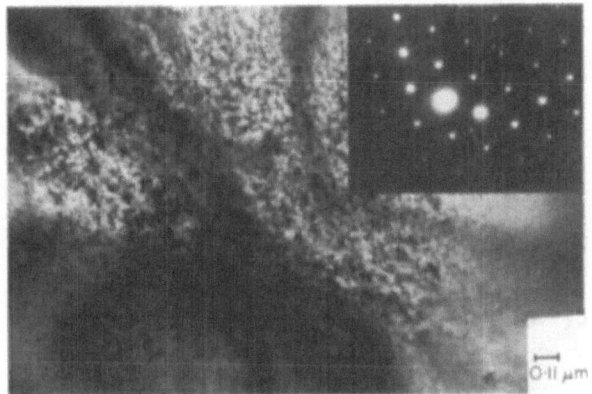

Fig.28 : Low temperature HVEM irradiation of Si at 1 MeV, showing a fine dislocation structure with no evidence of an amorphous phase, dose ∿ 10 displacements/atom.

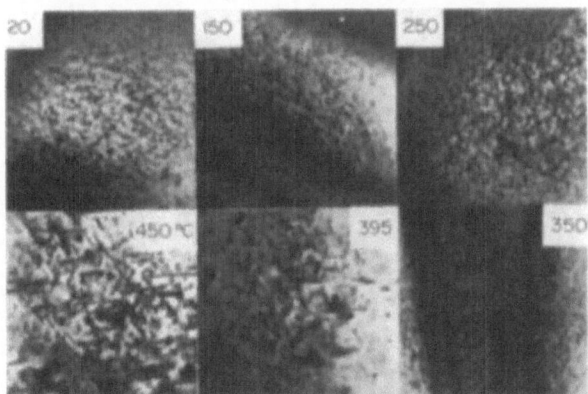

Fig.29 : The variation of defects in HVEM irradiated p type Si as a function of irradiation temperature, dose 2 displacements/atom.

300

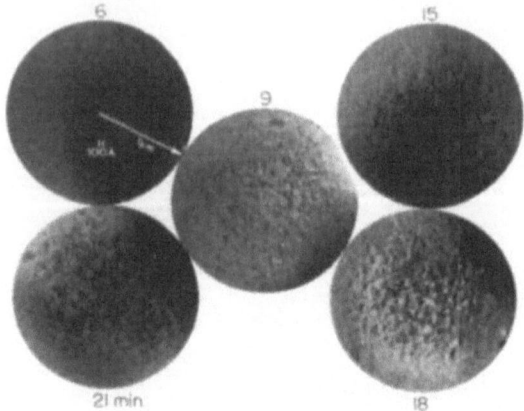

Fig.30 : The dose dependence of defects produced in Si using the HVEM at room temperature (10 min. = 2 displacements/atom).

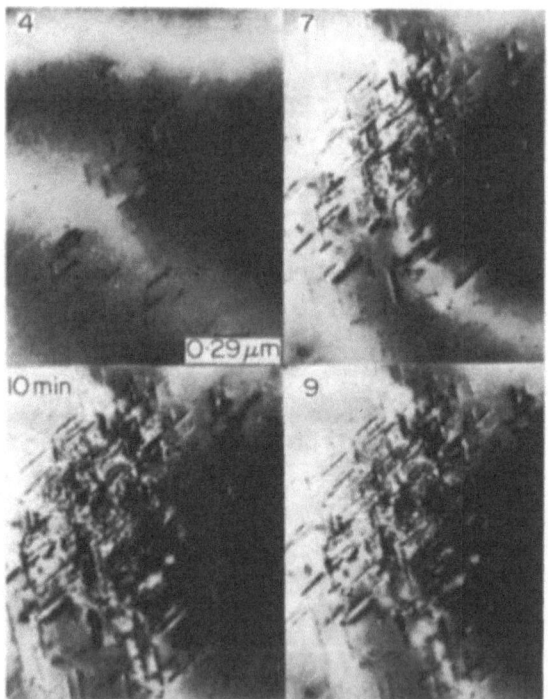

Fig.31 : The dose dependence of defects produced in Si using the HVEM at 450°C. (10 min. = 2 displacements/atom).

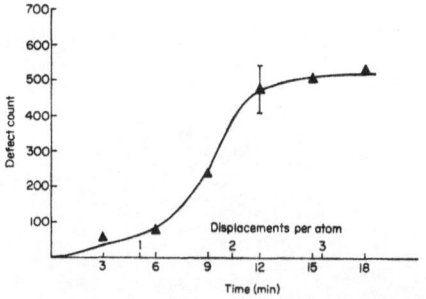

Fig.32 : Number of defects as a function of dose for 1 MeV electron irradiation of Si at room temperature.

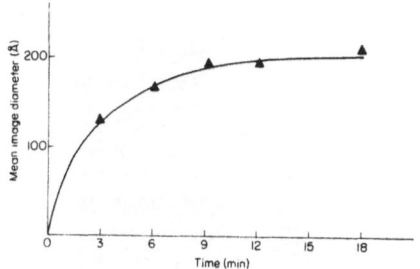

Fig.33 : Mean size of defects as a function of dose for 1 MeV electron irradiation of Si at room temperature.

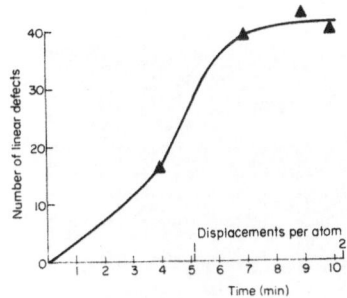

Fig.34 : Rod density as a function of dose for 1 MeV electron irradiation of Si at 450°C.

302

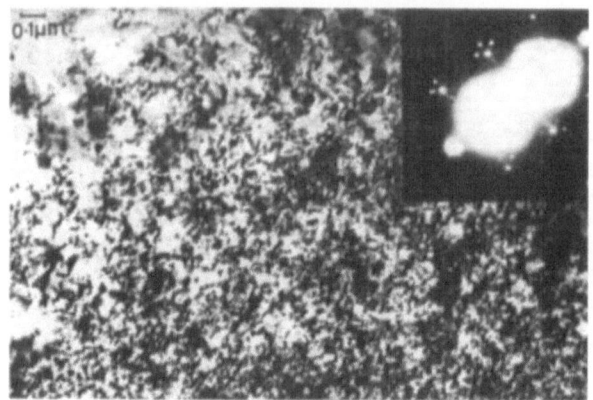

Fig.35 : Imperfect epitaxial recrystallization of Si, with diffraction pattern (annealed at 900°C for 1 hr).

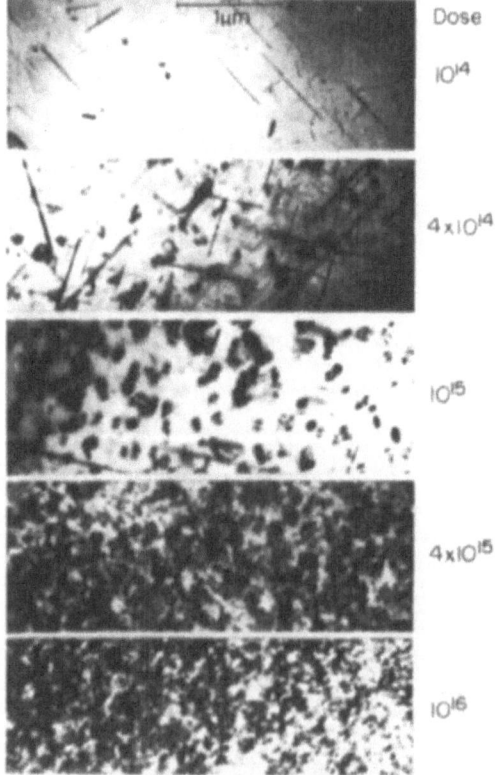

Fig.36 : Si irradiated with 80 keV B^+ ions to doses between 10^{14} and 10^{16} ions/cm^2 and annealed at 800°C for 1 hr.

damage. However, we have demonstrated that in the case
of heavy ion bombardment disordered zones are created
in the wake of energetic collision cascades. Furthermore, with a sufficiently large dose the damage is manifest as a continuous amorphous phase in the bombarded
region. On heating to temperatures above $\sim 600°C$ the
amorphous phase recrystallises (Mazey et al[18]) and it
is of interest to study the residual defect structures
which remain in the bombarded area after such annealing.

Early work by Mazey et al[19] and Large and Bicknell[12] has been followed by more detailed studies by
Davidson[7]. Two basically different regimes can be identified depending on whether the damaged region consti-
tutes a completely amorphous phase or is only partially
non-crystalline. In the first case, the recrystallisa-
tion follows a uniform behaviour virtually independent
of the ion type. After heating to about 600°C the a-
morphous layer recrystallises essentially epitaxially
onto the underlying crystalline material. However, re-
crystallisation is by no means perfect as can be seen
from the micrograph shown in fig. 35. Detailed diffraction analysis of this recrystallised phase (see diffraction pattern shown inset) suggests that the diffraction
contrast is, in fact, caused by microtwins having mi-
sorientations between 1-2° about the main orientation.
Once formed this phase appears quite stable to temperatures at least as high as 1020°C.

In the case when the damaged region remains essentially crystalline the detailed evolution of defect
clusters upon annealing depends intimately on the particular bombarding ion and on the ion dose. It is not
possible to detail the results for every situation in
this review and so we will highlight just a few notable examples. In general, all the amorphous diffuse
rings disappear from the diffraction pattern in the
temperature range 500-600°C when the pattern is returned to that typical of a good single crystal with very
little evidence for twinning. At 600°C the transmission
electron micrographs show a rather fine structure within which it is very difficult to resolve the individual defects. It is not until about 800°C that the microstructure has grown into an easily resolvable form.
Figure 36 illustrates the form of the microstructure
which is observed in Si after a room temperature B^+ implantation to various doses and annealing at 800°C. At
low doses the predominant defects are long rods, similar to those already described for proton and electron
irradiation, but as the ion dose increases the rods gi-

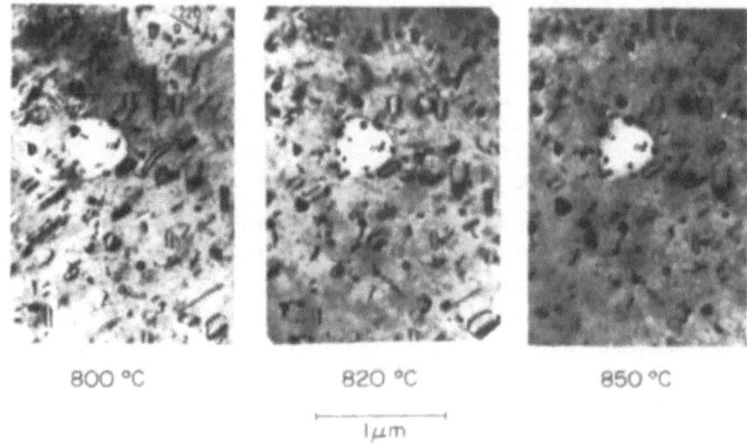

800 °C   820 °C   850 °C

1 μm

Fig.37 : Si irradiated with 10^{15} 40 keV B$^+$ ions/cm^2, and annealed between 800° and 850°C. This sequence clearly shows a number of rods either disappearing or becoming shorter with increasing temperature.

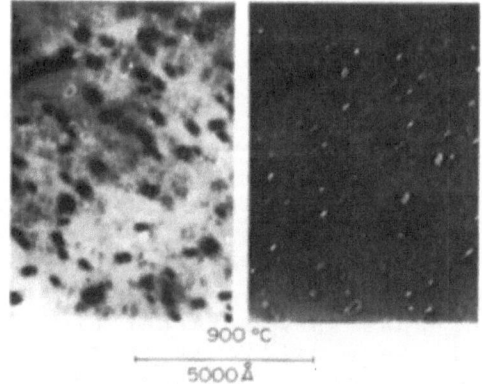

900 °C

5000 Å

Fig.38 : Si irradiated with 3.10^{14} 80 keV P$^+$ ions/cm^2, and annealed at 900°C for 30 min., showing a mixture of small loops.

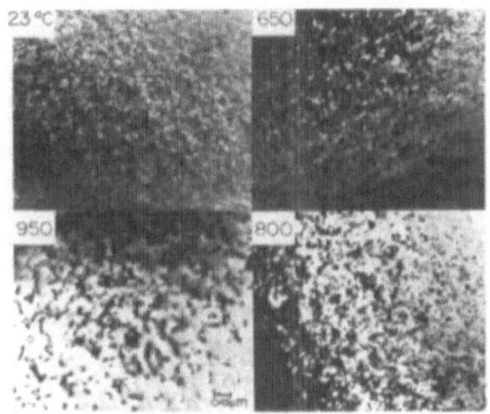

<u>Fig.39</u> : Annealing sequence of a room temperature HVEM irradiated Si specimen showing that very little changes occur until about 800°C (dose 7 displacements/atom).

ve way to more conventional dislocation loop structures which ultimately grow into a rather complex network. On heating to 900°C the configuration of these defects changes, the changes being more pronounced for the lower doses where the predominant defects are rods. Some rods are still present but many appear to have transformed into elongated dislocation loops, see fig. 37. It is interesting to note that after a liquid nitrogen B^+ ion implantation, the rod-like defects are relatively rare and the microstructure generally takes the form of dislocation loops. In the case of P^+ ion implantation, the situation is somewhat different. Rod-type defects are rarely, if ever, seen and the predominant defect structure is that of loops, see for instance fig.38.

It is also of interest to anneal an electron irradiated sample, where no disordered zones are created during the irradiation. Fig.39 shows a composite of an annealing sequence of a sample irradiated to 7 displacements/atom at room temperature. As we can see after annealing at 650°C very little change has occurred and it is necessary to anneal to ∿ 800°C before any major coarsening of the dislocation network takes place.

307

REFERENCES

1. Banbury P.C. and Haddard I.N., Phil.Mag. 14, 841 (1966)

2. Beevers C.J. and Nelson R.S.,Phil.Mag., 8, 1189 (1963)

3. Bullough R., IAEA Symp.on Rad.Dam.in Reactor Materials - Vienna, (1969)

4. Bullough R. and Perrin R.C., BNES Conf.on Voids-Reading, p.79, (1971)

5. Bullough R. and Nelson R.S., AERE R 7353, (1973)

6. Chadderton L.T. and Eisen F.H., Proc.Conf. "Ion Implantation", p.445, Thousand Oaks, California (Gordon and Breach) (1970)

7. Davidson S.M., D.Phil.Thesis, University of Oxford (1972)

8. Dienes G.T. and Damask A.C., J.App.Phys. 29, 1713 (1958)

9. Erginsoy C., Englert A. and Vineyard G., Phys.Rev. 133, 595 (1964)

10. Ham F.S., J.Phys.Chem.Solids, 6, 335 (1958)

11. Hudson J.A., Thesis, University of Cambridge (1969)

12. Large L. and Bicknell R.W., J.Mat.Sci., 2, 589 (1967)

13. Lehmann C and Liebfried G., Z.Physik, 162, (2), 203 (1961)

14. Leibfried G. and Dederichs P.H., Z.Physik, 120, 320 (1962)

15. Lomer J.N. and Pepper M., Phil.Mag. 16, 1119 (1967)

16. Matthews M.D., Private communication to be published (1971)

17. Matthews M.D., to be published (1972)

18. Mazey D.J., Nelson R.S. and Barnes R.S., Phil.Mag. 17, 1145 (1968)

19. Mazey D.J., Barnes R.S. and Nelson R.S., Int.Conf. Electron Microscopy, Kyoto p.363 (1966)

20. Mazey D.J. and Barnes R.S., Phil.Mag. 17, 387 (1968)

21. Nelson R.S., Hudson J.A. and Mazey D.J., J.Nucl. Mat., 44, 318 (1972)

22. Nelson R.S., Int.Conf.on Defects in Semiconductors Reading - (1962)

23. Nelson R.S., Phil.Mag. 11, 291 (1965)

24. Nelson R.S. and Thompson M.W., Proc.Roy.Soc. A259, 458, London - (1961)

25. Nelson R.S., "The Observation of Atomic Collisions in Crystalline Solids" (North Holland Pub.Co.) (1968)

26. Norgett M., Robinson M.T. and Torrens I., AERE R-TP 494, (1972)

27. Parsons J.R., Phil.Mag., 12, 1159 (1965)

28. Seeger A., Schumacher D., Schilling W. and Diehl J. "Vacancies and Interstitials in metals" - Jülich (North Holland) (1969)

29. Sharp J.V., Radiation Effects (1972)

30. Sigler J.A. and Kuhlmann-Wilsdorf D., Phys.Stat. Sol. 21, 545 (1967)

31. Silsbee R.H., J.App.Phys., 28, 1246 (1957)

32. Sigmund P., Ris (Denmark) Report n° 103, (1965)

33. Thompson M.W. and Nelson R.S., Phil.Mag. 7, 2015, (1965)

34. Turnstall W.J., Eriksson L. and Mazey D.J., Phil. Mag. 21, 617 (1970)

35. Vineyard G.H., Gibson J.B., Goland A.N. and Milgram M., Phys.Rev. 130, 1129 (1960)

36. Wilson M.M., Radiation Effects 1, 207 (1970)

COLLISION DAMAGE PROCESSES IN TRANSPARENT MATERIALS

D.POOLEY
Materials Physics Division
AERE - Harwell
Great Britain

INTRODUCTION

In alkali halides and in organic materials radiation damage occurs almost exclusively through the excitation of the electrons of the solid by the radiation,
and as a result even low energy electrons and x-rays
cause atomic displacements. The defects which are produced in these materials can usually be studied in microscopic detail, because the defect electron levels
generally lie between the valence band and the empty
conduction band, ans as a result optical or magnetic
resonance spectroscopy is possible. The situation in
metals is very different ; here radiation damage occurs
only by collision processes and only high energy electrons or heavy particles can cause atomic displacement.
Moreover the basic point defects cannot be studied in
any detail in metals because the conduction band electrons make defect spectroscopy impossible.

Some transparent materials fall into an interesting intermediate class, where damage occurs only by
collision processes but detailes defect structure and
behaviour can nevertheless be studied. The best known
examples are alkaline earth oxides, and in the limited
time available I will concentrate on them, but diamond
and other oxides could also be classified in this way.
Limiting the field to alkaline earth oxides still leaves a daunting volume of work from which to draw information, and I will, therefore, concentrate on the
damage process in these oxides, rather than on the de-

310

Table 1. Magnetic Resonance and Optical Data for F$^+$ Centres in Alkaline Earth Oxides

Crystal	g	Magnetic Resonance parameters			Optical Absorption parameters	
		Nuclei for hyperfine interaction	a(MHz)	b(MHz)	Peak energy at 4K(eV)	Width at 4K(eV)
MgO	2.0023	^{25}Mg	-11.03	-1.33	4.95	~ 0.6
CaO	2.0001	^{43}Ca	-25.66	-2.71	3.7	0.30
SrO	1.9845	^{87}Sr	-40.77	-4.28	3.0	0.28
BaO	1.9355	^{135}Ba	184.7	18.4	2.0	0.47
		^{137}Ba	206.9	20.3		

fects themselves.

However, I must first say a little about the structure and properties of defects in alkaline earth oxides.

THE NATURE OF DEFECTS IN ALKALINE EARTH OXIDES

The basic phenomenology of defects in the alkaline earth oxides has been well reviewed by Henserson and Wertz[1] and a more recent detailed review of the spectroscopy and the structure of the more important defects has been given by Hughes and Henderson[2]. It will be clear from these reviews that much remains to be done even in defect characterisation in the oxides, particularly for interstitial centres. Nevertheless the F^+ centre, an oxygen ion vacancy which has trapped only one electron and has therefore effectively one positive charge, has been identified as the primary defect in MgO, and also in the other alkaline earth oxides as well as in BeO and ZnO. Both EPR and optical measurements have been made on these centres, with the optical band often being identified through its correlation with the magnetic resonance spectrum. Some of the basic optical and magnetic resonance data for F^+ centres in the alkaline earth oxides are given in table 1. The magnetic resonance data are consistent with the simple atomic model for the centre, particularly the near spin g-value and the primarily isotropic hyperfine interaction. The optical data too seems unexceptional, although the behaviour of the F^+ band in CaO is by no means as simple as the data of table 1 suggest. In fact vibrational structure and a double humped shape is apparent in this band (figure 1), and the centre provides a possibly unique example of the two-mode dynamical John-Teller effect. (Hughes[3] and Escribe and Hughes[4]). The F centre too has been studied in MgO and CaO, and can usually be produced by irradiation or additive coloration. Anion vacancies then are the primary defects, just as in alkali halides.

However, let us now look at the collision damage process. As we do this I will try to make some very general points at the risk of some loss of rigor, in order to build a picture of what one might expect in oxides from what is known to happen in metals and alkali halides. I will also briefly survey the experimental measurements which have been made, and we will see how they compare with the general predictions.

312

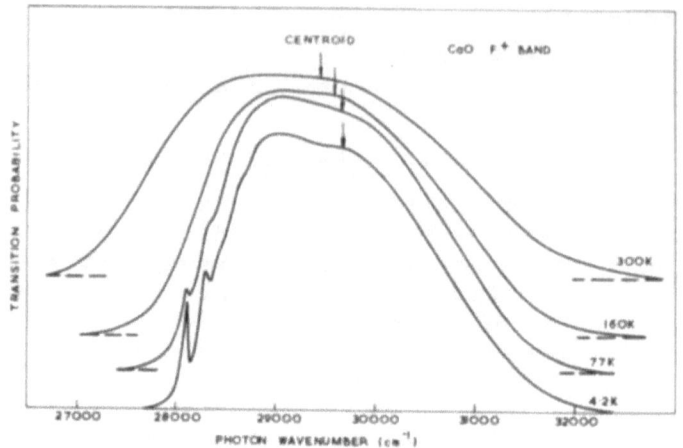

Fig.1 : Vibrational structure of CaO F⁺ band

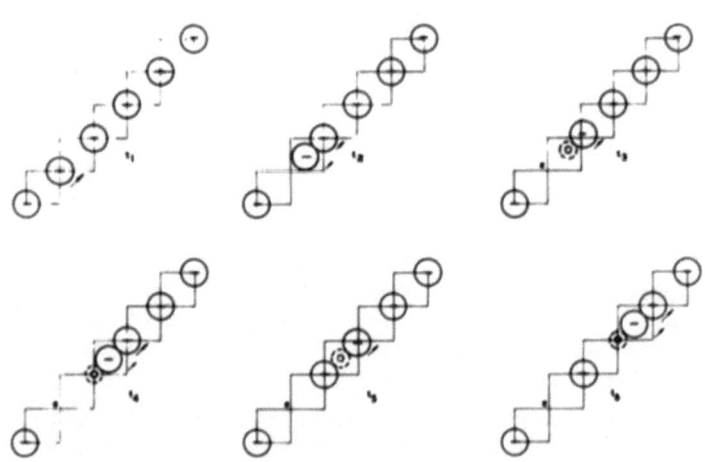

Fig.2 : A schematic representation of a <110> collision sequence in KCl including hole tunnelling.

THE ABSENCE OF IONISATION DAMAGE

We have been reminded by Sigmund that fast charged particles tend to lose most of their energy to the electrons of a solid. In alkali halides and organic materials (as Kabler and Charlesby have described) the excitation of electron leads to atomic displacement events, but in oxides and metals it does not. The reason is perhaps not immediately obvious but is really not difficult to find. The basic criterion is whether any excited electronic state which has a lifetime long enough for it to cause atomic displacement (say 10^{-13} or 10^{-12}S) also has enough energy to do the job. In a metal the answer is fairly obviously no ; any electronic excitation will be dissipated among the conduction electrons in a time of the order of 10^{-15} S, and there is no conceivable way that a large number of slightly excited conduction electrons can get together to move an atom into an interstitial position.

In insulators and semi-conductors the answer is less obvious. The degredation of a very energetic electronic excitation to a number of electron-hole pairs probably does occur too quickly, so that a Varley like displacement process is rather unlikely. On the other hand the electron hole pairs themselves have lifetimes of up to 10^{-6}S or more, and store energies of several electron volts. Whether or not they can create defects then depends on their energy relative to the required displacement energy. In an ionic crystal we might expect the energy required to displace an anion from its lattice site to be roughly equal to the lattice binding energy, that is

$$E_b = \alpha Z^2 e^2 / 4\pi E r_o$$

where α is the Madelung constant for the structure, Z is the range on the ion and r_o is the distance between neighbouring ions. The energy of an electron hole pair, on the other hand, is roughly

$$E_p = \alpha Z^2 e^2 / 4\pi E r_o + E-I$$

where E is the electron affinity of the anion and I is the ionisation potential of the cation. In our simple model we would expect to have ionisation damage in insulators if $E_p \gtrsim E_b$ but not otherwise. Table 2 lists these energies for KBr, MgO and CaO, and shows that E_p and E_b are comparable in KBr. Bearing in mind the crudity of our arguments we might therefore expect ionisa-

Table 2. Lattice electrostatic energies and
electron-hole pair energies in so e
ionic insulators

	Lattice binding energy $E_b = \frac{\alpha Z^2 e^2}{4\pi E r_o}$ (eV)	Anion electron affinity E(eV)	Cation ionisation potentiel I(eV)	Electron hole pair energy $E_p = E_b + E - I$
KBr	7.7	3.4	4.3	6.8
MgO	46.5	-9.5	15.0	22.0
CaO	42.5	-9.5	11.8	21.2

tion damage in KBr. For the oxides on the other hand E_p is much smaller than E_b, so that ionisation damage there would be very unlikely.

The experimental evidence confirms this ; it is overwhelming that ionisation damage does not occur in the oxides. The defects created in oxides by electronic excitations are all associated with impurities or other defects present before irradiation, and these have been reviewed by Sonder and Sibley.[5]

THE DISPLACEMENT THRESHOLD ENERGIES

We convinced ourselves that displacement threshold energies for oxygen in oxides were too high ro ionisation damage to occur by using very simple arguments, and it is probably worthwhile trying to refine somewhat the value of the threshold energy. In defining the F centre production process in alkali halides it has been argued by Pooley and Hatcher[6] and by Smoluchowski et al[7] that the ejected halide ion loses an electron in the ejection process (figure 2) and as a result the energy requirements are reduced. The minimum energy requirements are reduced. The minimum energy requirements will then be

$$E_d = E_b + E - E_F - E_H$$

as can be seen in figure 3. The energy required to remove an anion from the lattice is E_b as before and that required to remove its electrons is E. The energy gain in placing the electron back in the vacancy is the F (or F^+) centre binding energy E_F, and E_H is the binding energy of the ionised anion with lattice anions. With these arguments we find for KBr $E_d \simeq 6$ eV and for MgO we have $E_d \simeq 27$ eV.

In an actual collision damage event, however, dynamic effect might be expected to change these thresholds very considerably. Unfortunately, almost no computer simulations of radiation damage events in oxides have been carried out. Beeler did make some two body simulations of high energy events but they are not relevant to the threshold situation. Pooley has carried out many body simulations near threshold, but using a simple one dimensional crystal model which later proved totally inadequate for alkali halides, in that displacement energy thresholds calculated in the linear model (Pooley[8]) were about a factor of 5 'smaller than those calculated in a three dimensional model (Hughes et al[9] and Torrens and Chadderton[10]). Even in the linear model

$$
\begin{array}{ccc}
-\ +\ - & & -\ +\ - \\
+\ -\ + & \xrightarrow{\ E_b\ } & +\quad\ +\ \text{and}\ X^{-} \\
-\ +\ - & & -\ +\ -
\end{array}
$$

$$
\Big\downarrow \quad
\begin{array}{l}
E_d = E_b + E \\
\quad\ -E_F - E_M
\end{array}
\qquad
\begin{array}{c}
\big\downarrow\ E \\[4pt]
-\ +\ - \\
+\quad\ +\ \text{and}\ X^{\circ}\ \text{and}\ e^{-} \\
-\ +\ -
\end{array}
$$

$$
\begin{array}{ccc}
+\ X_i^{-}\ + & & \\
-\ +\ - & & \big\downarrow\ -E_F \\
+\ \bigcirc\ + & \xleftarrow{\ -E_M\ } & +\ \bigcirc\ +\ \text{and}\ X^{\circ} \\
-\ +\ - & & -\ +\ -
\end{array}
$$

Fig.3 : Born-Haber cycle for anion ejection with ionisation.

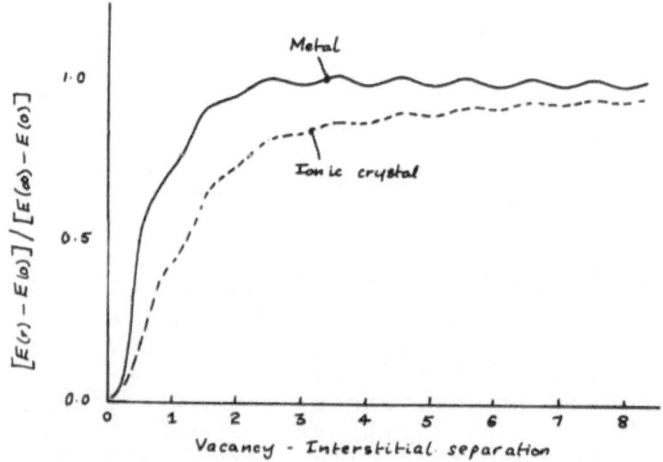

Fig.4 : Energy versus defect separation in metals and ionic solids.

the threshold energies calculated for MgO and CaO were in the region 30-50 eV, primarily because of the large separation between the oppositely charged interstitial ion and vacancy which is necessary for stability. This is in fact a major difference we might expect between ionic solids and metals, where no long range Coulomb attraction between vacancy and interstitial can occur, and the point is illustrated in figure 4.

In the oxides then we would expect that the thresholds will be at least $\sim$ 30 eV and probably significantly higher because of large separation of interstitial and vacancy which is required for stability. The threshold in MgO has been measured by Chen et al[11] using electron irradiation and optical monitoring of the F^+ and F bands. Their results are shown in figure 5 and they indicate a displacement threshold for oxygen in MgO of about 60 eV.

Unequivocal evidence for the displacement of cations in bulk oxides by irradiation is not available. Under certain conditions it appears that the V^- centre, a positive ion vacancy with one trapped hole, may be produced by irradiation at liquid nitrogen temperature but this is not certain. Displacement of cations in bulk crystals has often been inferred from the production of interstitial loops (Bowen and Clarke[12]) but as we shall see later the work of Hobbs et al[13] on alkali halides has cast doubt on that. A simple examination of the microscopic processes which would have to occur in the displacement of cations does lead to the view that the displacement threshold might be significantly higher than for anions. For example ionisation of the ejected ion will not reduce the energy requirements as it will for an anion. Tench and Duck[14] do have strong evidence for both F^+ and V^- centre production in MgO powder, but the production rates are higher in powders with a high surface area an in any case very energetic particles were used (20 MeV protons), so that nothing can be inferred about thresholds.

DISPLACEMENT EFFICIENCIES ABOVE THRESHOLD

Chen et al found a displacement threshold energy of $\sim$ 60 eV for oxygen in MgO, and in the same experiments found that the defect production rate at room temperature by electrons well above threshold (1.7 MeV compared with 0.33 MeV) was about ten times smaller than predicted by displacement models such as the Kin chin and Pease model. Their initial explanation was that MgO was behaving like many metals, where the damage rate is very high at very low temperatures but includes the production of close interstitial pairs which recombine at higher temperatures. However, I think we would expect the oxides to differ from metals in that close pairs would be very unlikely to be stable (figure 4).The situation shown in figure 4 is likely to lead to different displacement probabilities for oxides and metals as a function of the energy given to the displaced ion. This is shown schematically in figure 6. Here we are assuming that the minimum displacement threshold energies are the same for the metal and the oxide, and then arguing that if figure 4 is valid it follows that a displaced oxygen ion at a given separation from its vacancy is less likely to have retained the energy required to leave the vacancy altogether that is a metal atom under similar circumstances. It is, therefore, not surprising that displacement efficiencies in oxides are much smaller than one would expect from the measured threshold energies.

This picture has been confirmed by the recent work of Hugues[15] . He has found displacement rates in MgO which are again about an order of magnitude lower than those calculated from the thresholds (table 3). He has also established that the situation is not as in metals ; theedefect production rates at low temperatures are not vastly greater than at room temperature, and no significant defect annealing occurs when a crystal which has been irradiated at low temperature is warmed up and irradiated further (figure 7). The factor of two increase in the production rate in cooling to 15 K could easily be due to changes in the displacement probability curve (figure 6) caused by changing atomic vibration amplitudes ; Hughes points out that the change cannot be due to an increase in the distance at which interstitials and vacancies recombine spontaneously since this is inconsistent with no defect annealing when the temperature is warmed from 15K. This thermal ma annealing distance has been estimated to be 18 nearest neighbour distances in MgO by Hensersen and Bowen[16] from a measurement of the saturation concentration of F^+ centres under neutron irradiation, and this too is consistent with the general picture.

INTERSTITIAL CLUSTERS

As in alkali halides, in oxides the anion vacancies tend to remain isolated even at fairly high doses, and as in alkali halides there is evidence for the formation of clusters of interstitials. Henderson and Bowen

319

Table 3. The measured and calculated numbers
of oxygen vacancies produced in MgO
by proton irradiation

Proton Energy	Number of oxygen vacancies per incident proton			
	Calculated assuming $E_d \sim 60eV$	Measured at		
		15K	80K	300K
0.4 MeV	3.5	0.25	0.15	0.10
3 MeV	8.1	0.77	0.50	0.45

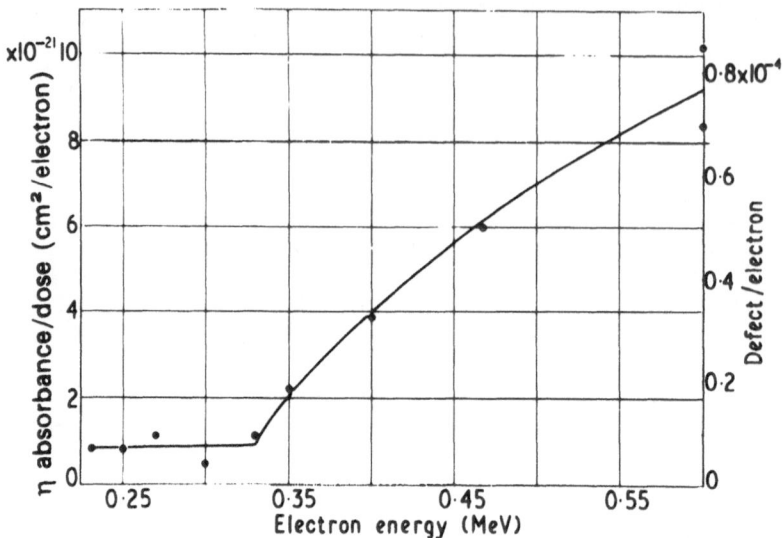

Fig.5 : Defect production rate, η, as a function of irradiation energy in MgO - Chen et al (1970)

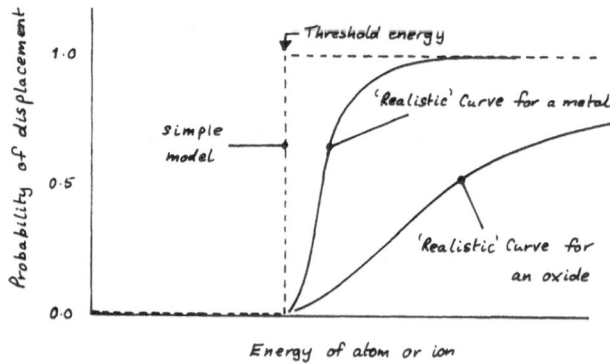

Fig.6 : Displacement probabilities for real solids

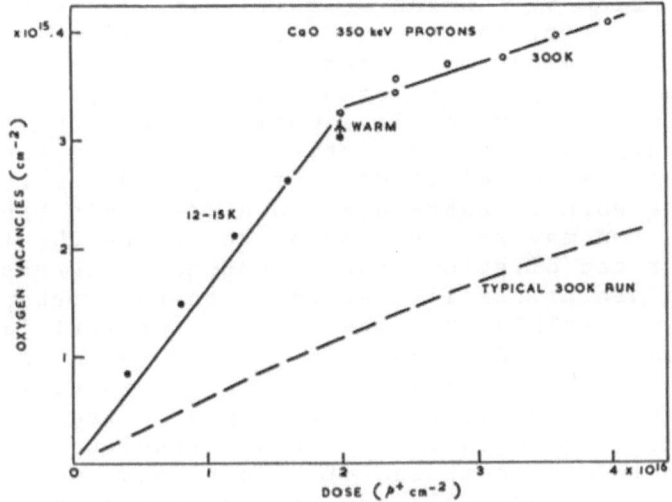

Fig.7 : Defect production in CaO (Hugues 1973)

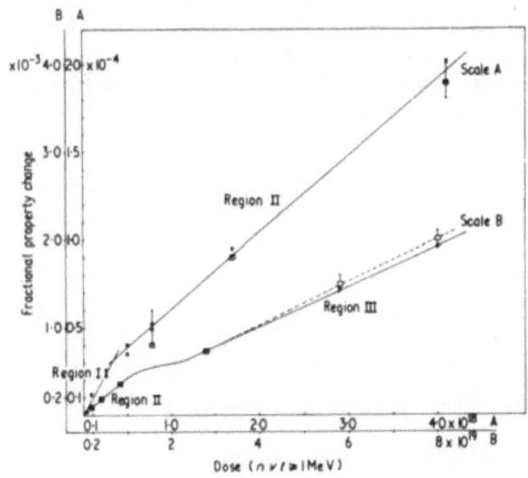

Fig.8 : The fractional change in lattice parameter $\Delta a/a$ and crystal density $\Delta\rho/\rho$ plotted as a function of dose up to 4×10^{18} nvt (scale A) and 8×10^{19} nvt (scale B). The scales for both ordinate and abscissa have been changed by the same factor x20, in converting scale A to scale B. Full curve with full squares, $\Delta a/a$; broken curve with open squares, $1/3\ \Delta\rho/\rho$. Menderson & Bowen (1971)

measured changes in lattice parameter and in density as a function of neutron dose (figure 8). They found that the density change was slightly larger than that of the lattice parameter and inferred that interstitial dislocation loops were formed. This interpretation is consistent with the earlier work of Bowen and Clarke[17] where small loops of interstitial character are seen with transmission electron microscopy. However, once again the work of Hobbs has shown that this kind of interpretation may well be too naive, since the lattice parameter and dilation changes caused by interstitial clusters are critically dependent on their shape. It is certainly possible that the kind of interstitial loop formation proposed by Hobbs et al for the alkali halides occurs more generally in ionic crystals. In the halides the evidence points strongly to the displacement of only halogen ions, but nevertheless perfect interstitial loops composed of both anions and cations are formed. We have already argued that displacement rates for anions in the oxides are probably larger than for cations, but it may be that perfect interstitial loops are formed nevertheless.

It would probably be fair to conclude that collision damage in oxides is providing a field where interesting confirmation and extension of the ideas formed during work on metals and alkali halides is possible, but that no very new or unexpected radiation damage phenomena have yet been found.

REFERENCES

1. Henderson and Wertz, Adv.in Phys. <u>17</u>, 749 (1968)

2. Hugues and Henerson, Chapter in "Point Defects in Solids" Eds. Crawford J.W. Jnr., Slifkin L.M., Plenum Press, 1972

3. Hugues, J.Phys. C., <u>3</u>, 627, (1970)

4. Escribe and Hughes, J.Phys. C., <u>4</u>, 2537, (1971)

5. Sonder and Sibley, Chapter in "Point Defects in Solids" Plenum Press, (1972)

6. Pooley and Hatcher, Bull.Am.Phys.Soc.<u>15</u>, 340 (1970)

7. Smoluchowski et al., Phys.Rev.Lett., <u>27</u>, 1288 (1970)

8. Pooley, Proc.Phys.Soc., <u>87</u>, 257 (1966)

9. Hugues et al., A.E.R.E.-5604 (1967)

10. Torrens and Chadderton, Phys.Rev. <u>159</u>, 671 (1967)

11. Chen et al., J.Phys. C., <u>3</u>, 2501 (1970)

12. Bowen and Clarke, Phil.Mag. <u>9</u>, 413 (1964)

13. Hobbs et al., Proc.Roy.Soc., <u>A</u> <u>332</u>, 167 (1973)

14. Tench and Duck, J.Phys. C., <u>6</u>, 1134 (1973)

15. Hughes, EPS Conference at Marseille-Luminy 1973

16. Henderson and Bowen, J.Phys.C., <u>4</u>, 1487 (1971)

COMBINED IONISATION AND COLLISION DAMAGE PROCESSES IN IONIC SOLIDS

D.POOLEY
Materials Physics Division,
AERE - Harwell
Great Britain

INTRODUCTION

It is probably reasonable to estimate that more than 95 % of radiation damage studies have been concerned with either collision or ionisation damage, but not with both at the same time. The reason is, of course, that where ionisation effects and photochemical damage do occur they usually provide extremely efficient ways of converting radiation energy into defects, whereas collision damage processes are generally far less efficient. If MgO is irradiated with 1 MeV electrons only one F or F^+ centre is produced for every 100 electrons incident on the crystal, the conversion factor is about 100 MeV/defect. In alkali halides, on the other hand, typical defect production rates give 1000 eV/defect and more sophisticated measurements (Sonder[1]) indicate even higher efficiencies of around 100 eV/defect. With photochemical damage up to a million times more efficient than collision damage it is small wonder that collision effects are impossible to measure when photochemical damage occurs.

Nevertheless, there are a number of ionic materials where, under suitable conditions, defects are created by both ionisation and collision damage effects. In some ways their rarity makes them curiosities, away from the main stream of radiation damage science, but it also makes them very interesting. I will discuss effects in MgO, LiF and MgF_2. Only in the last case is there evidence that the same defect, in this case the

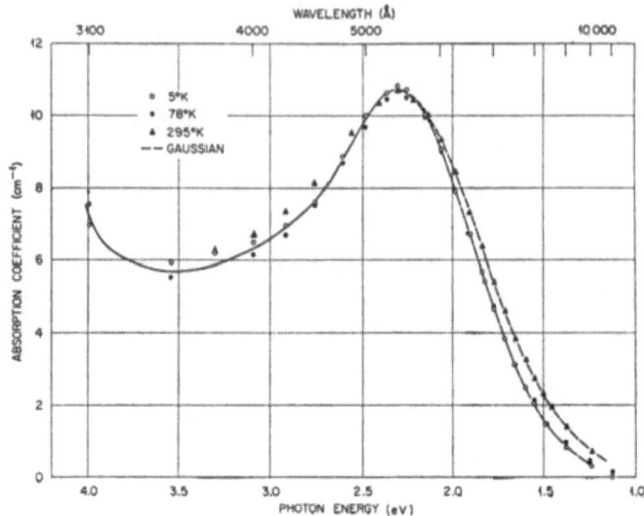

<u>Fig.1</u> : Absorption spectrum of γ ray irradiated MgO crystal.

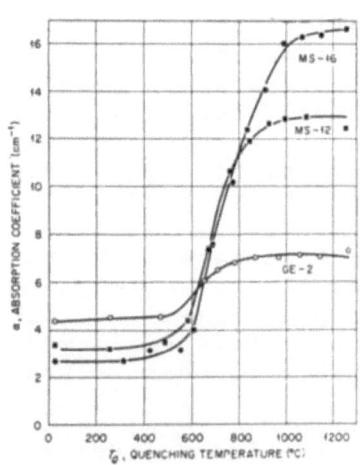

<u>Fig.2</u> : Absorption coefficient of the 2.3 eV band resulting from gamma irradiation versus the quenching temperature for three samples from different sources. Each sample had previously been slow-cooled to decrease the radiation-induced absorption coefficient.

F centre, is actually produced by both photochemical and collision damage process. In the other cases either different defects are produced by the two processes or the ionisation effects modify the end products of the collision damage.

V^-, F AND F^+ CENTRES IN MgO

Ionisation effects do not cause atomic displacements in alkaline earth oxides (we discussed this in the session on collision damage in oxides) bu they do change the electronic configuration of pre-existing defects and appear to create new ones. If typical MgO crystals are irradiated with γ rays to doses of about 1M Rad a band is produced at 2.3 eV (figure 1) which is due to a centre now called the V^- centre. (Henderson and Hughes[2]). The V^- centre was originally discovered by Wertz et al[3] and identified as a positive ion vacancy with a trapped hole using ESR techniques, the 2.3 eV band being associated with it later (Wertz et al[4]).

There is now very good evidence (Chen and Sibley[5]) that the positive ion vacancies must be present before irradiation and that the creation of V^- centres by irradiation consists of adding holes to these. Certainly the creation of V^- centres by 1 MeV gamma rays is far too efficient ($\sim$ 1000 defects per photon) for a collision damage process, saturates far too soon and is far too sensitive to impurities. For example if Mn, Ti or Co are incorporated into the crystal V^- formation is suppressed. On the other hand quenching from a high temperature, which is a standard method of generating lattice defects in crystals, raises the saturation V^- concentration considerably (figure 2).

In MgO then, one important lattice defect, the positive ion vacancy with a trapped hole, is created by an ionisation process. Its antimorph, the negative ion vacancy with a trapped electron (the F^+ centre) is, in contrast, created only by collision damage. Again the F^+ centre was first studied by ESR, and later on an optical band, at 4.9 eV, was assigned to it (Handerson and King[6]). Unlike the V^- centre F^+ production has all the characteristics of collision damage. Electrons below 0.33 MeV create no defects at all and even at 1.6 MeV nearly 100 electrons are required to create each F^+ centre. Reactor neutrons, which can transfer much more energy to the oxygen ions than can electrons are much more effective. Sibley and Chen[7] found that a given area dose of neutrons produced about 6 times as many F^+ centres in a thin sample as the same area dose

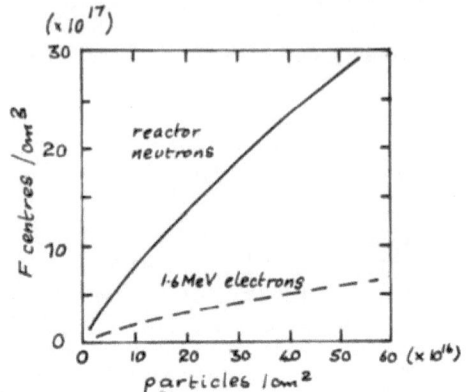

Fig.3 : Radiation damage in MgO

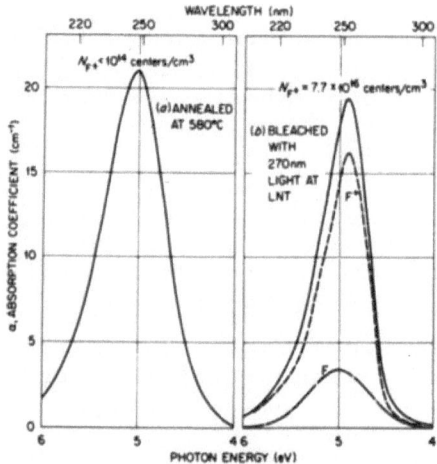

Fig.4 : F and F⁺ centers in electron irradiated MgO crystal.

of 1.6 MeV electrons (figure 3) and we should also remember that the neutrons are capable of irradiating
much thicker samples. Formally then important defects
in MgO are created by ionisation and collision damage
processes, although the ionisation involves only changing the electronic state of an existing defect.

In talking about F^+ centre production in MgO I
have deliberately avoided distinguishing between F and
F^+ centres. In fact both centres are now known to have
strong optical absorption bands at approximately the
same place, which caused considerable confusion until
the fact was fully realised (Chen et al[8]). Whereas only F centres are produced by additive coloration and
essentially only F^+ centres by neutron irradiation, in
electron irradiated crystals a mixture of F and F^+ centres are formed (figure 4). The production of the anion
vacancies is clearly by a collision damage process but
their occupancy by electrons is determined by ionisation effects.

DEFECT PRODUCTION IN LiF

When irradiated with x-rays LiF behaves superficially as the other alkali halides do, F centres are
created with high efficiency, M centres form later on
and eventually interstitial clusters can be seen
(Hobbs et al[9]). However, the F centre concentration
does not grow linearly with time for long, and at high
doses it rises like $t^{1/2}$ (Durand et al[10] and Farge[11]).
The reason for the $t^{1/2}$ growth is not absolutely certain although the explanation given by Farge and his
collaborators, namely that as the F centre concentration grows it becomes even more likely that interstitial atoms created by the radiation will recombine with
an F centre before finding an interstitial cluster and
being stabilised, seems the most likely. Whatever the
reason, at large doses the growth of anion defects has
considerably slowed.

Because of this, Farge et al[12] were able to see
a new band grow in LiF which had been heavily neutron
irradiated at 77K (figure 5). They attributed this
band to an interstitial Li centre created by a collision damage process. Their evidence that the centre is
an interstitial lithium atom, in the form of a <110>
centre analogous to the H centre, is very strong and
includes the observation that the I centre becomes unstable at about 400 K, at which temperature lithium
platelets are produced in neutron irradiated LiF (Lambert and Guinier[13]).

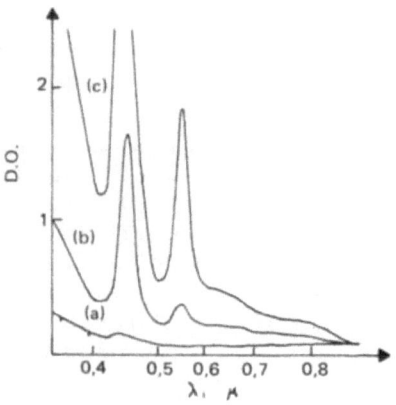

Fig.5 : Spectres d'absorption à 77°K de cristaux de LiF irradiés aux neutrons à 77°K

        a) 5×10^{14} n/cm^2
        b) $\qquad 10^{16}$ n/cm^2
        c) 2×10^{17} n/cm^2

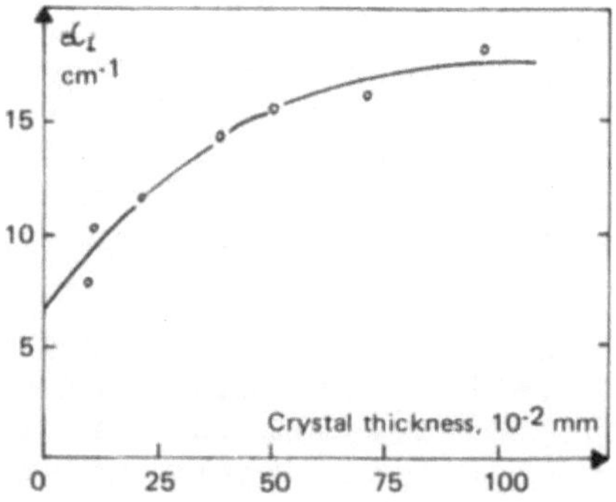

Fig.6 : Variation of the density of I centers formed with sample thickness.

Farge[14] has also collected fairly convincing evidence that the I centre is created by a knock-on process and not by ionisation. This is partly from the complete lack of correlation between anion defect production (F and M centres) and the formation of I centres, and partly from the energy dependence of I centre production by fast electrons. Farge inferred this energy dependence from the way I centre formation varied with crystal thickness (figure 6). The I centre concentrations formed were smaller in thin samples than thick ones by about a factor of three, whereas M centre concentrations did not vary significantly with thickness. This effect is consistent with knock on creation of I centres ; as the electrons pass through the crystal they lose energy and actually become more effective at creating collision damage because the Rutherford scattering cross section varies with $1/E^2$. However, this would not happen if the displacement threshold of the Li atoms were so high that relativistic electrons were needed for collision damage, and the data can be fitted satisfactorily only if the threshold energy is about 10 eV or below. A displacement energy of 10 eV for Li^+ seems rather unlikely, although not impossible in view of the strange behaviour of Li^+ in other alkali halides. However, it may well be that the formation of I centres is due to combined ionisation and collision effects.

Dupuy and his co-workers (Thevenard et al[15]) have also studied LiF irradiated under conditions where knock on damage is likely. They have irradiated with 56 MeV α particles and 28 MeV deuterons, which have penetration depths of 1-2 mm and allow them to measure defect distribution over the range of the particles. The particles create F and F aggregate centres, of course, but their charge states and annealing behaviour are critically determined by the type of irradiation used. They also create I centres and, after annealing, new bands at 290 nm and 435 nm (figure 7). These latter bands are not due to simple F aggregate centres, since the bands remain after annealing at 620 K for 30 mins, and are probably colloids. These centres and the I centres are created primarily at the end of the particle tracks which would be consistent with a collision damage process, but might also be an ionisation intensity effect.

In LiF then, there are strong grounds for believing that anion sub-lattice defects are created by photochemical processes, whereas cation defects are gene-

332

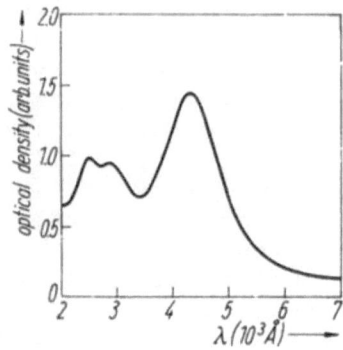

<u>Fig.7</u> : Optical absorption spectrum of a LiF sample irradiated with 4 x 10^{14} α-particles/cm^2 at RT and annealed at 620°K for 30 min.

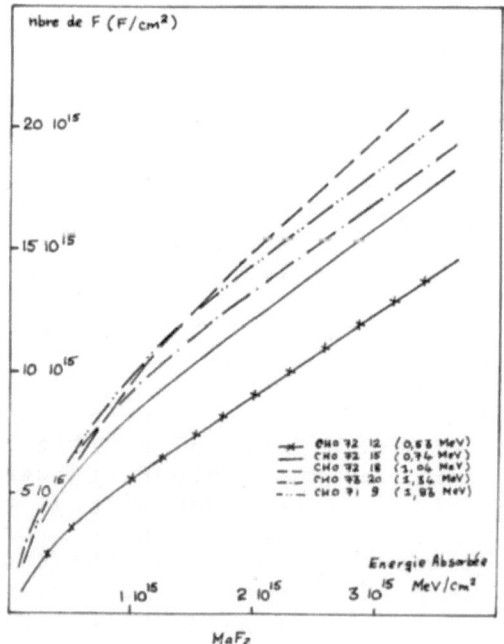

<u>Fig.8</u> : Courbes de dégradation pour différentes énergies des électrons incidents.

rated in direct collisions. In looking for cation defect generation in alkali halides materials like NaBr
and NaI should be better candidates than LiF, but to
my knowledge no evidence for Na$^+$ displacement in them
has been found.

F CENTRES IN MgF$_2$

It has long been known that F centres are created
in MgF$_2$ at room temperature by x-rays, and we would
therefore expect that the major defect production process would be photochemical. This has now been confirmed by Romero[16] who has found a complete lack of energy
dependence under electron irradiation (figure 8). At
room temperature the efficiency of F centre creation
in MgF$_2$ is about 10^{13} F centres/J (equivalent to 4 10^5
eV/F). Although this represents a far less efficient
process than in alkali halides it is still much more
effective at creating F centres than direct collisions
would be.

However radiation damage in MgF$_2$ is rather strange,
in that at lower temperatures F centre production appears to saturate (Sibley and Facey[17] and Buckton and
Pooley[18]). This is shown in figure 9, and has been attributed to the fact that close pairs are formed as
the initial step in photochemical damage. At low temperatures the close pairs are faily stable and large concentrations can be achieved, while at high temperatures
a large fraction can dissociate to give stable F centres
and separated interstitials. At intermediate temperatures, as in figure 9, the close pairs can recombine fairly quickly but cannot dissociate, so that only small
saturation concentrations can be built up. Under these
conditions the net production of defects by ionisation
processes is rather small, and we might expect to see
evidence of collision damage.

Figure 10 shows the F band growth curves in MgF$_2$
at 80 K under 400 keV proton irradiation ; there is a
continuous steady growth of the F band, in contrast
with the x-ray behaviour and consistent with collision
damage. What could well be happening here is that collision damage events can move the fluorine ion further
from the vacancy than can the photochemical process,
probably because it can give the ion more energy. The
fact that the momentum of the protons, rather than the
high ionisation density they cause, is important for
the steady growth of F centres is confirmed by comparing
electron and proton irradiation at 10 K (figure 11) at
approximately the same ionisation density. No F band

334

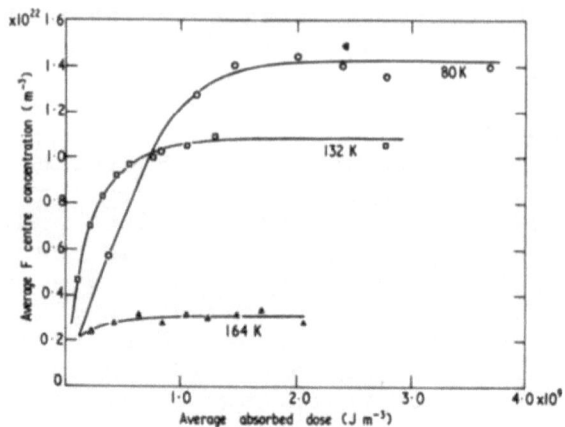

Fig.9 : The growth of the F band in MgF$_2$ at 80K, 132K and 164K under X-irradiation at 40 kW m^{-3}.

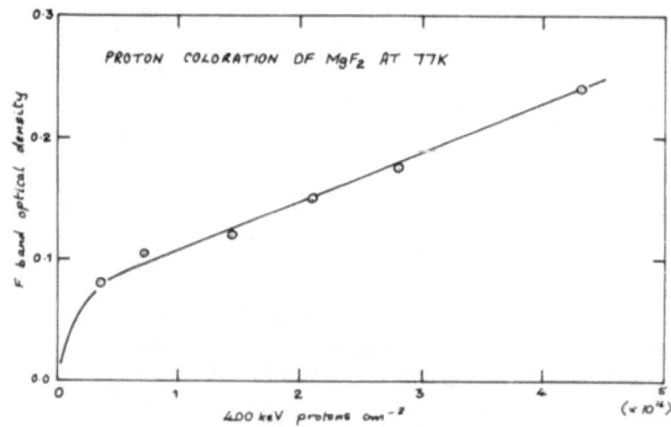

Fig.10 : Proton coloration of MgF$_2$ at 77°K

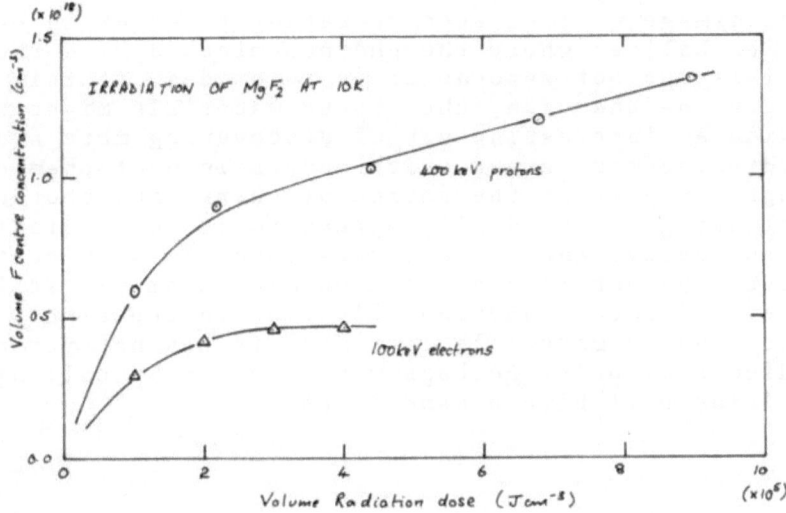

Fig.11 : Irradiation of MgF$_2$ at 10°K

growth can be measured under electron irradiation after 3-4 hrs, whereas under proton irradiation the F band is growing steadily, at a rate comparable with F^+ band growth in the oxides under proton irradiation.

It is very probable that collision and photochemical damage will be seen operating together in other complex halides where the photochemical damage rate is small but not zero as it is in oxides. If this does prove to be the case, then these materials might well provide an interesting way of discovering more about the displacement event which occurs in photochemical damage. At present the source of energy for photochemical damage is generally agreed to be non-radiative exciton decay, and the products to be F and H centres. However the detailed separation mechanism has still to be confirmed experimentally even in the alkali halides, and in materials like MgF_2 it has not yet been studied seriously. Perhaps what I began by calling curiosities will have a hand in this.

REFERENCES

1. Sonder, Phys.Rev., B5, 3259 (1972)

2. Henderson and Hugues, Chapter in "Point Defects in Solids" Plenum Press, 1972

3. Wertz et al., Disc.Faraday Soc. $\underline{28}$, 136 (1959)

4. Wertz et al., J.Phys.Soc.Japan, $\underline{18}$, Suppl. II, 305 (1963)

5. Chen and Sibley, Phys.Rev. $\underline{154}$, 842, (1967)

6. Henderson and King, Phil.Mag., $\underline{13}$, 1149 (1966)

7. Sibley and Chen, Phys.Rev., $\underline{160}$, 712 (1967)

8. Chen et al., Phys.Rev. $\underline{186}$, 865 (1969)

9. Hobbs et al., Proc.Roy.Soc. $\underline{A}$ $\underline{332}$, 167 (1973)

10. Durand et al., J.Phys.Chem.Solids, $\underline{30}$, 1353, (1969)

11. Farge, Ibid. $\underline{30}$, 1375 (1969)

12. Farge et al., J.Phys.Chem.Solids, 27, 499 (1969)

13. Lambert and Guinier, Compt.Rend. $\underline{246}$, 1678, (1958)

14. Farge, Phys.Rev. $\underline{B1}$, 4797 (1970)

15. Thevenard P. et al., Phys.Stat.Solidi ($\underline{a}$) $\underline{9}$, 517 (1972)

16. Romero, Report NT-02-29, CERT, DERTS, Feb. 1972

17. Sibley and Facey, Phys.Rev. $\underline{174}$, 1076 (1968)

18. Buckton and Pooley, J.Phys. C. $\underline{5}$, 1553 (1972)

IONIZATION EFFECTS ON DAMAGE PRODUCTION IN SEMI-CONDUCTORS

J.C. BOURGOIN

Groupe de Physique des Solides de l'E.N.S.
Université PARIS VII, Tour 23
2, place Jussieu

75221 PARIS Cedex 05 /France

INTRODUCTION

The production of defects by energetic particles can be divided in three stages : the energetic parti-cles transmit an energy to the atoms of the lattice ; when the transmitted energy is larger than a value characteristic of the irradiated material, called the threshold energy, the atoms are displaced from their substitutional site in the lattice and vacancy-inters-titial pairs are created ; the vacancies and the in-terstitials, when mobile, interact with each others or with impurities to annihilate or to form complex de-fects. While the defect production rate depends on the energy and on the type of the energetic particles (through the transmitted energy and the cross-section for the transmission of this energy), the nature of the defects produced is mainly dependent on the mobili-ty of the primary defects, vacancies and interstitials; indeed their mobility governs their interaction with each others or with impurities.

The mobility of defects (or of impurities) depends of course on the ˌtemperature but also on other parame-ters some of which are sensitive to the irradiation. An irradiation can enhance or induce the diffusion of defects through one of the following mechanisms[1] :

Defect mechanism
The diffusion of defects produced by irradiation can both enhance or retard diffusion processes[2] : the diffusion of an impurity atom via a vacancy mechanism

can be enhanced by the creation of additional vacancies ; the diffusion of a substitutional atom can be enhanced by converting it to an interstitial atom and permitting its diffusion by the usually faster interstitial mechanism ; the diffusion of an interstitial impurity can be retarded by trapping it on a vacancy, etc.

Recoil mechanism

Energetic particles can enhance diffusion by imparting a recoil energy to the defect either by a direct collision with the defect or by an indirect collision (for instance the energetic particle can put the defect in a high energy excited state which, when it decays, result in a recoil energy ; or the energetic particle can create energetic photoelectrons which in turn impart energy through coulomb collision with the defect).

Energy release mechanism

The capture of an electron or of a hole or the recombination of an exciton (produced by the elastic collision of the energetic particles with the electrons of the solid) at a defect site, all result in the release of an energy. When the release of this energy results in the creation of one or more phonons, i.e. is not radiative, it can contribute to the enhancement of the diffusion.

Ionization mechanisms

Under this denomination we consider the processes which, because of a change of the charge state or because of successive changes of the charge state of the defect, modify or induce the mobility of this defect.

The aim of this course is to develop the effects of the ionization processes due to the irradiation on the mobility of defects (or impurities) in semiconductors, i.e. the ionization and the energy release mechanisms. (We will not consider the direct creation of defects by ionization which could happen through processes similar to the Varley[3] and the Pooley[4] mechanisms in semiconductors whose ionicity[5] is sufficient). In the second section we will consider the processes which determine the electronic state of a defect ; in the third we will describe the different ionization enhanced diffusion (I.E.D) mechanisms, provide a theoretical justification for their existence and develop their phenomenology ; in the last section we will show how such mechanisms can provide an explanation for the behaviour of defects and of impurities in semiconductors under irradiation.

TRANSITIONS BETWEEN THE ELECTRONIC STATES OF A DEFECT

Consider a defect which can exist in two electronic states S and B, the B state having one more electronic charge than the S state, for instance. An energy level E_T in the band gap is associated with the transition from the S to the B state ; this transition can occur (figure 1) through the capture o the emission of an electron or of a hole :

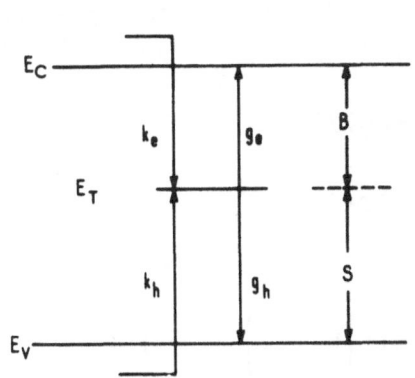

$$S+e^- \underset{G_e}{\overset{K_e}{\rightleftarrows}} B$$

$$B+e^+ \underset{G_h}{\overset{K_h}{\rightleftarrows}} S \qquad (1)$$

Fig.1 : Energy diagram of a defect.

Let K_e, K_h, G_e and G_h be the probabilities for the capture (K) and the emission (G) of an electron (e) or of a hole (h). If N is the defect population and α the fraction of time during which the defect is in the B state (i.e. the population of the B state is $b = \alpha N$ and the population of the S state is : $s = (1-\alpha)N$), the capture and emission probabilities, which are written as the product of the initial state by the final state (the coefficient of proportionality is $V\sigma$, where V is the thermal velocity of an electron or a hole and σ the cross section for electron or hole trapping on S and B states respectively), are :

$$K_e = V_n \sigma^n (1-\alpha) N \; n$$

$$K_h = V_p \sigma^p \alpha N \; p$$

$$G_e = V_n \sigma^n N_c \; \exp\left[-(E_c - E_T)/kT\right] \alpha N$$

$$G_h = V_p \sigma^p N_v \; \exp\left[-(E_v - E_T)/kT\right] (1-\alpha) N$$

(2)

with E_c and E_v the conduction and valence band energies, N_c and N_v the densities of states in the conduction and valence bands.

The dynamics of the rate of change of charge state can be written :

$$\frac{db}{dt} = -\frac{ds}{dt} \quad \text{and :}$$

$$\frac{ds}{dt} = -(K_e + G_h) + (K_h + G_e) \qquad (3)$$

Introducing the rate constants :

$$k_e = V_n \sigma^n n$$

$$k_h = V_p \sigma^p p \qquad (4)$$

$$g_e = V_n \sigma^n N_c \exp\left[-(E_c - E_T)/kT\right]$$

$$g_h = V_p \sigma^p N_v \exp\left[-(E_v - E_T)/kT\right]$$

equation (3) becomes :

$$\frac{d(1-\alpha)}{dt} = -\left[k_e + g_h\right](1-\alpha) + \left[k_h + g_e\right]\alpha \qquad (5)$$

At equilibrium

$$\frac{ds}{dt} = 0 \quad \text{and :}$$

$$\alpha = (k_e + g_h)/(k_h + g_e + k_e + g_h) \qquad (6)$$

α defines the ensemble average value of the occupation of the B state by a continual succession of transitions; if we consider a particular defect of the population, it changes its state with an average rate $1/\tau$ which can be written as the sum of the probabilities of change divided by the total population of the defects :

$$\frac{1}{\tau} = (K_e + G_h + K_h + G_e)/N \qquad (7)$$

or

$$\frac{1}{\tau} = \left[k_e + g_h\right](1-\alpha) + \left[k_h + g_e\right]\alpha$$

replacing α by its value :

$$\frac{1}{\tau} = 2/\left(\frac{1}{k_e + g_h} + \frac{1}{k_h + g_e}\right) \qquad (8)$$

THE DIFFUSION MECHANISMS DUE TO ELECTRONIC PROCESSES

A. Normal Ionization Enhanced Mechanism

The probability that a defect will jump from one equilibrium position to another over the barrier E_M presented by the host atoms is :

$$\nu = \nu_0 \exp (-E_M/kT) \qquad (4)$$

where ν_0 is a characteristic vibrational frequency ($\nu_0 \simeq kT/h$, h Planck's constant) typically on the order of 10^{13} s^{-1}. The equilibrium position of the defect

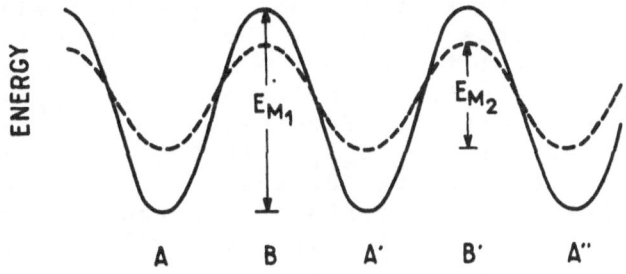

Fig.2 : Change of a defect barrier height.

in the lattice and the saddle point are characteristic of the detailed mechanism by which the diffusion takes place. The barrier height E_M describes the interaction of the defect with the surrounding atoms, which interaction may include repulsive or attractive electronic forces. Therefore this barrier may change with the charge state of the migrating defect[6] giving rise, when E_M is decreased (figure 2), to an enhancement of the defect mobility.

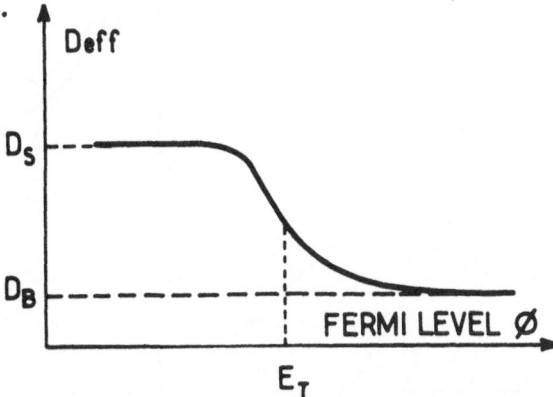

Fig.3 : Variation of the effective diffusion coefficient with the Fermi level.

Let D_B and D_S be the diffusion coefficients of a defect in the B and S states respectively ; the effective diffusion coefficient D_{eff} for the defect is the sum of the diffusion coefficients of the S and B states balanced by their respective occupation[9] :

$$D_{eff} = (1-\alpha)D_S + \alpha D_B \qquad (5)$$

D_{eff} varies from D_S to D_B with α, that is with the Fermi level. Let assume that the material is n type and that E_T is deep in the upper half of the forbidden gap; then at low temperature the Fermi level is well above E_T, the defect is in the B state and $D_{eff} = D_B$; when the temperature increases the Fermi level goes through E_T (when the Fermi level is on E_T : $\alpha = 1-\alpha = 0,5$ and $D_{eff} = (D_S+D_B)/2$) ; when it is well above E_T, $D_{eff} = D_S$. The variation of D_{eff} with the Fermi level is sketched on figure 3. The variation of D_{eff} with T depends on the respective variation of D_S, D_B and α with the temperature ; we illustrate the influence on the variation of D_{eff} with $1/T$ of the position of the level E_T (figure 4 : (1) $E_T = 0.1$ eV and (2) $E_T = 0.6$ eV, with $N_D - N_A = 10^{18}$ cm^{-3}) and of the doping and compensation (figure 5 : (1) $N_D - N_A = 10^{18}$ cm^{-3} and (2) $N_D - N_A = 10^{15}$ cm^{-3} with $E_T = 0.1$ eV) for the vacancy in silicon ($D_S = 10^7$ exp$(-0.18/kT)$ and $D_B = 10^{12}$ exp$(-0.33/kT)$; N_D and N_A represent respectively the donor and the acceptor concentrations.

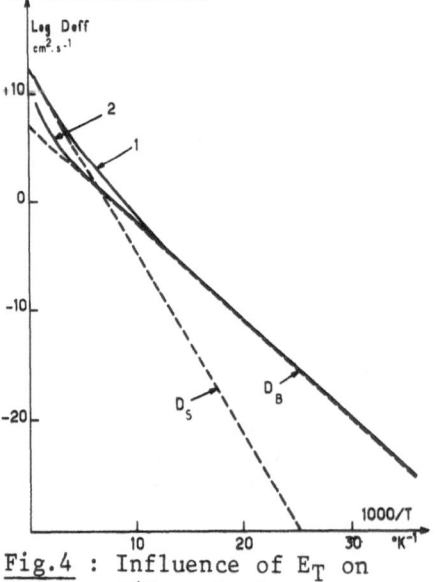

Fig.4 : Influence of E_T on D_{eff} vs $1/T$ variation.

Fig.5 : Influence of $N_D - N_A$ on D_{eff} vs $1/T$ variation.

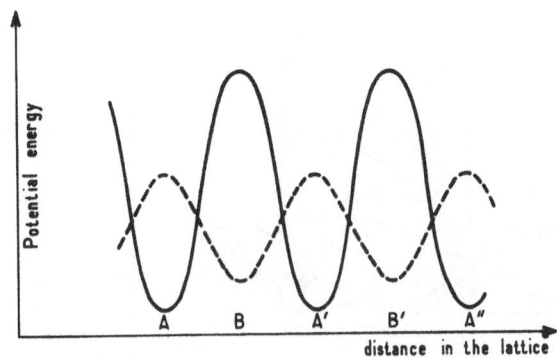

<u>Fig.6</u> : Potential energy of the defect vs its position in the
lattice.

B. <u>Bourgoin Mechanism</u>

Such mechanism[10] occurs when the equilibrium configuration of a defect is substantially changed when
its charge state changes. Assume that the potential
energy of the defect versus its position in the lattice
is as shown in figure 6. The defect, in a given charge
state, is in equilibrium at the bottom of the well at
site A. For that charge state the defect could move to
equivalent sites A', A", etc. by thermal excitation
over the saddle-point barrier at B, B', etc. Upon changing charge state, however, the defect, still located
at A, finds itself at the top of the saddle-point potential for that charge state and, say, relaxes to B.
Upon a subsequent charge change back, it changes potential energy curves again and can, say relax to A' (or
A). These two changes can move a defect through the
lattice <u>athermally</u>. For this mechanism we have to assume a change in the potential energy with charge state
such that, upon successive changes of the charge state,
the defect can proceed athermally from one equilibrium
site to another. (We do not of course require to precise the contours in figure 6).

We illustrate now how suitable potential curves
can arise for an impurity interstitial in the diamond
lattice. Following Weiser[11], we consider the tetrahedral (T) and hexagonal (H) sites for the equilibrium
site and the saddle point. The T site is characterized
by four nearest neighbors at distance $0.433 \ a_0$ from
the center and six next nearest neighbors at a distance
$0.5 \ a_0$ (figure 7) ; a_0 is the unit cell edge length.
The H site is at $0.415 \ a_0$ from 6 nearest neighbors and
$0.649 \ a_0$ from eight next nearest neighbors (figure 8).

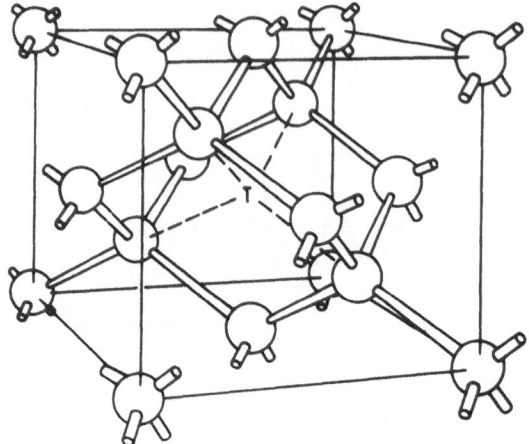

Fig.7 : Tetahedral site for an impurity interstitial in the diamond lattice.

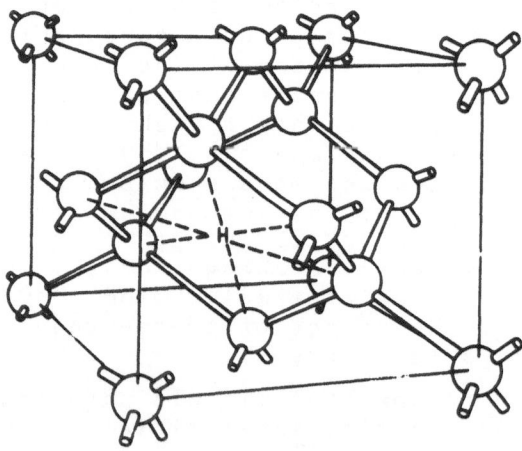

Fig.8 : hexagonal site for an impurity interstitial in the diamond lattice.

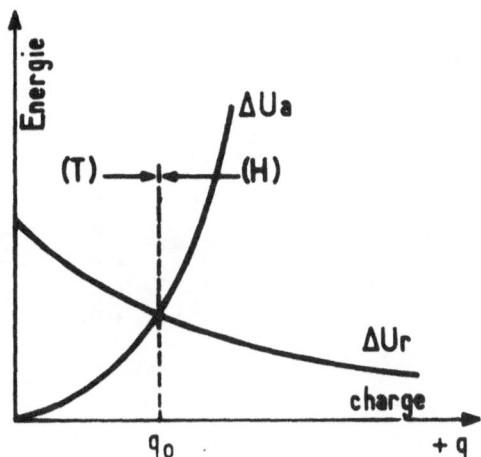

<u>Fig.9</u> : Variation of the equilibrium site of an interstitial
with q.

As Weiser argued, the interaction of a charged interstitial with its environment consists of an attractive
potential produced by the interaction of the charged
interstitial with the dipoles it has induced in the
host atoms and of a repulsive potential characteristic
of the interaction of the interstitial with closed
shells. Calculations based on such considerations have
been performed by Weiser[11] for Li^+ and Cu^+ ions in germanium and silicon. He was concerned with which site
was the equilibrium site for these ions ; hence he compared the total energy difference between the two sites.
He showed that the stability could be determined by
treating the differences in the attractive energy for
the two sites and separately the differences in the repulsive energy for the two sites and comparing these
differences.

We can use such considerations to study the question of the charge-state dependence of the stability.
The difference in attractive energies scales as q^2 (q
charge of the interstitial) since the potential created
by a dipole is linear with the charge of the elements
of this dipole ; the difference in repulsive energies
goes approximately exponentially with the ionic radius
and therefore decreases with increasing positive charge q. Hence, as shown on figure 9, the equilibrium site of the interstitial changes with q ; precisely where the change occurs depends on the details of the calculation, but the physical origin of the change does
not. At the point of the cross-over the migration ener-

348

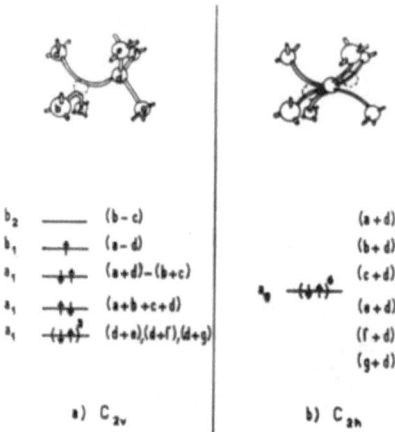

<pre>
b₂ ——   (b-c)   (a+d)
b₁ ——↑——   (a-d)   (b+d)
a₁ ——↑↓——   (a+d)-(b+c)   (c+d)
a₁ ——↑↓——   (a+b+c+d)   a ——↑↓↑↓—   (e+d)
a₁ ——↑↓↑↓—   (d+e),(d+f),(d+g)   (f+d)
                                     (g+d)
</pre>

a) C₂ᵥ   b) C₂ₕ

<u>Fig.10</u> : LCAO bonding scheme for different vacancy configuration.

gy vanishes (this point does not correspond necessarily to a possible equilibrium configuration). Of course
the Weiser's approach has a number of deficiencies :
multipoles are neglected, the effective charge of the
interstitial is assumed to be well localized which is
not necessarily the case, the interstitial configurations considered could be different from the tetrahedral and hexagonal ones[12].

   Change in equilibrium configuration as a consequence of a change of the charge state could also happen for vacancies[14]. Vacancy migration is conventional-
ly visualized as consisting of jumps from substitutional sites to substitutional sites with the split vacancy as saddle point. The configuration of the split-
vacancy (in which an atom, neighbour of the vacancy,
is displaced at equal distances between its original
position and the center of the vacancy as shown on figure 10a) could be the equilibrium configuration for a
particular charge state of the vacancy. We argued[14],
following Watkins[15] who showed that the electronic states of the vacancy are well described by a one electron
LCAO treatment with the states composed of orbitals
from the nearest neighbors (figure 10a), that the addition of an electron on a V⁻ vacancy, localized on this
vacancy, would result in such a substantial electron-
electron interaction that the V⁻⁻ assumes the split-
vacancy configuration with a LCAO bonding scheme as
described on figure 10b.

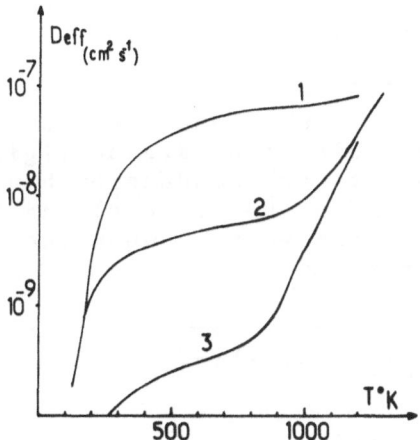

<u>Fig.11</u> : Variation of D_{eff} vs temperature for different values
of the impurity concentration.

To summarize, the potential curve required for
the Bourgoin mechanism (as shown on figure 6) can exist
in covalent materials for interstitial or vacancy type
defects. If the system has some ionic character, the
interaction of the defect with the host atoms would
make the potential more sensitive to the charge state
and the mechanism is expected to be even more important
for more ionic materials[5]. It will therefore apply to
III-V and II-VI semiconducting compounds and to ionic
crystals[16].

We will consider the effective diffusion coefficient D_{eff} according to this mechanism in the case
where no thermal activated diffusion of either S or B
states occurs. Since $1/\tau$ is the rate of jump, D_{eff} can
be written :

$$D_{eff} = \frac{a^2}{4} \frac{1}{\tau} \qquad (6)$$

with a the lattice parameter. What will be the influence on D_{eff} of the different parameters which define
the material and the irradiation ? Let us consider the
case of an n-type material in which E_T is deep compared to the donor level (the case of a p-type material

would be treated in a similar fashion and therefore will not be done here). We have to consider two distinct cases : the material is in thermal equilibrium or out of equilibrium, under the external electronic excitation due to the irradiation.

In thermal equilibrium

At low temperature the band-to-band and level E_T-to-band thermal recombinations are negligible, the Fermi level is close to the conduction band, E_T is occupied (the defect is in the state B) and $D_{eff} \simeq 0$. Only at high enough temperature, when the transitions from the level to the bands can occur :

$$D_{eff} = \frac{a^2}{2} \left(\frac{1}{\sigma^n V_n n + \sigma^p V_p N_v \ \exp\{-(E_T - E_v)/kT\}} + \frac{1}{\sigma^p V_p p + \sigma^n V_n N_c \ \exp\{-(E_T - E_c)/kT\}} \right)^{-1} \quad (7)$$

has a non negligible balue.

Figures 11 and 12 give as illustration the variation of D_{eff} vs temperature for different values of the impurity concentration and of the level E_T, respectively. The cross-sections σ^n and σ^p are arbitrarily chosen : $\sigma^n = 10^{-16} \ cm^2$

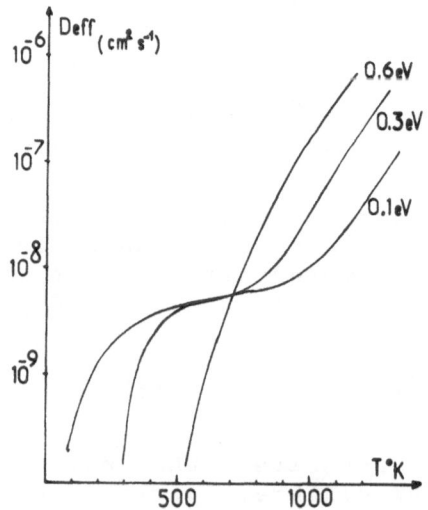

Fig.12 : Variation of D_{eff} vs temperature for different values of the level E_T.

and $\sigma^P = 10^{-15}$ cm^2 ; the value E_T is given with the
origin of the energy at the bottom of the conduction
band ; all the electrons and holes present are assumed
to recombine through the level E_T. Figure 11 shows the
influence of the doping concentration ((1) : $N_D=10^{18}$
cm^{-3} $N_A = 10^{14}$ cm^{-3}, (2) $N_D = 10^{16}$ cm^{-3} $N_A = 10^{14}$ cm^{-3},
(3) $N_D = 2.10^{15}$ cm^{-3} $N_A = 10^{15}$ cm^{-3}, with $E_T = 0.1$ eV)
and figure 12 the influence of the position of E_T
(with $N_D = 10^{16}$ cm^{-3} and $N_A = 10^{14}$ cm^{-3}).

Under external excitation

In the case of irradiation with ionizing particles
of illumination with light or of carrier injection, e-
lectron-hole pairs are created in the material. The
concentration C of these pairs is : $C = g\theta$ (g, generation rate of electron-hole pairs by the excitation and
θ minority carrier lifetime). Under such excitation
the Bourgoin mechanism can occur ; in the case where
the temperature is low enough to prevent thermal excitation of the carriers from E_T (kT low compared to
E_T-E_V and E_C-E_T), τ is reduced to :

$$\tau = \frac{1}{2C} \left(\frac{1}{\sigma^n v_n} + \frac{1}{\sigma^P v_p} \right) \tag{8}$$

when C is large compared to n and p. D_{eff} varies linearly with the concentration C of the electron-hole
pairs injected in the material.

C. Energy Release Mechanism

Any mechanism which releases energy in the locali-
ty of a defect may cause defect migration.Thus the act
of electron-hole pair recombination, carrier trapping,
excitation recombination, exciton recombination, etc...
at a defect site could create localized energy released
and defect motion[17] when this energy is released in a
cascade of phonons and when it is larger than the thermal migration energy E_M.

The probability of diffusion of a defect by this
mechanism is given by the product of the several probabilities : 1) that the energy source be created ; 2)
that the energy source reach the defect ; 3) that the
energy be released at the defect in an usable form ;
and 4) that the released energy result in a defect
jump. The first and second probabilities can be calculated in principle, using standard ways. The third and
fourth probabilities may be more by a problem. There
may be a number of competing ways which the energy may
be released, each may variously contribute to defect
mobility and in most instances the nature of the

energy release mechanism will have to be examined in
detail for each defect considered. There is one form
of energy release which we can describe and which probably has wide applicability : the "temperature spike".
Lax[18] argued that carrier capture results in a cascade
of single, acoustic phonons being released locally ;
this probably occurs in a number of processes. These
phonons represent a localized increase in lattice temperature. The theory of temperature spikes was developed by Seitz and Koehler[19] for the case of a very high
energy release by a collision between a high-energy
particle and a lattice atom ; in that application the
theory has the drawback that the very high energy lattice vibrations are not describable by thermal diffu-
sivities. In our application, however, we are dealing
with much smaller amounts of released energy, with lattice phonons, and the mathematics should apply directly. Briefly, if the energy release E_R is described as
a temperature pulse $T(\tau)$ which dissipates due to thermal diffusivity on a defect site then the number of
jumps experienced by this defect in this heat pulse
is :

$$n_j = \int_0^\infty \nu_0 \exp\left(-E_M/kT(t)\right)dt \qquad (9)$$

which Seitz and Koehler show to be equal to :

$$n_j = 0.093 \; \rho(E_R/E_M)^{2/3} \qquad (10)$$

with ρ-a dimensionless parameter which gives the ratio
of ν_0 and of the frequency with which the pertinent
vibrational energy is transferred from atom to atom
(when dealing with thermal phonons, $\rho \simeq 1$).

EVIDENCE FOR DEFECT AND IMPURITY MIGRATION UNDER IONIZATION

Electron and γ-ray irradiation are one of the ways
for creating electron-hole pairs in a semiconductor ;
such irradiations cause a change in the Fermi level
position and successive changes of the charge state of
the defects (or of the impurities) due to the recombination of electrons and holes on the defects (or on
the impurities) sites ; migration of defects (or of
impurities) is therefore expected to occur through one
of the mechanisms previously described. Since in this
lecture we are dealing with the processes produced by
irradiation we will consider the effect of ionization
on the behaviour of defects (and of impurities) only in
the case where this ionization is created by irradia-

tion (with electrons or γ-rays).

A. Defects

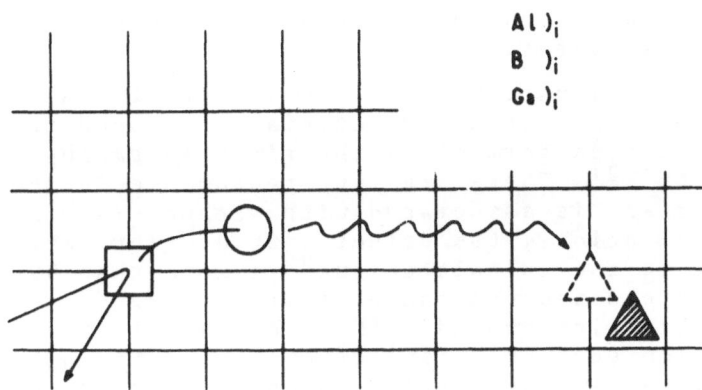

Fig.13 : Replacement sequence for group III atom doped silicon.

We will limit the discussion to defects whose behaviour can be clearly interpreted in terms of I.E.D.;
these defects are vacancies and interstitials, created
by electron irradiation at low temperature, in silicon
and germanium. In silicon, it is observed that :

1-aluminum interstitial atoms appear in aluminum
doped samples. Watkins[20] has shown that such appearance can only be due to the following replacement sequence, sketched on figure 13 : silicon interstitial atoms,
created by the irradiation, migrate to substitutional
aluminum atoms (or any group III atom) and replace
these atoms, giving rise to aluminum interstitial
atoms.

2-the vacancy anneals at different temperatures,
depending on the type (n or p) of the sample[15] : around
80°K in n-type material (with an activation energy of
0.18 eV)   and around 165°K in p-type material (with
an activation energy of 0.33 eV).

Such behaviours are characteristic of I.E.D. mechanisms. The activation energy for the migration of
the vacancy changes with its charge state since the
vacancy is neutral in p-type material and doubly negative in n-type material ; the migration of the vacancy is enhanced or retarded following a change of its
charge state : the corresponding mechanism is the normal ionization enhanced diffusion mechanism. As for

the interstitial, the only possible explanation for its migration, without activation energy (the migration energy should be on the order of $kT \simeq 0.3 \times 10^{-3}$ eV at 4°K), under electron irradiation, is that an I.E.D. mechanism, such as the Bourgoin or the energy release mechanism, is operative.

A better illustration of the effect of ionization on the behaviour of the interstitials is found in germanium. In n-type germanium, the first annealing stage occurs at 65°K[21]. There are many reasonable indications that this stage is associated with vacancy-interstitial pairs annihilation : the defect introduction rate is comparable to the calculated one[22] ; more than 99 % of the defects created can anneal in this stage[22] ; the annealing kinetics around 65°K is characteristic of the recombination through diffusion of correlated pairs[23] ; the energy release per defect in this stage is comparable to the estimated formation energy of an interstitial and a vacancy. One of the defects associated with this stage is a double acceptor[22] which can be easily ionized by γ-ray or electron irradiation or by light illumination since, the cross-section for the trapping of a hole being large, it has a large probability for trapping a hole. It is observed that under irradiation or light illumination the 65°K stage shifts to 27°K[23] ; the annealing can occur at even a lower temperature (4°K[24]) when the electron-hole pair concentration injected by the irradiation or the illumination is on the order of, or larger than, the electron concentration at equilibrium. We proposed[25] that the shift of the annealing stage from 65°K to 27°K is a manifestation of the normal ionization enhanced diffusion mechanism : the interstitial, whose charge state is I_2 under equilibrium condition, traps a hole under irradiation : $I_2 \rightarrow I_1 + e^-$; with its new charge state I_1 it anneals with a lower activation energy at 27°K. We also suggested[25] that the annealing at 4°K is due to the successive changes in charge state from I_2 to I_1 because of the Bourgoin mechanism (it is also possible that the Bourgoin mechanism is operative between I_1 and I_0, the doubly ionized interstitial, since the annealing at 4°K seems to occur only when a high injection condition is fulfilled). Actually an energy release mechanism could be operative instead of the Bourgoin mechanism.

B. Impurities

A change in configuration as a consequence of a change in the charge state is expected with impurities on which electrons are strongly bound since the electrostatic interactions are important only for such impurities ; these impurities will be associated to a deep level (E_T) through which the recombination of the electrons and the holes created by an excitation (an irradiation for instance) will occur. Metallic impurities which exist, at least partially, as interstitials such as Au, Cu, Fe, Zn, etc. are good examples of such impurities[26].

Indications for the migration of metallic impurities can be found in the litterature[27]. Here we will only consider the case of gold migration in silicon for two reasons ; the first reason is that the parameters, such as the cross-sections for electron and hole trapping and the associated energy levels, are well known for gold in silicon[28]. The second reason is that experiments reporting the migration of gold in silicon at room temperature under X-ray irradiation have been described in the litterature[27] ; but because the technique used in these experiments has been seriously questionned, we will describe another type of experiment.

Gold is deposited by evaporation on the surface of a silicon (10 Ω.cm, n-type) sample which has been previously cleaned by sputtering ; a Schottky diode is formed which is irradiated at room temperature with 200 keV electrons (the energy of which is too low to create defects). The gold profile -i.e. the concentration of gold vs depth- is determined after different doses of irradiation using capacitance -voltage measurements.

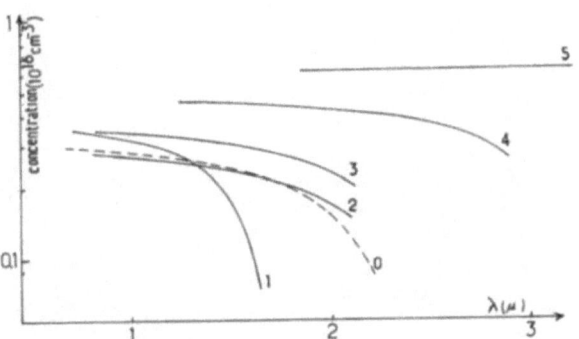

Fig.14 : Gold profiles of an irradiated schottky diode for increasing doses.

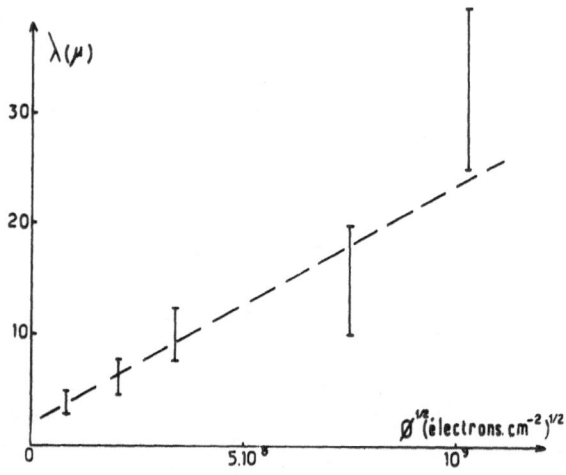

Fig.15 : Variation of the diffusion lenght λ with the dose φ of irradiation.

Typical profiles are shown on figure 14 for the following doses of irradiation : 0) before irradiation; 1) 7.15×10^{15} cm^{-2} ; 2) 4.26×10^{16} cm^{-2} ; 3) 1.15×10^{17} cm^{-2} ; 4) 5.65×10^{17} cm^{-2} and 5) 1.05×10^{18} cm^{-2}. It is observed (figure 15) that the diffusion length λ associated with a profile varies approximately linearly with the square root of the dose ϕ of irradiation, as expected if the diffusion is due to an electronic process. Indeed, in such a process :

$$\lambda = \sqrt{D_{eff} \cdot t} \qquad (11)$$

t : time of irradiation ; using formula (8)

$$\lambda = a \left[C.V \left(\frac{\sigma^n \sigma^p}{\sigma^n + \sigma^p} \right) t \right]^{1/2} \qquad (12)$$

with : $V_n = V_p = V$; Ct is proportional to ϕ.

Calculating the electron-hole pair generation rate g, a fit with experimental data is obtained for the transition $Au^{\circ} \rightleftarrows Au^{-}$ ($\sigma^n = 10^{-16}$ cm^{-2}, $\sigma^p = 10^{-15}$ cm^{-2}) using the value : $\theta \simeq 10$ µs for the minority carrier lifetime, in good agreement with the experimental value.

The linear variation of the diffusion length of gold in silicon vs $\sqrt{\phi}$ is a strong indication for the existence of an I.E.D. mechanism ; the magnitude of the diffusion length suggests that the Bourgoin mechanism is operative in this case.

CONCLUSION

The concept of ionization enhanced diffusion is, in many respects, a new concept which appears lately in the litterature ; up to now, it was recognized that defect migration could occur through the normal ionization enhanced mechanism but the fact that defects or impurities could diffuse athermally was not considered. It is not possible to assert that all the aspects of the I.E.D. mechanisms we described here really occur in the nature and that the observations we reported for gold migration in silicon are valid for all interstitials impurities with which a deep level is associated. As for defects, the indications we described for interstitial defects are less direct than for impurities and the experiments do not allow to distinguish whether an energy release mechanism or the Bourgoin mechanism is operative ; as for vacancy type defects no reasona-

358

ble indication of the existence of I.E.D. has been
found yet.

ACKNOWLEDGEMENT

The author is indebted to J.W.Corbett for his co-
operation in the work described in this course.

REFERENCES

1. Bourgoin J.C. and Corbett J.W., Rad.Effects (to be published)

2. Damask A.C., Studies in Radiation Effects in Solids, ed. Dienes G.J. Gordon and Breach, New-York, 1967 , Vol.2, p.1

3. Varley J.H.O., J.Phys.Chem.Solids, 23, 985, 1962

4. Pooley D., Proc.Phys.Soc., 87, 245+247, 1966

5. For a definition and an evaluation of the ionicity see Phillips J.C., Rev.Mod.Phys., 42, 317, 1970

6. The calculation of the variation of E_M with the charge state has been attempted in the case of the vacancy migration by Kiv and Umarova[7] using the Morse function treatment developed by Swalin[8] ; the calculation described in the next paragraph gives an idea how to treat the problem in the case of the interstitial.

7. Kiv A.E. and Umarova F.T., Sov.Phys.Semicond. 4, 474, 1970

8. Swalin R.A., J.Phys.Chem.Solids 18, 290, 1961

9. A rigourous demonstration of expression (5) can be found in Bourgoin J.C., Corbett J.W. and Frisch H.L., to be published

10. Bourgoin J.C. and Corbett J.W., Phys.Letters 38 A, 135, 1972

11. Weiser K., Phys.Rev. 126, 1427, 1962

12. Watkins G.D. et al[13] have shown, using a semi-empirical LCAO method, that such change could occur between the "bond centered" and the "split" interstitial configuration.

13. Watkins G.D., Messmer R.P., Weigel C., Peak D. and Corbett J.W., Phys.Rev.Letters 27, 1573, 1971

14. Corbett J.W. and Bourgoin J.C., Point Defects in Solids, ed. Crawford J.H. and Slifkin L.M. Plenum Press, New-York , Vol.2, to be published.

15. Watkins G.D., Radiation Damage in Semiconductors ed. P.Baruch Dunod, Paris, 1965 , p.97

16. In ionic crystals, Varley[3] has proposed a mechanism generically related to this mechanism to account for the production of Frenkel defects by ionizing radiation. In Varley's mechanism, atomic displace-

ments are produced by radiation as a consequence of multiple ionization : multiple (at least double) ionization of an anion leaves it positively charged in its substitutional site - where normally it is negatively charged. The interaction with the Coulomb field produced at this site by the neighbouring ions will tend to expel this multiply ionized anion, now positively charged, from its normal substitutional site to an interstitial site.

17. An example of such a mechanism is found in the alkali halides where the so-called radiolytic mechanism[4] is proposed to account for the creation of defects by ionizing radiation ; in this radiolytic mechanism the energy required for the formation of a Frenkel pair is given to an anion by the recombination of an exciton, the separation of the interstitial from the vacant site being achieved by a replacement sequence along a <110> direction.

18. Lax M., Phys.Rev., 119, 1502, 1960

19. Seitz F. and Koehler J.S., Solid State Physics, ed. Seitz F. and Turnbull P. Academic Press, New-York, 1956 , Vol.2, p.307

20. Watkins G.D., Symposium on Radiation Effects in Semiconductors Components, ed. Cambou F. Journées d'Electronique, Toulouse, 1968, p.A 1.

21. Mac Kay J.W. and Klontz E.E., J.Appl.Phys. 30, 1269, 1959

22. Calcott R.A. and Mac Kay J.W., Phys.Rev., 161, 3, 1967

23. Zizine J., Radiation Effects in Semiconductors, ed. Vook F.L. Plenum Press, New-York, 1968 , p.186

24. Arimura I. and Mac Kay J.W., Radiation Effects in Semiconductors, ed. Vook F.L. Plenum Press, New-York, 1968 , p.204

25. Corbett J.W., Bourgoin J.C. and Weigel C., Radiation Damage and Defects in Semiconductors, The Institute of Physics, London, 1973 , p.1

26. The levels associated with the different charge states of impurities in germanium and silicon can be found, for instance, in Hannay N.B. Semiconductors, Reinhold Pub.Corp., New-York, 1960 p.341

27. This question is reviewed in ref.1.

28. Tash A.F. and Sah C.T., Phys.Rev. B1, 800, 1970

TRACK PROCESSES

M.MONNIN

Laboratoire de Physique Corpusculaire
Université de Clermont
B.P. 45
63170 - AUBIERE   /France

INTRODUCTION

During the last decade, a considerable amount of effort has gone into the development and understanding of track processes.

Before proceeding further, one must define what, in this context, is the meaning of the word "track". In the following text, "track" will refer to the sucessive phenomena taking place along the trajectory of an ionizing particle in matter. This track will result in a continuous trail of damage. Therefore, it is understood that only heavy ionizing particles such as heavy ions ($Z > 2$, $M > 2$) are producing such tracks. In other words, this paper will be devoted to the study of the effect of heavy ions on matter.

Physicists, chemists and biologists have studied such effects. But, most of the work in this field of research has been initiated by the development of the so-called "solid state nuclear track detectors". Thus, it is through the study of those detectors that we are going to approach the study of heavy ions effect on matter.

HISTORY

Apparently, the first paper related to this topic is due to Young in 1959[1]. Lithium fluoride monocrystals were irradiated by fission fragments. Then the sample were etched by a mixture of hydrofluoric acid, acetic acid and ferric fluoride. After this treatment, the samples exhibited square shaped etched pits.

The next step is due to Silk and Barnes in 1959[2]. These authors used muscovite mica also irradiated with fission fragments[2]. The samples were viewed under a transmission electron microscope. Silk and Barnes have been able to show that every dark line observed under such conditions was due to the path of a fission fragment. During the following years, several authors observed such tracks and etched pits in mica[3,4], molybdenite[5,4], uranium dioxyde[6], silver cyanide[3], graphite[7] and metals[8,9].

The electron microscope is a powerful tool but not very easy to use. It is probable that without the discovery of Price and Walker[4] in 1962 the previous observations would not have lead to a great development (10). During their pionnering work, these authors found that fission fragments irradiated mica could be etched by hydrofluoric acid. The etching process took place along the fission fragments trajectories and consequently they could observe, under optical microscope, the track of every fission fragment. The latent tracks observed with the electron microscope were of the order of 100 Å wide. The etching process discovered by Price and Walker amplified the diameter of the track up to several microns. When the diameter was of the order of the light wavelength, the track acted as a strong diffusion center and was made visible.

The phenomenon, discovered with irradiated mica, has been extended to many different materials, inorganic as well as organic. In 1965, a handful of laboratories were working in this field. In 1969, a conference held in Clermont-Ferrand, France, gathered more 100 scientists from 20 different countries. At the present time the development and use of these detectors are the aim of even more people. The 1972 survey on track registration undergone by R.V.Griffith at the Lawrence Livermore Laboratory[11] gives the list of 118 different laboratories interested in track registration throughout the world. The study of developped tracks has provided interesting results in different fields of science especially in space physics -where very beautiful results were obtained- in nuclear physics, geology, radiography, trace analysis, radiation chemistry and biophysics. This list is obviously not exhaustive and to cite all the works performed with the aid of such detectors would take more than 21 pages -the imposed length of this paper.

SOME EXPERIMENTAL FACTS

1. Nature of the recording media

Not all the irradiated materials are able to exhibit tracks after proper etching. Only electrical insulators are suitable to be used for this track recording. Tracks have been obtained in inorganic materials such as glasses, mica, rocks (olivine, anorthosite, zircon...) and organic materials such as high polymers.

In table I[12] a list of different materials is given together with their electrical resistivity and their ability to record etched tracks.

Table I

	Materials	Resistivity (ohm.cm)
TRACKS	Silicates Glasses Polymers Molybdenite V_2O_5 glass	10^6 to 10^{20} 3,000 to 25,000 2,000 to 20,000
NO TRACKS	Silicon Germanium	10 to 2000
	Tungsten Zinc Copper Aluminium Gold	10^{-6} to 10^{-4}

It can be seen that below a 2000 Ω.cm value for the electrical resistivity no tracks are obtained. This fact may be explained by noticing that in good conductors irradiation created defects do not remain in the bulk of the sample whereas they subsist in insulators. Izui[5] claimed that he obtained fission fragments tracks in graphite but Fleischer et al[12] were unable to reproduce that experiment.

2. The basic phenomenon

When a heavily ionizing particle enters an insulating material it loses its energy through different processes mainly :

a) - generation of impurity atoms as a result of nuclear reactions

b) - atomic displacement

c) - electron displacement (ionization and excitation).

For non relativistic particles the probability of
energy transfer through a nuclear reaction is small.
Except at very low velocities of the incident particles,
the energy transfer through atomic displacement can
also be neglected. Thus, the primary mechanism for e-
nergy absorption is electronic displacement.

This energy absorption process will last for $\sim$
10^{-15} sec.[13]. Whatever are the following steps (diffusion, thermal equilibrium, chemical equilibrium), the
net resulting effect is the creation of a highly perturbated region along the particle's path. This region,
also called "latent track" will be more sensitive to
chemical reagents than the bulk material. If the irradiated sample is then immersed in a suitable reagent,
chemicals will attack both the latent track and the
bulk material. Fig.1 shows a schematic picture of the
etching process. After some time of treatment, the initial surface S_o of the sample is etched out and becomes
S_t, whereas the latent track becomes an empty channel
of conical shape. As previously said, when the channel
diameter reaches values comparable to visible light wavelengths, the light is strongly scattered and the
track is visible under optical microscope.

Two essential facts will result from the previous
description :

a) - tracks will be obtained only if the incident
particle enters the detector. In other words, no tracks
are observed if the particle trajectory begins and stops
within the sample. Indeed, in this case, there is no
way either for the etchant to reach the latent track,
nor for the reaction products to get out (unless the
medium is especially permeable, which is not generally
the case)

b) - One can define v_T as the etching speed along
the track, $v_T = dL/dt$, L being the track length and t
the time ; and v_B the etching speed of the bulk material ($v_B = 1/2 \ dE/dt$, E being the thickness of the sample
if its two faces are identical). It is obvious that a
particle whose incoming angle with respect to the sample surface is less than the critical angle θ_c such as
$\theta_c = \sin^{-1} v_B/v_T$ will not be registered. The bulk mate-

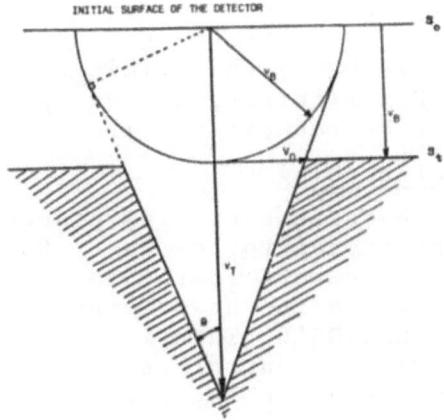

Fig.1 : Schematic picture of the etching process

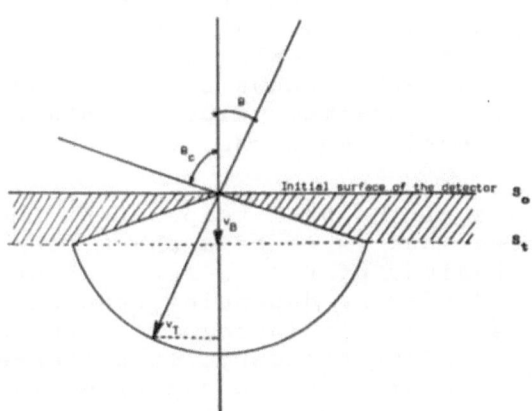

Fig.2 : No track registered when incoming particule angle is less than θ_c .

rial is destroyed faster than the track is etched. For instance, in cellulose nitrate, alpha-particles are not recorded when the incident angle is less than 35°ref.14. This kind of detector presents the greatest angular limitation. On the other hand, this limitation is only 1° for polycarbonate. For minerals, depending of the cleavage plane, it might be reduced to zero. (fig.2).

3. The chemical etching

In almost all cases, the etching process is performed in a temperature controlled bath. Sometimes, for very careful measurements (particle identification) it is required to use a bath whose temperature does not change by more than .03°C. This is achieved by using two precisely temperature controlled inner and outer baths[15]. The etching solution is stirred in such a way that no concentration gradient will appear during the etching process. The stirring allows the reaction products to be removed from the vicinity of the sample.

The ability for a detector and an etchant to form an etched track is defined by the ratio $R = v_T/v_B$. In order to obtain an etched track, it is necessary to find a reagent that gives an R value at least equal to 1. The greatest the R value, the better is the recording ability.

The increase of the reagent concentration, increases the etching speed. This statement must be considered as a naive picture. The plot of R as a function of the reagent concentration shows that for some materials the curves present a maximum value. In other case, R is continuously decreasing when the concentration is increasing[17].

The increase of the bath temperature increases the etching speed. But as we said about reagent concentration, the behaviour of R as a function of temperature is extremely different depending upon the nature of the recording material. It increases in the case of mica[18], glasses[19]. On the contrary, it decreases in the case of cellulose nitrate[20]. In addition, in polymeric materials, it is important that the temperature is above the first glass transition temperature. If it is lower, it is sometimes impossible to get a track[21,22].

Obviously, the three etching parameters are not independent. To a given set of concentration and temperature corresponds a given time for the track of a given particle to etch completely.

In table II the reader will find a limited list
of some insulators in which tracks have been obtained
together with the corresponding etchant. It should be
pointed out that for a given detector the user has to
set the etching parameters depending upon the purpose
of the experiment and the nature of the recorded particle.

The detailed mechanism of chemical etching is still
rather unknown. Since the latent track is much beyond
the sensitivity of the optical microscope, no information has been obtained by this technique. At least in
organic media, the use of the transmission electron microscope has provided a view of the etching process[16].
In a first step, the duration of which is a function
of the linear energy transfer of the particle (LET),
nothing happens. The greater the LET, the longer is
this first step. Then a dark line (50 to 100 Å wide)
appears. It corresponds to the diffusion of the reagent
in the latent track. Eventually, the channel i-e the
etched track begins to form.

Since diffusion processes seem to be involved in
the etching mechanism, it is expected that untrasonic
beam applied during the etching will increase the R value together with the etching speed. This behaviour has
been observed with different substances, inorganic[23]
and organic[20,24,25] as well. A 50 % increase of R can
be achieved.

4. Detection sensitivity

As was previously pointed out in the first section
there is no hope, at the present state of our knowledge, to get any track of particles lighter than alpha
particle or, at extremum, proton. In general, inorganic
materials are less sensitive than organic ones. Even-
though alpha particles have been recorded in some glasses and mica, usually only very heavy ions such as Argon or Iron are registered. On the other hand, on average, high polymers are able to record from alphas to
fission fragments ; but, inorganic materials are much
less sensitive to thermal annealing of tracks[10,26,27,28]

5. Detection energetic threshold

During the early period of study of solid state
nuclear track detectors, it was found that the particles
were registered only if their kinetic energy was below
a critical value (fig.3). This critical value was called the detection threshold. Recent work showed that
this threshold was probably an artifact[29,30]. In the

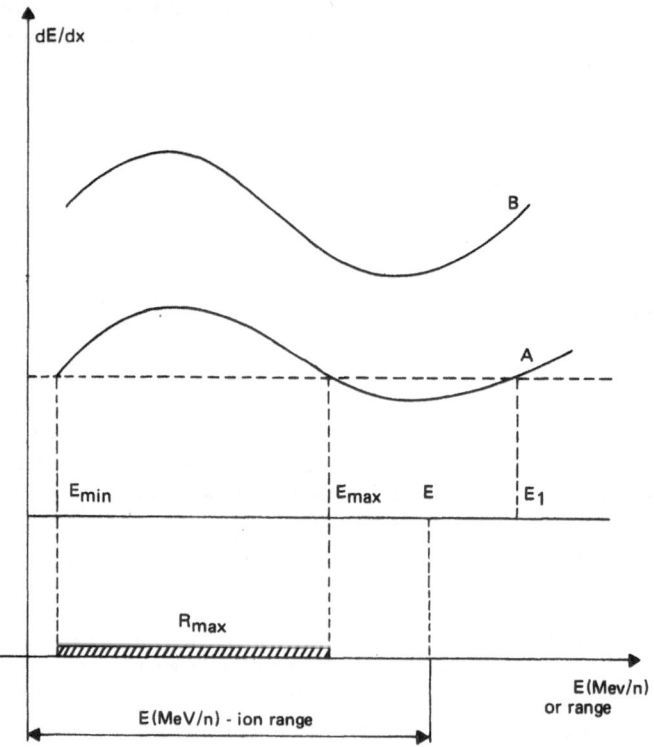

Fig.3 : "Detection threshold" for
particules.

following sections we will give more details about this
behaviour while discussing the various models of heavy
ions track formation.

Table II

Detectors	Etchant
Feldspar	NaOH
Mica	HF
Apatite	HNO_3
Glass	HF
Polystyrene	$K_2Cr_2O_7$
PVC	$KMnO_4$
Cellulose compounds	NaOH
Polyimide	NaClO
Polycarbonate	NaOH
PMMA	$KMnO_4$
Polyethylene	$K_2Cr_2O_7$

TRACKS FORMATION PRIMARY MECHANISMS

Almost since the discovery of solid state nuclear
track detectors, different scientists have tried to
build a model of track formation mechanism. At least
one common peculiarity of these models is that they
are expected to give a satisfactory explanation and to
be able to make correct prediction by looking only at
the energy loss mechanisms. It is a very questionnable
point of view. We would like to point out here that
between the moment the track is obtained, the medium
suffers a lot of different perturbations and rearrage-
ments. The etched track is the result of several successive steps :

a) - the energy loss by the impinging particle
(10^{-15} sec)
b) - transport and deposition of energy within the
medium (10^{-12} to 10^{-9} sec.)
c) - chemical equilibrium. This stage may be extended by diffusion processes and reaction rates
from a minimum of about 10^{-8} sec. up to years
in certain solids[13].
d) - the chemical etching process.

Table III shows an outline of the fundamental processes in latent track formation in organic media[13].
The same kind of speculation could be offered about i-
norganic media. This table is an evidence of the great
number of possible final products resulting from the
heavy ion irradiation. Therefore, it is not surprising

$(t\sim 10^{-15}\,\text{sec})$ ENERGY DEPOSITION
Electron displacement

M^* M M^+ e^-

EXCITATION

M^* — Radiation transition $M^* \to M + h\nu$

Radiationless transition $M^* \to M + KT$

Energy transfer $M^* + A \to M + A^*$

Bond Rupture — heterolytic cleavage $M^* \to A^+ + B^-$
Stable molecule $M^* \to A + B$

free radicals $M^* \to R_1^\bullet + R_2^\bullet$

Abstraction $R_1^\bullet + M \to A + R_2^\bullet$

Combination $R_1^\bullet + R_2^\bullet \to R_3$

Molecule reaction $M^* + M \to MM^*$?

IONIZATION

M^+

charge transfer $M^+ + A \to A^+ + M$

electron capture $M^+ + e \to M^*$

Bond Rupture:
Radical + ion $M^+ \to R^\bullet + N^+$

Radical ion + stable molecule $M^+ \to R^{+\bullet} + N$

Hot molecule ion reaction $M^+ + M \to MM^+$

Neutralization $M^+ + e^- \to M^*$

e^-

trapping $e^- + \square \to e^-_\square$

Neg. ion $e^- + M \to M^-$

10^{-12} to 10^{-9} sec

Luminescence, Heat

Pot. En. of reactive species

$(t > 10^{-8}\,\text{sec})$

Potential Energy dissipated-Establishement of Chemical Equilibrium

untrapping $e^-_\square + KT \to e^- + \square$

Table III

that tentative explanation of track formation mechanism based upon energy loss of the particle were not completely satisfactory. On the contrary, the amazing fact is that they were so much successful and provided a very fruitful guideline. In the following sections we are going to review the several models that have been proposed. Since inorganic materials and organic materials are very different in their structure some models are especially devoted to one kind or to the other kind of material.

1. Inorganic media

A. Thermal spike

This hypothesis is probably the first attempt of explanation of latent track formation[31,32]. When a target atom receives an energy E_0 less than the displacement threshold, it is not displaced. The energy is converted into vibrational motion. Then the surrounding atoms undergo vibration also from the energy transferred to them by the initial target atom. This phenomenon has a short duration because the energy is rapidly diluted in the medium. Everything happens as if a small sphere was suddenly heated up. More precisely, the electrons of the target atom are displaced through coulombian interaction. In turn, these electrons transfer their energy to the surrounding atoms through electron-phonon interaction. The thermal equilibrium does not exist. The temperature variation ΔT as a function of time t can be written :

$$\Delta T = D^{-1} \frac{\partial T}{\partial t}$$

D being the thermal diffusion constant.

One can obtain the temperature T (r,t) at distance r from the target atom :

$$T(r,t) = \frac{E_0}{(4\pi)^{3/2}} \frac{D}{C} (Dt)^{-3/2} \exp\left(-\frac{r^2}{4Dt}\right)$$

where C is the thermal conductivity of the medium.

When the target atom receives an amount of energy greater than E_0, it is displaced. Then a cascade reaction occurs. Several thermal spikes as previously described are created. An approximate calculation is possible. It can be shown[33] that an energy transfer of 300 eV yields a temperature raise of about 1000°C in a 30 Å diameter sphere within 5×10^{12} sec., in metal.

372

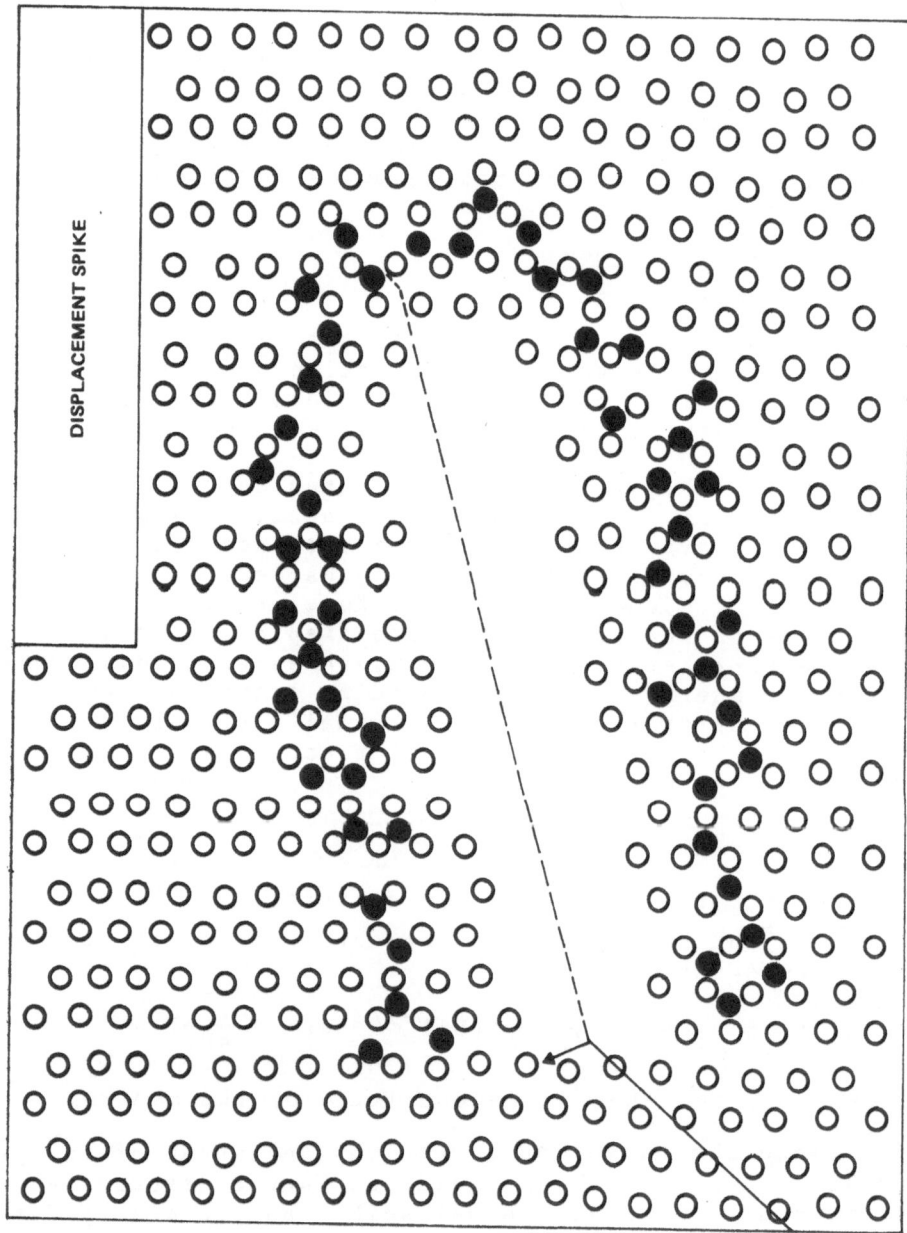

Fig.4 : Displacement spike along incident particule
trajectory.

The cooling time is very short (10^{-11} sec).

If one considers now a particle travelling through the medium, the heated spheres overlap : it is a cylindrical thermal spike. The temperature is :

$$R(r,t) = \frac{LET}{4\pi Ct} \exp(-\frac{r^2}{4Dt})$$

It is possible to define the radius of the latent track as the radius r_0 of the cylindrical area around the ion path where the temperature is greater than the thermal decomposition temperature T_D of the material[3]. For different materials, r_0 is proportional to $T_D^{-1/2}$.

According to the thermal spike theory, semi-conductors that suffer a greater damage density -since r_0 is smaller- than insulators should record tracks : they do not[12]. In addition, it has not been possible to find any relation[10] between T_D and the latent track radii as it is predicted by the thermal spike model. Hence, it does not seem that this model applies correctly to explain track formation.

B. Displacement spike

An attempt has been made to apply the displacement spike mechanism previously proposed by Brinkman[34,35] about the creation of defects. Along the incident particle trajectory some atomic collisions occur. Each collided atom initiate a cascade process. Consequently 3 or 4 times more atoms are concerned than the number of primarily collided atoms. The mean free path of the primary atoms is less than the distance between atoms in the atomic layer. Hence, only atoms very close to the primary atom are displaced. These atoms are placed in interstitial positions and a lacuna is created. See figure 4.

Unfortunately this simple mechanism is contradictory to experiments. If elastic atomic collisions were responsible for the track formation, tracks should be obtained in metals as well ; they are not. The atomic collision cross section increases when the energy of the incident particle decreases. Displacement spike should be more important at the end of the range. This has not been observed. In mica, an alpha particle dose 20 times higher a given 2 MeV/nuc-Ne ion dose should produce the same effect. This is wrong. Even up to 500 times, alphas do not produce the same effect as Ne-ion do. Therefore, this model is not convenient either to explain latent track formation.

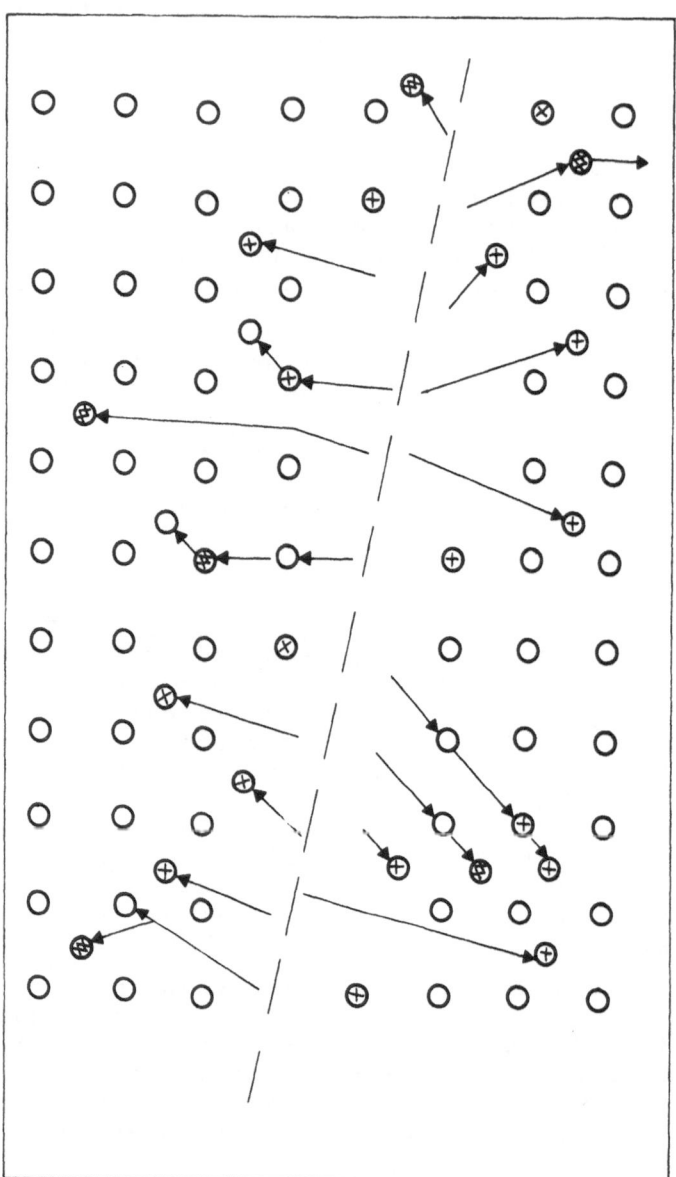

Fig.5 : Ion explosion spike.

C. Ion explosion spike

A more sophisticated theory has been proposed by Fleischer, Price and Walker[12,36]. Tracks are supposed to be "the result of defects created by the removal of the electrons. A narrow cylinder which is densely filled with excess positive ions is thereby created. These positive ions strongly repel one another and are ejected into interstitial positions surrounding a now depleted core region"[12]. Fig.5 is a schematic diagram indicating the resultant ion displacements but not the possible subsequent rearrangements. This model is based upon four necessary conditions.

a) - the electrostatic repulsive force has to be greater than mechanical strength (or bonding strength). The force between two multiple charged ions (charge number n) at an interaction distance a from each other is :

$$(1/\varepsilon)(n^2 e^2 a^{-2})$$

where e is the electron charge and ε the dielectric constant. Then, the local force per unit area is

$$(1/\varepsilon)(n^2 e^2 a^{-4})$$

On the other hand, the mechanical tensile strength σ of a material of Young's modulus E is approximately $(1/10)E$. The first necessary condition can be written as

$$n^2 > R \equiv E \, \varepsilon \, a^4/10 \, e^2$$

R is called "stress ratio" by the authors.

b) - in order to have a continuous trail of damage one needs at least 1 ionization per atom plane crossed by the bombarding particle. This condition is written

$$n > 1$$

c) - the time for electrons to diffuse into the ionized region must be greater than the time required by the ions to get into interstitial position. Indeed, if other electrons where able to replace those ejected by the primary particle the ion explosion spike could not take place. Let n_e be the density of free electrons and n_a the number of ionizations per atomic plane. If the ionized region is considered as a cylinder of radius r, this radius is given by

$$\pi r^2 \ a \ n_e = n_a$$

The time for electrons to diffuse a distance
r is r^2/D, where the diffusion constant D is
given by the Einstein relation $\mu kT/e$, k being
Boltzman's constant and μ the electron mobility. Thus, the third condition is written

$$n < n_a e/\pi \ a \ \mu \ k \ T \ t$$

d) - the hole mobility μ_p must be small enough so
that they do not move away and suppress permanent track formation. To be inefficient this
move must be less than one inter atoms distance. Therefore

$$\mu_p < a^2 e/k \ T \ t$$

will be the way to write the 4th condition.
This relation requires that at room temperature
track will not appear in materials whose hole
mobility is more than about $.2 cm^2.V^{-1}.sec^{-1}$
such as metals and most of the semi-conductors.

According to the authors, this model is "greatly
over simplified". However it is striking to note the
good agreement between experiences and predictions.
The chemical reactivity of the perturbated area is the
same as observed in regions formed with holes and ions
in interstitial positions. Metals and semiconductors
are not able to record tracks because of their great
hole mobility. The widths of latent track observed by
electron microscopy are compatible with the ion explosion spike model. The detection threshold varies in
the same way as the stress ratio.

However, the ion explosion spike is not convenient
to explain track formation in polymers. Alpha-particles
track have been obtained in several polymers. According
to the model, in this case, the distance between displaced atoms would be too large to produce an etchable
track. In other words, at the present moment, the Ion
Explosion Spike model is probably the best as far as
inorganic materials are concerned. On the other hand,
one has to look in a different direction for models
explaining track formation in organic media.

2. <u>Organic Media</u>

A. <u>The LET$_{crit.}$ model</u>

As described by Benton at the Clermont conference,
the first proposed criterion postulated that for etchable track formation the total rate of energy loss by

the bombarding particle must exceed a critical value which is characteristic of the detecting material[37]. This is a single adjustable parameter criterion. It was found to work over a narrow region of particle velocities. It failed when applied to relativistic particles; the reason being that energetic delta-rays tend to be multiply scattered away from the latent track region.

B. The primary ionization model

This model postulates that an etchable track will appear if the rate of primary ionization exceeds a critical value which is characteristic of the detecting material[38]. The primary ionization is calculated through the formula of Bethe :

$$\frac{dI}{dx} = \frac{aZ^{*2}}{I_0\beta^2} \left(Log \frac{2\ mc^2\beta^2}{(1-\beta^2)I_0} - \beta^2 + 3.04 \right)$$

where I_0 is the ionization potential of the outermost shell of the stopping atoms, m the electron mass, a a constant characteristic of the medium.

Two adjustable parameters are used : I_0 and the critical value PI_{crit} of primary ionization. After proper adjustments, the model fits the experimental data very nicely. The detection threshold as a function of energy is correctly predicted and in addition the track etching speed v_T as a function of PI allows to perform heavy particle identification. Unfortunately a value of $I_0 = 2$ eV must be utilized in order to fit the data. This value appears to be low to fit the physical meaning given to it by Bethe. We believe that the PI model should be considered more as a very convenient tool than as a realistic description of the latent track formation mechanism.

C. The delta-Rays models

Several authors have tried to present comprehensive models based upon the energy density distribution around the heavy ion trajectory. All these models consider that the secondary electrons (delta-rays) ejected by the primary particle play a prominent role in the heavy particle effect on matter. Some of these models have been inspired by track studies (Benton, Katz, Monnin), others have been initiated by radiation chemistry studies (Mozumder, Chatterjee). Many points are common to these models. Instead of reviewing them one after each other, we intend to take a look at the more recent works by Chatterjee et al[39] and by Faïn et et al[40],[41]. The other models are contained in the later. The two models have been independently developed. They start from different premises. They do not use any adjustable parameter. Thus, it is worth mentioning that the agreement between the two models is rather remarkable.

The first question that can be raised is to figure out what is the average distance of action of the primary particle. According to Bohr's impulse principle, the maximum distance at which an electron can be perturbated by the primary particle electric field is $r = \hbar v/2E$ where v is the primary particle velocity and E the first allowed electronic transition. This distance can be extremely large when the particle velocity is high. But very few electrons will be perturbated at this distance. More important is to know the average distance of primary energy deposition. A very rough estimate of this geometrical dimension can be obtained by using uncertainity principle and through the use of atomic excitation cross section. The differential excitation cross section for an atom raised from its ground state to the energy level n corresponding to a transfer of momentum q is given by the Bethe formula

$$d\sigma_n = \frac{4\pi z^2 e^4}{v^2} \; |\varepsilon_n(\vec{q})|^2 \; \frac{d(q^2)}{q^4}$$

where z and v are respectively the charge and velocity of the incident particle. $\varepsilon_n(\vec{q})$ is given by

$$\varepsilon_n(\vec{q}) = \int \bar{u}_n(\vec{r}) . \exp(i\vec{q}.\vec{r}/\hbar) \; u_o(\vec{r}) \; d\vec{r}$$

u_o and u_n are the initial and the final wave function of the bounded electron in the atom ; r is the electron coordinate. The main value of q^2 is :

$$<q^2> = \int |\varepsilon_n(\vec{q})|^2 \; \frac{dq}{q} / \int |\varepsilon_n(\vec{q})|^2 \; \frac{dq}{q^3}$$

Let d be the average atomic radius. The wave functions u_n and u_o decrease suddenly when $r > d$. Thus $\varepsilon(\vec{q})$ becomes vanishingly small for value of $q > \hbar/d$. The contribution of momentum transfer for distances greater than $\hbar/d$ to the total cross section is small. In view of the smallness of the term $q \times r/\hbar$, one can write :

$$|\varepsilon_n(\vec{q})|^2 = |\int \bar{u}_n \, (1+i\vec{q}\vec{r}/\hbar) \, u_0 \, d\vec{r}$$

$$= q^2 \, |\int \bar{u}_n \vec{k}\vec{r}/\hbar \, u_0 \, d\vec{r}|^2$$

$$= q^2 \cdot g_n$$

In the above expression k is an arbitrary unit factor whose choice cannot modify the value of g_n. Since g_n is only related to the atomic structure, the average value of q is given by

$$<q^2> \int_{q_m}^{\hbar/d} q \, dq / \int_{q_m}^{\hbar/d} dq/q = (\frac{\hbar^2}{2d^2} - q_m^2) \, Log \, \frac{\hbar}{dq_m} \quad (1)$$

where q_m is the minimum value of q given by $\Delta E/\beta c$. ΔE is the excitation energy. For $\beta > .03$, $q_m^2 << (\hbar/d)^2$ and this results into

$$<q^2> = \frac{1}{2} \frac{\hbar^2}{d^2} / \, Log(\frac{\hbar}{d} \cdot \frac{v}{\Delta E})$$

The mean quadratic projection of q on an axis perpendicular to the direction of propagation of the incident particle is given by

$$<q_x^2>^{\frac{1}{2}} = 2^{-\frac{1}{2}} <q^2>^{\frac{1}{2}}$$

Applying Heisenberg's principle involving position and momentum, the root mean square of uncertainty of the site of excitation around the ion path is estimated as

$$<\delta_\rho^2>^{\frac{1}{2}} = 2^{+\frac{1}{2}} <\delta x^2>^{\frac{1}{2}} \approx \hbar/(2^{1/2} <q_x^2>^{1/2})$$

If $d \sim 1$ Å and $\Delta E \sim 5$ eV then for incident particle energy lying between. 1 MeV/nuc and 1000 MeV/nuc the variation in logarithmic term of equation I ranges between 1 and 5. Therefore the mean radius of the primary energy deposition is of the same order of magnitude as that of the atomic radius.

This result implies that secondary electrons must play an important role in the heavy track processes.

In the Chatterjee's model one starts from the equipartition between knock-on and glancing collisions energies. This is justified by experimental facts such.as high energy channeling experiments. In the

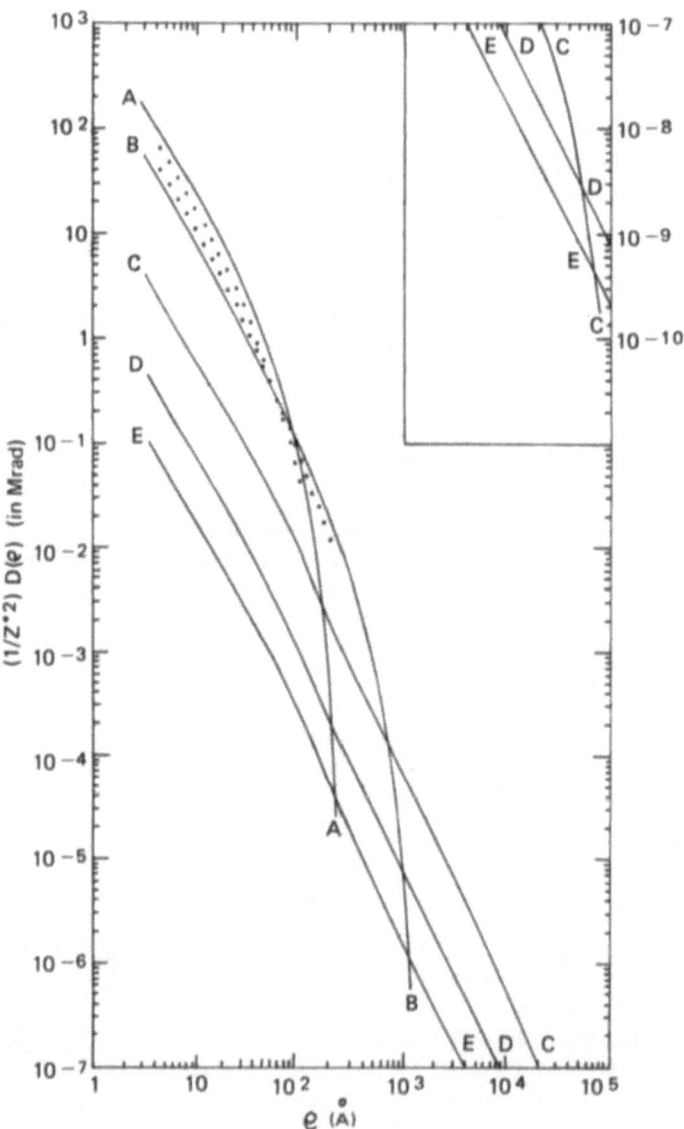

<u>Fig.6</u> : Distribution of energy density deposited by
secondary electrons.

Faïn et al.'s model one calculates the secondary electrons energy spectrum using Gryzinski's theory[42] refined by Gerjuoy and Vriens[43,44]. This means that electrons are not considered as free electrons but as binded electrons. Then, using the angular distribution of secondary electrons, their scattering cross sections and their range down to 10 eV electrons based upon experimental data one gets the distribution of energy density deposited by secondary electrons (fig.6). Experimental points from Wingate et al[45] data are also on the same plot, for 1 and 3 MeV alpha-particles.

The models allow to calculate the partition of energy deposition between excitation and ionization, and the W values (energy needed to create an ion pair) as a function of radial distance.

This kind of calculation gives a rather peculiar picture of heavy ion track. At very short distances from the ion path, the energy density is extremely high then decreases very fast at greater distances.

In his formulation known as "Restricted Energy Loss", Benton[46] chose an adjustable parameter w_0. w_0 is the maximum energy of secondary electrons that are efficient. Electrons of kinetic energy greater than w_0 are supposed to leave their energy to far from the ion path and hence do not contribute to the track formation.

Katz[47] supposes that a minimum energy density is needed in a minimum diameter core for an etchable track to be formed. But in his calculation this author has introduced binding energy as a variable parameter and has extrapolated the energy deposited by high energy electrons (greater than 10 keV) to low energy region.

It must be pointed out that these models intend to describe the track structure at the earliest step of its formation ($\sim 10^{-15}$ sec.). Therefore it would be surprising if they would be able to predict very accurately the experimental observations done much later. Particularly, these models ignore the rearrangements, excitons interactions, radical-radical reactions, diffusion phenomena, etc. ion recombination molecular dissociation processes. These models must be considered as the first steps of a complete theory of heavy ion track processes.

However, despite their uncompleteness, these models have lead to interesting developments and shown that heavy ions do not behave at all as low LET particles do. Consequently, heavy particles will turn into very valuable tools in different fields of research.

It should be remembered that :

- heavy ions exhibit very little or no scattering
- precises range-energy relationships (no straggling) are available
- linear energy transfer depends as Z^2 and hence is rapidly increasing with particle charge
- as heavy ion slows down and eventually comes to rest it gives a wide variation of LET
- the Bragg's peak to plateau ratio is very important
- the spatial distribution of energy density around heavy ion path is very high and covers a wide spectrum of energy densitites. Particularly, close to the ion path, energy densities not available by other means are deposited during a very short period.

We would like to describe a technique that is a direct consequence of the previous studies and several other applications either actual or potential.

POLYMER GRAFTING TRACK TECHNIQUE

According to the previous calculations, at distances from the ion path less than 300 Å, the energy density due to a 1 MeV-alpha particle is at least of 10,000 rad. On the other hand it is known[48,49] that free radicals and peroxydes functions are created by heavy ion irradiation. In addition, experiments have shown[50,51] that electron bulk irradiation of polymers at doses as low as 10,000 rad are sufficient to initiate polymer grafting using the radicals or the peroxydes groups. The similarity of the situation is striking. It must be possible to graft a monomer B in the region surrounding the ion path in a polymer A_n = A-A-A-A-A-.... .The experiment has been succesfully achieved using cellulose triacetate (polymer A_n) irradiated by fission fragments. The monomer employed was pure propenoic acid. The grafting reaction is performed at 55°C under vacuum. It results in polymer B chains grafted onto polymer A_n chain. Since the monomer B is acid, one can take advantage of its chemical properties to visualize the ion path. A basic fluorescent dye (rhodamine B) is reacted and fixed onto the B_n chains, and only there. Then the sample is viewed under microscope with ultraviolet light illumination. The tracks are seen as bright lines among a dark field of view (see picture). The efficiency of this technique is a 100 % for fission fragments. The principle of this method allows   to

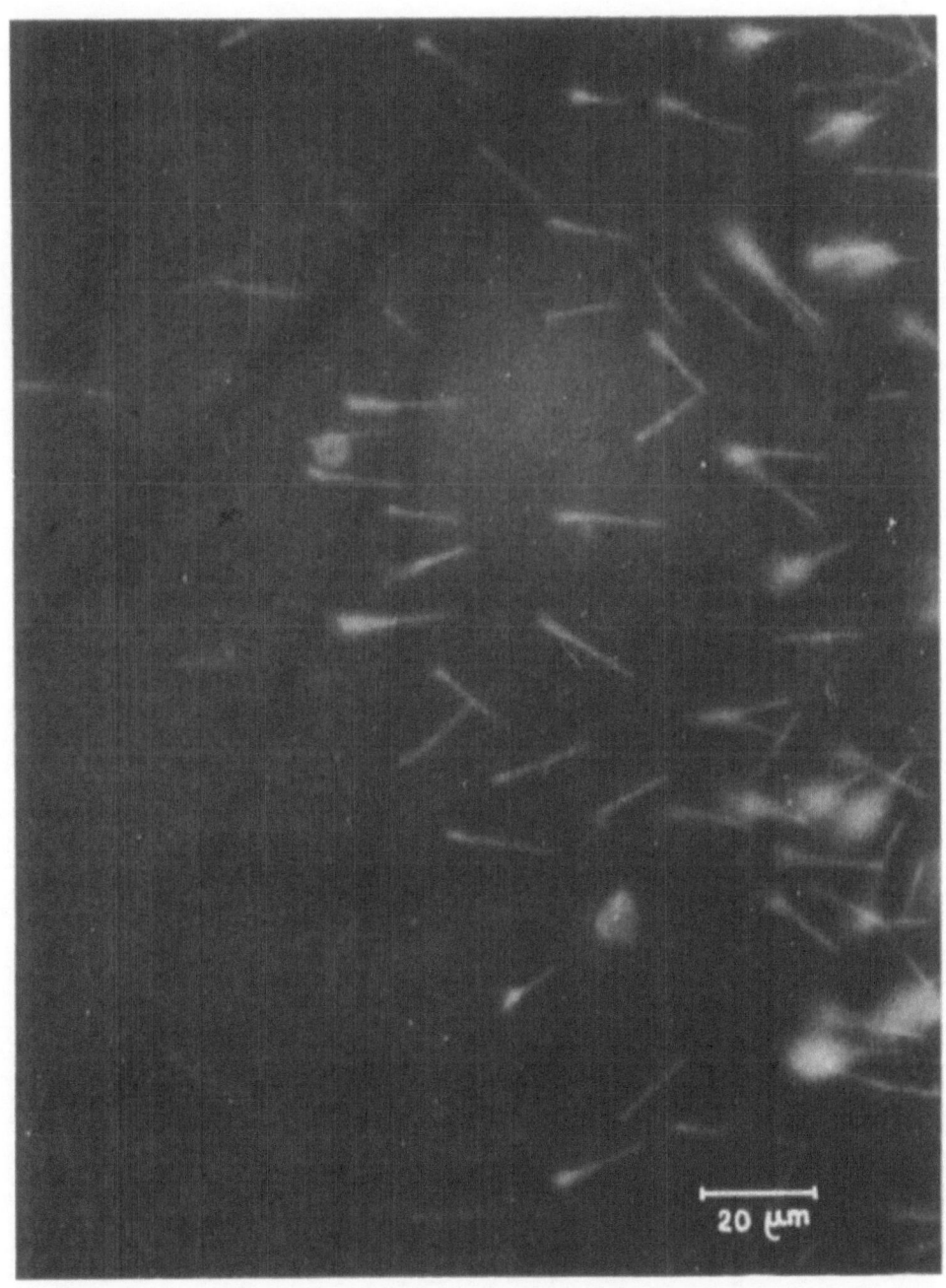

20 μm

Fig.7 : Fission fragment tracks obtained by the
polymer grafting technique.

expect a higher sensitivity than the now classical solid state nuclear track detectors. Before it can be developped into a valuable method much work has to be
done. But it is worth trying because this technique
does not destroy the surrounding material ; tracks can
be revealed in materials highly sensitive to radiation
damage such as polytetrafluorethylene (Teflon) ; no
angular limitation is expected ; automatic measurements
should be possible...

OTHER APPLICATIONS

- The technology of fast breeder reactors request
highly radiation damage resistant materials. Heavy ions
can be used for material testing because they are very
damaging and obtainable at high fluxes. Thus experimental irradiations which would take a year or two in a
classical reactor can be done in a day.

- Heavy ions have a large X-ray fluorescence cross
sections ($\sigma > 10,000$ barns). Analysis of trace elements
can be rapidly performed with a high sensitivity.

- Modifications of physical properties are also
possible. A great deal of work has been done in order
to increase the critical current of type II semiconduc-
tros like $Nb_3(Al,Ge)$. Semiconductors of n-type can be
converted into p-type and vice versa.

Alloys may undergo phase diffusion and therefore
get new interesting properties.

- Since heavy ions produce highly ionized regions
they can possibly be used as triggers for plasma studies.

- The classical chemistry may also benefit from
heavy ions tracks structure. The development of contemporary chemistry is partially due to utilization of
techniques under extreme conditions : very high or very
low temperatures, high pressures... Heavy ions are able
to produce deeply highly ionized atoms and therefore o-
pen a new way of research for the rare gases and non-
metals chemistry. The same is true since recent studies
have shown the existence of compounds containing cations coming from high ionization potential molecules
(O_2^+, I_2^+ ...)

The chemistry of excited species gives information about reaction processes. Indeed the reaction products are different from those obtained from species
in their ground state (singlet state carbon atoms, excited oxygen molecules...).

- Particularly at the Universtiy of California, very interesting work has been performed in the field of radiobiology. Dr Tobias's group has developped a valuable experimental method in tumor therapy. The heavy ion beam irradiated tumor in such a way that the Bragg peak occurs within the tumor. Therefore, the tumor receives a great amount of energy whereas the surrounding tissue is not or almost not irradiated. In addition the oxygen enhancement ratio is equal to zero, in the case of heavy ions. It is known that cells deprived of oxygen are much more radiation resistant than oxygenated cells. The core of a tumor is more resistant and hence is usually more difficult to destroy. This is no longer true with heavy ions that are as efficient with or without the presence of oxygen.

- The same group has succesfully explained the phosphene phenomenon observed by astronauts during interplanetary flights. While having eyes closed astronauts saw flash lights or strikes. It has been shown that this phenomenon is due to heavy cosmic ray striking the retina cells.

CONCLUSION

It is impossible to make reference to all the works that have been performed in this field. We ask authors whose interesting discoveries have not been mentionned to use their magnanimity.

Our aim was essentially to show how heavy ionizing particles are different from low LET particles. Several different models have been proposed. None of them are completely satisfactory but all of them bring a stone to build a complete theory. Various experiments are still needed to clarify the situation. It is undoubtful that heavy ions beams will open new unsuspected doors.

ACKNOWLEDGEMENT

We are indebted to Miss Janet COOK who unselfishly devoted much of her time to correct the manuscript.

REFERENCES

1. Young D.A., Nature, 182, 375, 1958

2. Silk E.C., Barnes R.S., Phil.Mag. 4, 970, 1959

3. Bonfiglioli G., Mojoni A., J.Appl.Phys., 32, 2499, 1961

4. Price P.B., Walker R.M., J.Appl.Phys., 33, 3400, 1962

5. Ifui K., Fujita F.E., J.Phys.Soc.Japan, 16, 1779, 1961

6. Bierlien T.K., Master B., J.Appl.Phys., 31, 2315, 1961

7. Ifui K., Fujitz F.E., J.Phys.Soc.Japan, 16, 1032, 1961

8. Kelsch J.S., Kzmmerer O.F., Buhl P.A., Appl.Phys. 11, 555, 1960

9. Silcon J., Hirsch P.B., Phil.Mag. 4, 1356, 1959

10. Price P.B., Walker R.M., J.Appl.Phys., 33, 3400, 1962

11. Griffith R.V., Survey on Track Registration, Lawrence Livermore Laboratory, University of California, 1972

12. Fleischer R.L., Price P.B., Walker R.M., J.Appl. Phys. 36, 3645, 1968

13. Benton E.V., Int.Conf.on Track Registration, Clermont-Fol, 1969 - Isabelle Monnin -publ.

14. Debeauvais M., Ve Conf. Int. Phot.Corp. Florence 1966

15. Blanford G., P.h.D, Thesis, Washington University St-Louis Missouri, 1971

16. Tripier J., Debeauvais M., Stein R., Relarosy J., J.Microscopie, 7, 811, 1968

17. Somogyi J., Varnagy M., Medvécfky L., Int.Conf. on Track Registration, Clermont-Ferrand 1969, and Radiation Effect.

18. Price P.B., Walker R.M., J.Appl.Phys., 33, 3407, 1962

19. Becker K., Rapport U.S., NRDL, TR 904, 1965

20. Benton E.V., Henke R.P., Nuel.Instr. and Methods, 70, 183, 1969

21. Monnin M., Radioprotection II, 2, 105, 1967

22. Monnin M., Parizet M.J., Int.Conf. Clermont-Ferrand and Rad.Effect, 1969

23. Maurette M., PhD Thesis University of Paris, 1965

24. Anno J., 3rd cycle thesis, University of Toulouse 1970

25. Monnin M., Sanzelle S., Chave A., Menigot M., Journées d'Electronique Toulouse, Tome III, 1968, University of Toulouse Publ.

26. Maurette M., Bull.Soc.Franc.Mineral.Crist., 89, 41, 1966

27. Maurette M., J.Phys. et le Radium, 27, 505, 1966

28. Pelelygin V.P., Tretiakova S.P., Shadieva N.H., Cieslak E., Intern.Conf. on track registration, Clermont-Ferrand, 1969

29. Price P.B., Lal D., Tamhane A.S., Perelygin V.P., Characteristic of tracks of ions in common rocks silicates, Sub itted for publication Jan. 5th., 1973

30. Bibring J., Borg J., Dran J., Maurette M., Meunier R., Peters J., Walker R., Détecteurs minéraux et traces d'ions lourds de très faible énergie, 1973. to be published

31. Chadertton L.T., Rad.Damage in cristals. Methnel & Co London, 1965

32. Chaderton L.T., Mc Torrens I., Fission damage in cristals. Methnen & Co. London, 1965

33. Dienes G.J., Vineyard G.H., Radiation Effects in Solids. Intern.Science Pub. New-York, 1957

34. Brinkman J.A., J.Appl.Phys. 25, 961, 1954

35. Brinkman J.A., Amer.J.Phys. 24, 246, 1956

36. Fleischer R.L., Price P.B., Walker R.M.,Ann.Rev.Sc. USA 15, 1, 1965

37. Fleischer R.L., Price P.B., Walker R.M., Hubbard E, Phys.Rev. 133, A, 1443, 1964

38. Fleischer R.L., Price P.B., Walker R.M., Phys.Rev. 156, 353, 1967

39. Chatterjee A., Maccabee H., Tobias C.A., To be published in Radiation Res., 1973

40. Fain J., Monnin M., Montret M. Paper submitted to

Rad.Res., 1973

41. Fain J., Monnin M., Montret M., Int.Conf. on Micro-dosimetry Stresa, 1972

42. Grysinski M., Phys.Rev. 138, A, 336, 1969

43. Gerjuoy E., Phys.Rev. 148, 54, 1966

44. Vriens L., Proc.Phys.Soc. London, 90, 935, 1967

45. Wingate C.L., Baum J.W., Brookhaven Nat.Lab. Repp. 14767. 1971

46. Benton E.V., PhD Thesis, Stanford California 1968

47. Katz R., 7th.Int.Conf. on corpuscular Photography Barcelona 1970, P.Cuer Pub.

48. Henke R.P., Benton E.V., Int.Conf. on track registration. Clermont-Ferrant 1969

49. Chambaudet A., Roulet H., C.R.Acad. Sc. Paris, 274 B, 145, 1972

50. Chapiro A., Radiation Chemistry of Polymeric Systems. John Wiley & sons, New-York 1972

51. Bonnefis J.C., Thesis Paris 1969

52. Monnin M., Blanford G., Science 181, 743, 1973

FORMATION OF AMORPHOUS MATERIAL

E.DARTYGE

Laboratoire de Physique des Solides[*]
Bâtiment 510 - Univèrsité PARIS-Sud

91405 - ORSAY /France

PRELIMINARY : DEFINITION OF AN AMORPHOUS STATE - DIRECT METHODS OF OBSERVATION

This lecture is restricted to inorganic materials.

When melts or solutions of compounds are cooled to form a solid state, generally the atoms or ions which compose the solid are regularly packed in the three directions of space. The electronic density of the crystal is a triply periodic function of space, defined by three vectors a, b, c on which a unit cell is built. Real crystals possess different types of defects (e.g. point defects, clusters, precipitates, dislocations) and may have small sizes but in any of these cases, the long range order is preserved. When a long range order does not exist in a material, i.e. it can't be defined a unit cell, the material is amorphous.

The direct methods of observation of crystalline or amorphous states are X rays, electrons or neutrons diffraction. With electrons, it can be observed materials of $\sim 1\mu$ dimension, with X rays, of ~ 0.5 mm and with neutrons ~ 1 cm.

By diffraction experiments, it is possible to make a difference between very little crystals and amorphous materials[1] : a powder of crystals of ~ 100 Å di-

---

[*] Laboratoire associé au CNRS

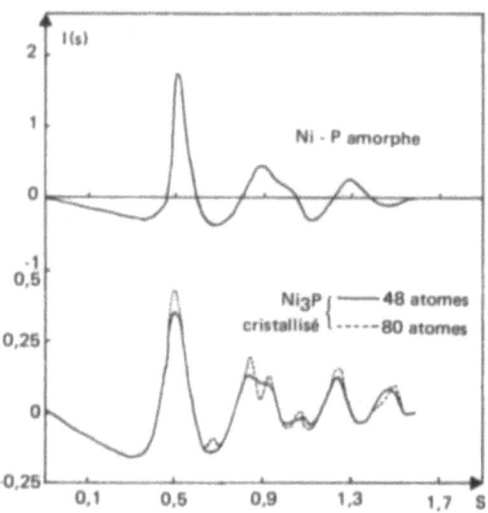

<u>Fig.1</u> : Interference function I(s) of 2Ni$_3$P models made respectively of 48 and 80 atoms, compared with experimental interference function of Ni-P amorphous.

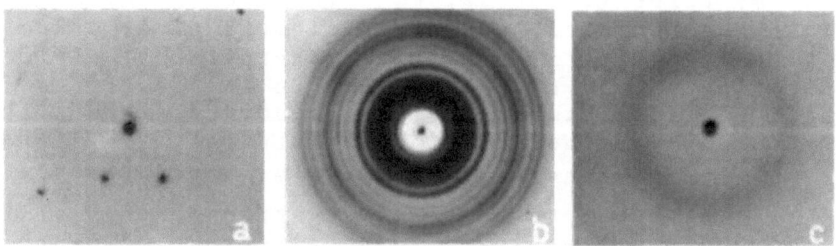

<u>Fig.2</u> : X ray diffraction photograph of 3 samples of muscovite mica :

  a) monocrystal without treatment

  b) powder of fine crystals of decomposed mica, obtained by heating the specimen to 1400°C

  c) radiation - amorphised mica.

mension gives lines of X ray diffraction broader than
that expected by the definition of the device, whereas crystals of greater size (1000 Å) give fine diffraction lines. And even for dimensions smaller than 100 Å
there exists a difference between the diffraction pattern given by a powder of 50 atoms crystallites and
that given by a completely amorphous material. Fig. 1
illustrates this difference in the case of NiP alloys.

It can be seen that a 50 atoms crystal gives more
diffraction lines than an amorphous material. Fig. 2
is an example of the different kinds of X rays photograph obtained with a single crystal, a powder of little crystals and an amorphous material.

COMPARISON OF DIFFERENT MATERIALS

    1. Comparison of Radiation Damage in Different
       Materials by High Doses

    a) - Metals

Metals and metallic alloys can't become amorphous
by irradiation. The main effects produced are creation
of voids[2] and swelling of metal. Only alloys composed
with transition metal plus ∿ 20 % metalloïd become a-
morphous[3] by irradiation with fission fragments. But
the introduction of Si or any covalent metalloïd adds
covalent bonds to the metal.

The defects produced by irradiation are somewhat
similar to those produced by a rapid quenching from
liquid state. If it is possible to obtain an amorphous
alloy by rapid quenching, it should be possible to obtain amorphisation of the same alloy by irradiation[3].

    b) - Ionic crystals

The defects produced by irradiation in ionic crystals present radiation defects similar to those obtained in metals : with high doses there is a formation of
voids and precipitates. As an example, LiF irradiated
by thermal neutrons has been studied with small angle
scattering of X rays[4] : the α and H[3] particules produced by $Li^6(n_1, \alpha)H^3$ reaction are the origin of the observed defects ; different kinds of defects have been
identified : cavities, metallic lithium precipitates,
bidimensional lithium agglomerates. But despite high
irradiation doses, these ionic compounds never become
amorphous and the crystals progressively·turn into metallic lithium, loosing F_2 in the atmosphere. As for
metals, this result is connected to the fact that it is
not possible to obtain amorphous ionic crystals by

other methods.

c) - Covalent crystals

It seems that amorphisation can only be found in crystals with covalent or partially covalent bonds. Covalent bonds are very strong ones and prevent reordering of the disordered material. Besides, the same individual group of atoms can be found in the crystalline and amorphous state (e.g. tetrahedras).

2. Evidence of Amorphisation : Examples

a) - Natural irradiation

Metamict crystals have been studied since 1930[5,6]. Uranium and Thorium atoms embedded in certain radiactive minerals produce α particules and recoil atoms. Such materials present amorphous zones together with desordonned crystalline matrix and also a desorientation of the crystal which breaks into crystallites[7]. The most studied mineral was metamict zircon. Holland and Gottfried[6] by X ray diffraction found that the concentration of undamaged zircon is negligibly small when the dosis of α rays reached 10^6/mg. Fig. 3 shows the difference between undamaged and metamict zircon ; the pattern of metamict zircon presents an amorphous ring of zirconia, the diffraction spots are broader, and do not exist at large angles ; a desorientation of the crystal can also been observed.

Ultrathin amorphous coating have also been observed by high voltage electron microscopy by Bibring et al[8] on dust grains from Apollo 11, Apollo 12, Apollo 14 and Luna 16 missions. They showed that these coatings result from an "ancient" implantation of solar wind ions in the grains.

b) - Neutron irradiation

Single crystals of quartz become amorphous after a strong irradiation with fast neutrons (dose > 10^{20} n/cm^2). By X rays experiments, M.C. Wittels[9] and R. Comes et al[10] have studied the mechanism of this transformation : in the first stage (inferior to 2 x 10^{19} n/cm^2), the defects are homogeneously dispersed through the crystalline lattice thus giving a normal pattern diffraction, with increased lattice parameters. When the dose of neutrons increases, areas of highly disturbed matter are embedded in a still crystalline matrix. The disordered phase is not isotropic but keeps some relation with the matrix : the diffraction pattern is not the single ring of amorphous silica. (fig.4)

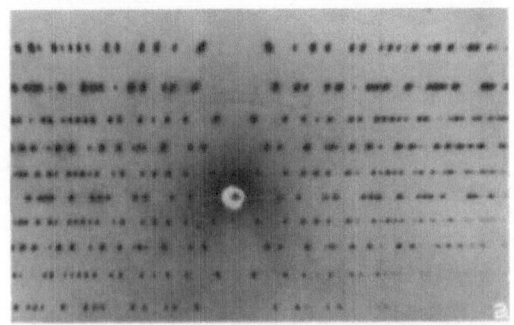
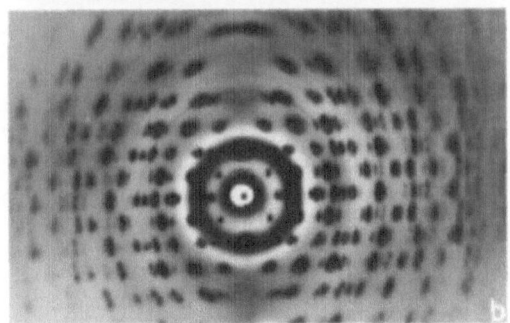

Fig.3 : Rotating X ray photograph of Ceylan zircon

a) non metamict

b) metamict

Mo Kα monochromatic 40 kV, 20 mA

394

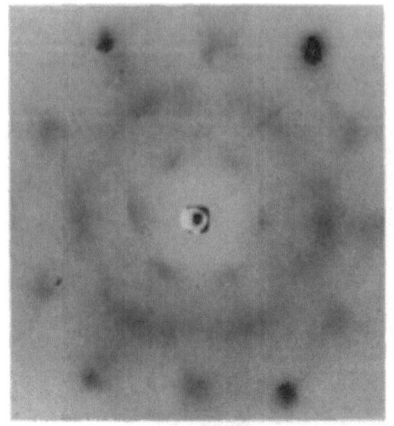

Fig.4a

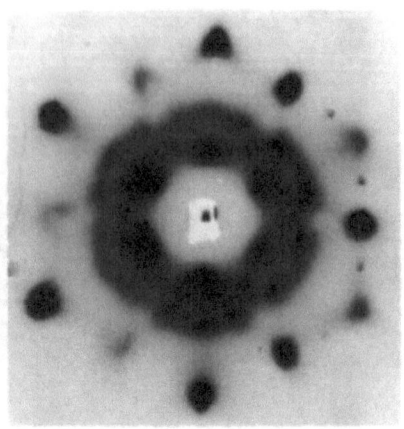

Fig.4b

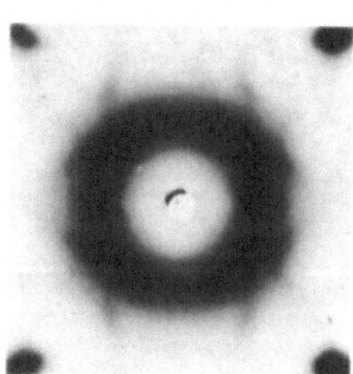

Fig.4c

Fig.4 : X ray diffraction photograph of quartz crystal (9)

   a) unirradiated exposure time : 16hrs

   b) 3.10^{19} exposure time : 5hrs

   c) 7.2×10^{19} exposure time : 3 hrs
   fixed crystal Mo Kα monochromatic.

At this stage, the crystalline matrix presents a disorder characterized by diffuse scattering localized on [101] and [100] planes of the reciprocal lattice.

It means that chains of atoms parallel to |100| and |110| axes are moved together so that the periodicity is preserved along the chain but disturbed between the chain and its neighbours. Above 6×10^{19} n/Cm^2, the crystal has an hexagonal symmetry which results from a mixture of domains of α_1 and α_2 structures in equal proportion. For every high doses (superior to 10^{20} n/cm^2), the crystalline pattern disappears completely and thus there only remains the amorphous ring, almost identical to the ring given by pure vitreous silica.

Neutron irradiation experiments have also been tried by M.L.Swanson et al[11] and Wittels[12] on Ge and Si up to 5×10^{20} fission'neutrons/cm^2. They could not obtain any amorphism in Ge or Si.

c) - Irradiation by ions

- ions of energy inferior to 1MeV/nucleon

High doses irradiation effects have been widely studied in semiconductors[11,13]. A description of the different stages of amorphisation is given for Si and Ge by M.L.Swanson et al[11] by means of electron microscopy and diffraction experiments. When Ge is bombarded with a dose superior to 10^{12} 100 keV O$^-$ ions/cm^2, an amorphous ring appears, coexistent with crystalline pattern. With a dark field photograph, it can be seen individual amorphous zones in the crystal. J.R.Parsons[13] measured the size of the amorphous zones : they are dependent upon the irradiation temperature : Germanium when bombarded at room temperature presents zones of diameters nearly equal to 68 Å, and when bombarded at temperatures inferior to 30°K, zones of diameters nearly to 89 Å. It is thought[12] that the individual damage regions produced by irradiation are crystalline, since the areas observed at low ion fluences anneal at lower temperatures than the amorphous zones. At high fluences, overlap of damage regions should make them amorphous. The temperature dependence of formation of amorphous zones can this way be explained[13]. The "primary damage zones" should contain defects whose concentration should vary until 0.02, yielding to amorphous zones.

It is generally noticed two types of defects : point defects and large defects ; in minerals, the same observation has been made by M.Lambert et al[14] in muscovite mica irradiated by Argon ions, the energy of

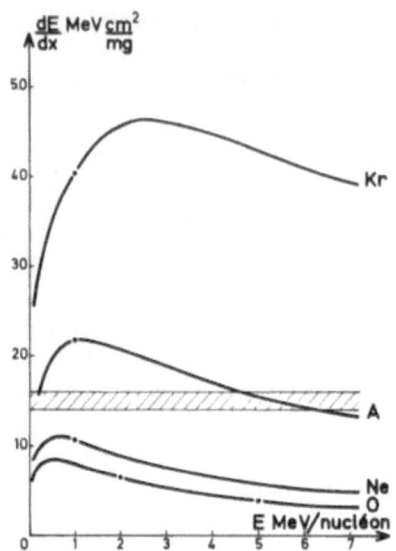

Fig.5 : Unitary energy loss $\frac{dE}{dx}$ of incident ions, in muscovite, varying with the energy of incidence. The different irradiations realized are represented by points on the curves.

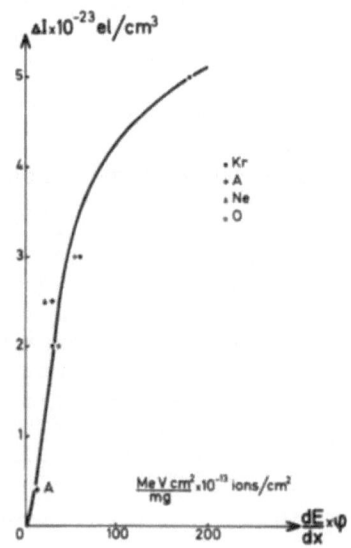

Fig.6 : Variation of point-defects concentration in mica with $\frac{dE}{dx} \times \phi$

which varies between 10 MeV/nucleon and 1 MeV/nucleon.

- ions of energy superior to 1 MeV/nucleon

They have been mainly used on minerals[15] in connection with astrophysics and geological experiments. The path of heavy ions of high energy in minerals is materialized by a "latent track" which can be chemically etched after irradiation and then seen through an optical microscope. The criterions and theories of registration of charged particule tracks in solids have been formulated by R.L.Fleischer et al[16] as "ions explosion spike mechanism". They also come to the hypothesis[17] that the "latent track" in minerals is composed of desordoned zones the size of which varies around 50 Å diameter. They could visualize such zones by electron microscopy.

The nature and their formation conditions and also progressive amorphisation will be defined in muscovite mica. Small angles scattering of X rays[18] were used to study the produced defects in muscovite mica when bombarded with oxygen, neon, argon and krypton ions. I(s) is measured, when I(s) is the absolute scattering power measured in electron units, and where s is the length of the scattering vector,; $s = 2 \sin\theta/\lambda$ when 2θ is the angle between incident and scattered beam and λ the wavelength of the incident beam. Point defects give a scattering power which does not depend on the angle, $I(s) = cst$. In this case, I(s) is proportional to the concentration of point defects. Large defects give a scattering power which increases when the scattering angle decreases. Then, one can analyze I(s) in gaussian in order to evaluate the dimension of the defects and their concentration. The experimental conditions are defined in fig.5 where is drawn the unitary energy loss of the ions, dE/dx in function of the energy of incidence. Table I gives an abstract of the results.

In every sample, point defects of a radius inferior to 3 Å are formed. If it is supposed that these defects are vacancies, with a volume of the order of 30 Å, then the concentration n of these point-defects can be calculated. It is observed that n is proportional to dE/dx and to fluence ϕ, in the first stages of irradiation, as it is shown in fig.6 .

When ϕ increases, a saturation of the concentration of point defects is noted. This saturation corresponds to the appearance of the amorphous state.

In the case of irradiations by Kr, Ar and Ne ions, larger defects are localized on the ion-path. The dimensions and concentrations of large defects are given in table I. For the sample irradiated by Kr ions, the size of defects is approximate because it is larger than can be measured by the experimental device. The concentration of defects can only be calculated if their electronic density is known. A minimum number of defects per incident ion is necessary in order to explain the selective etching along the ion-path. A suffisant number of large defects is calculated only in the hypothesis of a little variation of density between mica and large defects. Otherwise, at sufficiently high fluences (10^{13} ions/cm^2 = ϕ_c for irradiations by Kr ions and 10^{15} for Ne ions), mica sample becomes amorphous, as can be observed on a transmission X ray diffraction pattern, in fig.7.

It can thus be concluded that the structure of the studied zones is amorphous. The following schema illustrates the disposition of the defects in the crystal. (at the end of the text).

In the case of Ar ions, the average distance d between defects does not vary much with the fluence and is of the order of 60 Å.

The progressive amorphisation of mica can be followed this way. At low fluence, there exists point defects and amorphous zones along each ion-path. When the fluence is sufficiently high so that the damage regions overlap, there is superposition of amorphous ring and diffraction spots : at complete overlapping of damage regions, the mica becomes completely amorphous. But the question is : overlapping of which region is important ? If it is the overlapping of ion-paths in this case, for two different ions producing defects of radii r_1 and r_2, mica becomes amorphous when ϕ_{C_1} and ϕ_{C_2} are fluences necessary to the amorphisation of mica for ion 1 and 2 respectively. If it is the overlapping of individual amorphous zones, in this case $\phi_{C_1} r_1^3 = \phi_{C_2} r_2^3$.

Those two equations can be tried in the case of irradiation by Ar and Ne ions :

overlapping of ion-paths : $\phi_{C_1} \ r_1^2 = 1.2 \ 10^{17}$ (Ne)

$$\phi_{C_2} \ r_2^2 = 2.4 \ 10^{17} \quad \text{(Ar)}$$

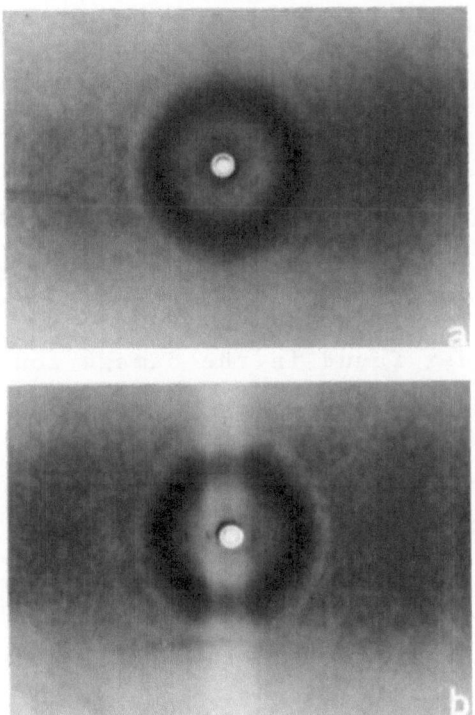

Fig.7 : a) rotating X ray photograph of mica irradiated with
1.2 10^{15} Ar Ions/cm^2

b) same specimen, 15° oscillation around 0 0 1 reflexion

Cu Kα 20 mA 40 kV, monochromatic.

overlapping of amorphous zones :

$$\phi c_1 r_1^3 = 1.3 \ 10^{18} \qquad (Ne)$$

$$\phi c_2 r_2^3 = 4.6 \ 10^{18} \qquad (Ar)$$

It seems that amorphisation is rather due to overlapping of ion paths.

Before complete disappearance of crystalline pattern a limit case can be obtained when crystalline pattern is limited to the first (ool) reflexions of mica. In this case, it seems that inside the amorphous mica little regions are left in which the only direction of order is perpendicular to the (001) planes.

In conclusion, the mechanism of amorphisation produced by irradiation seems very similar in all studied cases : first appear point defects, damage zones, then overlapping of damage zones and in the end complete amorphisation. The only differences lie in the state of disorder found in the damage zones.

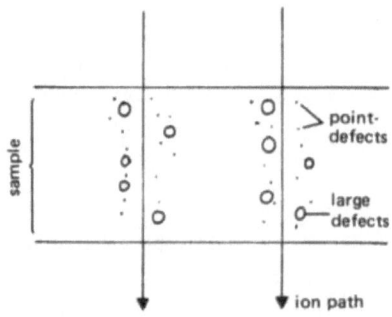

Disposition of the defects in a crystal.

Nature of ions Energy	Energy loss per unit length $\frac{dE}{dx}$($\frac{Mev\ cm^2}{mg}$) from	Dosis (ions/cm²)	$\frac{dE}{dx} \times \overline{10^{\ 3}}$	Concentration of point defects	Radius of large defects (Å)	Concentrations of large defects number per one incident ion and 10 μ
Kr 1 Mev/nucleon	40	7.5×10^{12}	30		30	
Kr k Mev/nucleon	40	1.5×10^{13}	60	1,5	30	
Kr 1 Mev/nucleon	40	4.5×10^{13}	180	2.6 coexistence of cristalline+ amorphous state		
Kr 1 Mev/nucleon	40	3×10^{14}	180	amorphous		
Ar 1 Mev/nucleon	15	2×10^{13}	30	1,3	18	1 500
Ar 1 Mev/nucleon	15	7×10^{12}	10.5	0,2	18	1 200
Ar 1 Mev/nucleon	15	8×10^{14}	10.5	amorphous		

TABLE I

Ne 1 Mev/nucleon	10.5	2×10^{13}	21	1,3	11	4 000
Ne 1 Mev/nucleon	10.5	10^{15}	105	amorphous		
O 2 Mev/nucleon	5.5	10^{14}	55	1,5	absents	absents
O 5 Mev/nucleon	3.5	10^{14}	35	1	absents	absents

Table I continued.

REFERENCES

1. Dixmier J., Thesis, Orsay, 1969

2. See for example a review-article about the subject by Norris D.I.R., Radiation Effects, 14, 1, 1972

3. Lesieur D., Thesis, Orsay 1973
   C.R.A.S. Paris, 226 B, 1038, 1968

4. Lambert M., Thesis, Paris, 1958
   Lambert M., Lefevre S., Guinier A., C.R.A.S., 255, 97, 1962
   Comes R., Lambert M., Perret R., Guinier A., Phys. Stat.Sol. 13, 265, 1966

5. Chudoba K.F., Stackelberg M., Viristall Z., Kristall., 95, 230, 1936
   Pellas P., Mémoires du Museum National d'Histoire Naturelle, C XII, 16, 1965

6. Holland H.D. and Gottefried D., Acta Cryst., 8, 291, 1955

7. Bursill L.A., Mc Laren A.C. and Phokey P.P., International Conference on Electron Diffraction and Crystal Defects, Melbourne, II B-3, 1965, Pergamon Press

8. Bibring J.P., Durand J.P., Durrieu L., Jouret C., Maurette M. and Meurier R., Science, 175, 753, 1972

9. Wittels M.C., Phil.Mag., 2, 1445, 1957

10. Comes R., Lambert M. and Guinier A., Interaction of Radiation with Solids, Plenum Press, 1967

11. Swanson M.L., Parsons J.R. and Hoelke C.W., Radiation Effects, 9, 249, 1971

12. Wittels M.C., J.Appl.Phys., 28, 921, 1957 and 40, 2909, 1969

13. Parsons J.R., Phil.Mag., 12, 1159, 1965

14. Lambert M., Levelut A.M., Maurette M. and Heckman H. Radiation Effects, 2, 1970

15. Price P.B. and Walker R.M., J.App.Phys. 33, 3407, 1962

16. Flescher R.L., Price P.B. and Walker R.M., J.App. Phys. 36, 3645, 1965

17. Price P.B. and Walker R.M., J.App.Phys. 33, 3400, 1962

404

18. Dartyge E., Lambert M., Radiation Effects, to be published.

THE PHOTOGRAPHIC PROCESS*

L. SLIFKIN

University of North Carolina
Chapel Hill, N.C.
U.S.A.

The silver halide photographic emulsion is a unique optical recording system. Although our technology employs a wide variety of methods for optical imaging --office copying devices, blueprinting, and television cameras are common examples-- none can compete with photographic film in that combination of virtues which make the film so useful as a camera-working medium : compactness, convenience, economy, resolution, permanence, sensitivity, and a satisfying rendition of intermediate grey tones. (Parenthetically, however, it may be remarked that with regard to degradation of signal-to-noise ratio, photographic emulsions are considerably poorer than television cameras ; in part, this is a price paid for the tone quality and latitude of the film).

The present discussion will focus on the underlying physics of this unique position of the silver halide system. It will emerge that we do not yet have a complete understanding of all of the microscopic details of the physical processes involved. The reason for these obscurities, in spite of many decades of energetic research, will be appreciated to result from the circumstance that many of the crucial steps in the process take place on an atomic scale, at a very few

*Preparation of this manuscript supported by the U.S. National Science Foundation (Grant n° GH-35794).

special sites. The very efficiency of the scheme makes
its functioning difficult to elucidate. In particular,
the surface of the silver halide will be seen to play
a predominant role ; thus, the understanding of the
physical properties of large single crystals is necessary to --but not sufficient for-- the understanding
of the phenomena which take place in the microcrystals
of the photographic emulsion.

A number of recent reviews of the photographic
process are listed as references 1-8. From them, the
interested reader will be able to discern a variety of
points of view and also an evolution of our ideas over
the past decade and a half. The remaining references
are to more recent research reports, primarily papers
which have not yet been discussed in the reviews.

THE SILVER HALIDE PROCESS

The photographic "emulsion" consists of a suspen-
sion of microcrystals of silver halide --AgBr or AgCl,
often with several percent AgI in solid solution-- in
gelatin, along with small amounts of additives which
have been found to improve performance in various ways.
The gelatin serves passively as a protective anti-flo-
cculation agent, a mechanical support, and a medium
which will be permeable to the aqueous solutions to be
used in the subsequent processing of the system. It
also plays an active role in the photographic process,
as an acceptor for halogen produced by photolytic decomposition of the silver halide and as a supplier of
a sensitizing agent, which we shall see is critical to
the efficient functioning of the scheme.

The microcrystals typically have dimensions in the
range 0.1 to 1 micron. Depending on the precipitation
conditions at the time of preparation, they may be ir-
regularly shaped, cubic, octahedral, or tabular ; for
example, precipitation of AgBr in the presence of excess bromide ion favors the survival of (111) surfaces.
The concentrations of divalent cation impurity may well
be as low as 10^{-6}. The microcrystals may each contain
none or several lattice dislocation, depending on how
they have been prepared. In particular, mixed crystals
of AgBr and AgI are strained because the AgI is concentrated mainly in the center (smaller solubility product), thereby causing a gradient in lattice parameter.

Now it has long been known that exposure to light,
especially near the violet region of the spectrum, will
cause photolytic decomposition of silver halides, leaving a residue of "print-out" silver which appears

black because of the irregularity of the particles. Although this phenomenon could be used to record an optical image, its sensitivity is many factors of ten too low for any technological application. The key to the modern photographic process is that formation only of a microscopic, invisible speck of silver metal can be made to serve the microcrystal as an adequate memory of a previous exposure to light. This speck, known as the "latent image", functions as a catalyst in a subsequent process, called "photographic development", in which the emulsion is immersed in a solution of an organic reducing agent.

The mixture of silver halide crystals and reducing agent is thermodynamically unstable relative to metallic silver plus an oxidized form of the developer molecule, but kinetically there is a substantial barrier to the transfer of electrons from the organic developer to the ionic silver halide crystal. The latent image speck virtually eliminates this barrier, thus catalyzing the complete reduction to metallic silver of those grains which carried one or more such specks, while the majority of the grains without a latent image are completely unaffected by the development. This autocatalytic functioning of the latent image results in an amplification of the optical signal by a factor of 10^8 - 10^9. It is also evident that since a given grain is either completely converted to silver or else not at all (i.e., it acts as a "yes" or "no" element), the production of intermediate grey tones on a photograph must arise from a <u>statistical</u> relation between exposure and formation of the latent image. This comes about not only because of the statistical distribution of photon absorption by the various grains but, more importantly in most cases, because the grains themselves have a distribution of sizes and perfections and therefore do not all have the same efficiency for the conversion of optical quanta into a viable latent image speck.

The microscopic mechanism of this development reaction is not known. One proposed model imagines a catalytic reaction to take place between the silver halide and adsorbed developer at the line of contact between these and the metallic silver of the growing image speck. A second model suggests that the latent image speck acts as an electrode in an extended electrochemical cell. Consistent with this are recent results of Pontius and Willis[9], in which the dependence of the rate of development on electrochemical potential of the solution and on the surface area of the growing development center shows that the process is limited by the transfer of electrons from solution to the silver speck, and not by the rate of diffusion of silver ion within the microcrystal. On the other hand, electron microscopic studies of Mueller[10] have shown that the filaments produced in the early stages of development are not of silver, but rather consist largely of some complex which only later is converted to the metal ; these results would seem to indicate that the development process may be rather complicated.

Phenomenologically, the response of an emulsion, after development, as a function of the exposure (a measure of the integrated incident flux of photons) is displayed on a plot of the optical density versus the logarithm of the exposure (the density is defined as

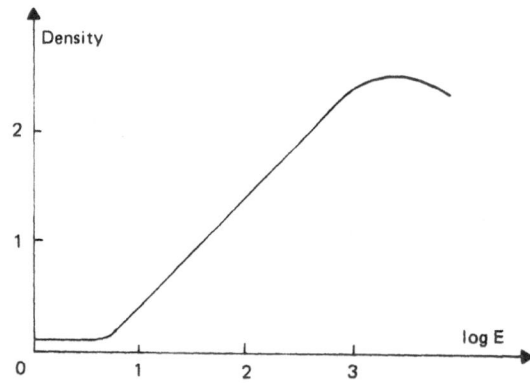

the negative of the logarithm of the fraction of incident light which is transmitted by the developed emulsion). Such a graph is known as a Hurter and Driffield, or H & D, plot ; it is shown schematically in the accompanying figure 1a. There is a non-zero optical density even for zero exposure ; this is called "fog" and is due to the prior presence on a few of the grains of sites or specks which will catalyze development. The "speed" of the emulsion is measured (inversely) by the value of the exposure at which the curve turns upward ; the contrast is determined by the slope of this rising region ; and the latitude by the extent of the horizontal projection of this portion of the curve. The ultimate decrease in density at high exposures is known as solarization, and results from an attack of the latent image by halogen.

Of particular interest is the shape of the H & D

plot at low exposures. The fact that there is no increase in density until after a threshold exposure has been exceeded (i.e., the slope at low exposures is zero) tells us that more than one photon must be absorbed by a grain to render it developable. Also, the shape of the curve after this threshold is consistent with the notion that at least several silver atoms are required in order to form a latent image speck. Since the latent image is the critical factor in the photographic process, we now focus our attention on its formation and its properties.

THE LATENT IMAGE

Our model for the formation of the latent image is due to Gurney and Mott (in 1938), and is schemati-

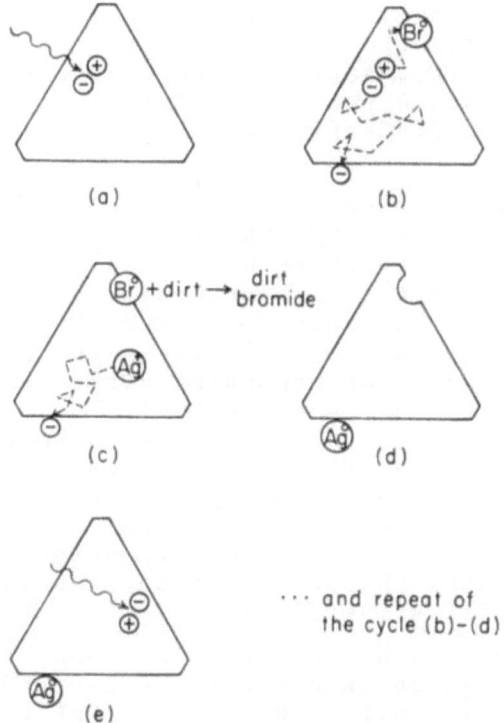

cally illustrated in the adjacent diagram. The absorption of an appropriate photon removes an electron from a bound state, produing a free photoelectron in the conduction band of the crystal, and leaving a hole in the valence band. These photocarriers wander through the crystal, and eventually become permanently trapped. If the electron is to contribute to the formation of the latent image, it is essential that this final trapping be at the surface of the grain. If the hole also reaches the surface (at a site different from that of the electron ; otherwise, recombination will destroy the effect of the illumination), it gives rise to an atom of halogen. Various substances --some in the gelatin, and some deliberately added to the emulsion during preparation-- are in contact with the grain to react with this halogen, and thereby eliminate the possibility of a subsequent attack on the growing latent image speck. (It is the exhaustion of these halogen acceptors that gives rise to the solarization effect shown on the H & D plot at high levels of exposure).

This concludes the electronic component of the Gurney-Mott cycle. The next step is ionic : the migration to the trapped electron of an excess, interstitial silver ion, resulting in the formation of an atom of metallic silver. The cycle is now complete, the original electron trap is reset, and the grain is ready for absorption of a second photon. Ultimately, the speck of silver grows sufficiently large to function as a latent image.

This two-stage, electronic/ionic cycle has been beautifully demonstrated by experiments using short flashes of light, synchronous with a pulsed electric field --first on large single crystals (by Haynes and Shockley) and later on emulsion grains (by Hamilton and Brady). The pulsed field displaces the photoelectrons, resulting in a corresponding displacement of the latent image specks that are eventually formed. An inverse to this experiment was performed by Klein and Matejec, who first electrically polarized the grains in the absence of illumination, producing a redistribution of the interstitial silver ions ; a subsequent illumination then results in a distribution of latent image which reflects the polarization-induced assymetry of interstitial silver ions.

The question now arises as to how many atoms a given speck must acquire in order to become a viable latent image. Heuristically, one might argue that the latent image speck functions because it is metallic, that a single atom is not metallic, that the smallest 3-dimensional array of points is a tetrahedron, and that therefore the minimum size of a latent image speck must be about four atoms. Amusingly enough, the experimental evidence indicates that this result is approximately correct : the probability that a grain will become reduced during development rises sharply toward u-

nity as the number of silver atoms in the largest --
perhaps the only-- speck increases from 3 to 5.

In the first place, we have already seen that the
details of the H & D curve at low exposures make it
clear that absorption of only one photon cannot render
a grain developable. Moreover, statistical studies of
the developable fraction as a function of photon flux
(i.e., the shape of the H & D curve at higher exposures) indicate that virtually no grains can produce a
latent image speck upon absorption of only one or two
photons.

A convincing demonstration of the necessity of
repeated cycling of Gurney-Mott mechanism in order to
produce a latent image is the existence of what are
known as high- and low-intensity reciprocity failures.
These terms refer to a decrease in efficiency under
illumination at either very high or very low intensity. High-intensity reciprocity failure occurs when a
second photon is absorbed by the grain before the completion of the first ionic step of the Gurney-Mott cycle, which typically requires about 1 to 10 μ sec.
Since the second photoelectron cannot be trapped at
the site of the first until the first is neutralized
by an interstitial silver ion, then this second photon
does not contribute to the growth of the latent image
speck. Low-intensity reciprocity failure, on the other
hand, arises because a single atom of silver is unstable against re-ionization back into the grain ; its
lifetime at room temperature is about 1sec. Unless a
second photoelectron is captured at the site of such
an atom within this lifetime, the atom will be lost
(the re-ionized electron apparently ultimately combines with adsorbed oxygen or a trapped hole in the grain). Once a two-atom silver speck is formed, however,
the lifetime increases to several days, and further
growth of the speck can proceed quite slowly without
loss of efficiency. Because Ag_2 is the smallest stable
speck, the rate of formation of latent image at low
intensity goes as I^2. Note that neither of these two
reciprocity failures would be observed if the latent
image did not require several silver atoms.

These stability properties were used by Berg and
Burton and by Webb and Evans to show that the minimum
number of silver atoms which will constitute a latent
image is very near to 4. After a moderate exposure to
produce Ag_2 specks on a substantial fraction of the
grains, the emulsion was further illuminated at very
low intensities -intensities so low that any new sin-

gle atoms that might form would not be stable. Thus, this second exposure served only to induce growth in the previously existing Ag_2 specks. It was found that the grains became developable only when the second exposure was at least as great (i.e., involved as many absorbed photons) as was the first, implying that the specks had doubled in size. Hence the smallest latent image speck must be approximately Ag_4.

A further confirmation is provided by computer simulation studies of Hamilton and Bayer[11], who found that they could fit experimental H & D curves if the critical size for the latent image were chosen between 3 and 5 atoms of silver. For minimum sizes outside this range, the fits rapidly became very poor.

We now examine the various steps of the Gurney-Mott cycle in detail. The solid state properties of the silver halide grains which are critical to the mechanism are each discussed as they enter the process.

ELECTRON ENERGY BAND STRUCTURE

The first step in the Gurney-Mott cycle is the absorption of a photon to produce a photopair. Untreated silver halide crystals are transparent over much of the visible range, with an absorption that begins at approximately 420 nm (3 eV) for AgCl at room temperature, 480 nm for AgBr, and 510 nm for AgBr containing 3 % AgI. This optical absorption produces photoconductivity, so that it must involve the excitation of electrons into the conduction band, but --unlike the usual band-to-band absorptions in other materials --the absorption coefficient rises only slowly with decreasing wavelength, reaching values of about 10^2 to 10^3 cm^{-1} at the violet end of the spectrum and about 5×10^5 cm^{-1} well into the ultraviolet (at approximately 5 eV). Moreover, this absorption "tail" is strongly dependent on temperature, shifting to longer wavelengths and larger absorption coefficients as the temperature is increased. Finally, as compared to other ionic materials, such as the alkali halides, the onset of a photoresponse in the visible region of the spectrum, rather than the UV, is somewhat unusual.

This peculiar behavior has its origin in a coincidental near-degeneracy of energy levels, which greatly affects the electronic energy band structure. As it turns out, the Madelung potential within the crystal so raises the energy of the outer 4d electrons of the silver ion and lowers that of the halide p-electrons

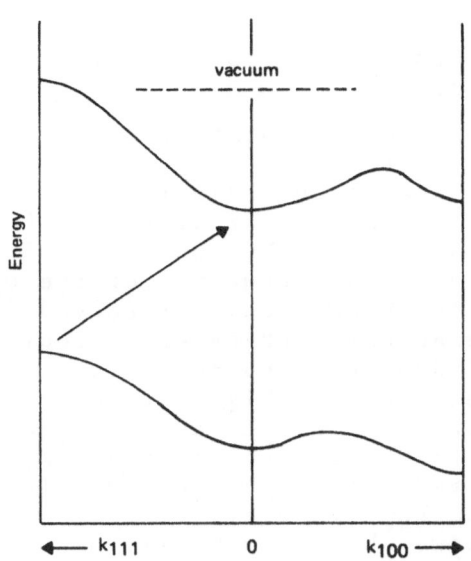

as to bring the two almost into coincidence. The result of the consequent mixing of these states is to bend the valence band upwards, producing a maximum at L, at the zone boundary in the (111) direction. A portion of the band structure as calculated by Scop and by Bassani, Knox and Fowler, is shown in the accompanying sketch. The symmetry of this structure has been verified by the piezo-optic studies of Ascarelli and magnetoconductivity measurements on photocarriers, by F.Brown. Also of interest is the fact that, for the silver halides, the bottom of the conduction band lies about midway between the vacuum level and the top of the valence band*.

It is the valence band maximum at L which causes the threshold for photoconductivity and the photographic process to appear in the visible, instead of well into the ultraviolet. Thus, whereas the direct transition from the top of the valence band at k = 0 would require some 5 eV, the indirect transition from L requires only 3.2 eV for AgCl and 2.7 for AgBr. Also these valence band states at L are predominantly d-like, so that it is perhaps more nearly correct to say that

---

* Contrast this with the alkali halides, for which the conduction band minimum lies nearer to the vacuum level, or with the semiconductors Ge and Si, for which the conduction band minimum is much closer to the valence band. The silver halides thus represent an intermediate between the two extremes.

the excited electron comes from a silver d-shell than
from a halide ion.

Now it is not sufficient that a band-to-band transition in the visible region be possible ; it must also be probable. Because the transition is indirect,
the simultaneous absorption or emission of a phonon is
required if momentum is to be conserved. At low temperature, where the phonon structure of the optical absorption and luminescence can be resolved, it has been
found that either an 8 meV transverse acoustic or a 12
meV longitudinal acoustic phonon is involved, in both
AgCl and AgBr. At the much higher temperatures at which the photographic process is employed, other phonons,
including those from the optical branches, are undoubtedly also involved. These phonons are plentifully
available at room temperature, because the Debye temperature is so low, about $150°K$ (12). Also, the rather
high ionicity of the silver halides assures a good
coupling between the phonons and the electronic system.
Moreover, in mixed crystals, such as the AgBr : I commonly used in emulsions, the selection rule on conservation of k-vector is relaxed, thereby decreasing the
need for phonon participation and further increasing
the optical absorption coefficient in the visible. Finally, because the large dielectric coefficient in the
silver halides causes the excitonic radius to be very
large and the binding energy quite negligible (compared to kT), the quantum efficiency for production of
free electrons by the absorption of photons is very
nearly unity.

Thus, absorption of light in the blue end of the
visible spectrum is reasonably efficient : with an absorption coefficient exceeding 10^{-2} cm^{-1}, a layer several thousandths of a centimeter thick would make use
of a substantial fraction of the incident blue light.
But what about the remainder of the visible spectrum ?

The spectral sensitivity can be extended throughout the visible region, and even into the near infra-
red (out to 1.3 microns), by providing the grains with
an adsorbed layer of an appropriate dye. The dye molecules are not randomly disposed, but form an ordered
layer, called a J-aggregate, in which the coupling between the dye molecules is strong. The excitation produced by an absorbed photon can thus be efficiently
transmitted along the layer to a suitable site on the
surface of the silver halide, where an electron is injected into the conduction band of the grain.

There is no violation of the law of conservation of energy in this phenomenon ; one is not obtaining a band-to-band excitation at a bargain price, because the resulting hole is left in a localized state somewhere above the top of the valence band. Such states undoubtedly exist at various sites on the silver halide surface ; they may also be provided by the layer of dye.

The mechanism of dye sensitization is not fully understood. Two possible alternatives have been extensively investigated. In the first, the excited dye is considered to transfer an electron directly into the conduction band of the silver halide. Consistent with this scheme is the evidence that those dyes which can sensitize photographic grains all seem to have their first excited states at energies above the bottom of the silver halide conduction band[13]   . On the other hand, it is known that the dye molecules are not exhausted during exposure --it is possible to inject many times more electrons than there are molecules-- so that one must also find a way to transfer a hole from the dye to the silver halide surface. Furthermore, there is some uncertainty in the absolute positions of the energy levels of the dyes when in the adsorbed, J-aggregated state. Thus, Levy and Lindsey[14] have proposed a variant --an inverse-- to this transfer mechanism, in which the AgBr/dye system is treated as a photodiode ; the dye is considered to be the negatively charge p-type component (just the opposite from the role usually assigned it), so that photoelectrons produced in the junction region will drift into the silver halide microcrystal. In support of this mechanism, Levy and Lindsey have demonstrated that an AgBr/PbO junction behaves under illumination as predicted.

The second mechanism for dye sensitization assumes that the excited dye transfers to the microcrystal only excitation energy, not charge, and that this energy then lifts electrons from surface states into the conduction band. The strongest experimental argument for such a process is the demonstration by Kuhn and colleagues that the interposition of a layer of inert fatty acid, 54 A thick, impairs neither the sensitization process nor the ability of the silver halide to quench the luminescence of the dye. Since the lifetime for electron tunnelling through such a layer is much greater than the luminescence time, it would appear that the dye is coupled to the grain by energy transfer, rather than by electron transfer. These results, however, are still somewhat controversial : O'Brien[15] has presented evidence that there are sufficient defects in the fatty acid film to explain the data in terms of possible electron transfer, while Wullschleger and Steiger[16], working with other systems, have corroborated Kuhn's conclusions.

Thus, the available evidence does not yet allow of a definite choice of mechanism (it is, of course, possible that both --or neither-- of these operate). In any event, the <u>net</u> role of the layer of sensitizing dye is that of a high oscillator strength absorber which can readily transport, without loss, the energy of the red or green photon to surface electronic states lying within the energy gap of the silver halide.

THE SUB-SURFACE SPACE CHARGE

Thus far, we have tacitly assumed that the microcrystal and its properties are, except for possible inhomogeneities in iodide concentration, uniform from the interior to the geometric surface. In actuality this is not so ; there exists a separation of ionic charge over a rather extensive region near every dislocation and surface, which results in significant potential differences and electric fields. Since these space charge effects are important to each of the remaining steps of the Gurney-Mott cycle, it is useful now to examine their origin and properties.

Every crystal has, in thermodynamic equilibrium, a number of point defects. In the interior of a pure, otherwise perfect crystal of silver halide, the dominant defects are equal numbers of interstitial silver ions and the complementary silver ion vacancies (this combination is known as the Frenkel defect). But whereas both components of a Frenkel pair must be created or annihilated simultaneously in the interior, this is not necessarily true at the surface or at a dislocation. Jogs on dislocations or at a surface ledge can, by the addition of a silver ion to a negatively charged site or the removal of Ag^+ from a positively charged site, separately create or destroy each component of the Frenkel defect. Thus, dislocations and surfaces must come to equilibrium separately with each type of point defect in the crystal, and if the free energies of formation of the two components are not the same, then an excess of one of them is "evaporated" from the source. This leaves the surface with a net charge, counterbalanced within the crystal by a space charge which consists of the excess quantity of the low-energy defect. The resulting potential difference between surface and interior can be shown to be $\Delta G/2e$, where ΔG is the difference between the two free energies of formation, and e is the electronic charge. It should be appreciated that, since the energy involved in reversing the sign of a jog should depend on the surroundings, therefore the quantity ΔG is expected to be different for surfaces and dislocations, and may well also be a function of orientation.

Real crystals, of course, are not perfectly pure. In the case of silver halide emulsion grains, the approximately 1 ppm of divalent cation impurity introduces an additional 1 ppm of vacancies, which -at room temperature- overshadows the number of intrinsic defects. The surface thus sees an excess of vacancies, with the result that there is a condensation onto jogs, converting the majority of them to negatively charged jogs. One thus expects that, at least near room temperature, the surfaces of the microcrystals carry a net negative charge, compensated by a sub-surface space charge which should consist of interstitial silver ions and divalent cations.

In addition to theoretical treatments of this situation, there have been a number of experiments on AgBr and AgCl which demonstrate the existence of both the surface charge and also the sub-surface layer, which possesses abnormally high ionic conductivity, presumably a result of the excess of interstitial silver ions. These investigations include such ingenious experiments as a measurement of the ionic conductivity of emulsion grains by Hamilton and Brady (the ionic relaxation time was observed by following the decay of the force exerted on photoelectrons), a measurement by Hamilton and Baetzold[17] of the surface component of conductivity in thin films, and a measurement of the force opposing a photoelectron which is drifting out toward the surface of a large single crystal (by Saunders, Tyler and West).

The present state of our knowledge of these surface charge effects in silver halides has recently been reviewed[18]. Although there are some conflicts and complications in the experimental data that have been reported the main features seem to be clear. The surface of a typical crystal of AgCl or AgBr is negative relative to the interior by approximately 0.15 volts, this potential difference being accomplished over a depth of the order of 0.1 micron. There is thus a substantial electric field within the space charge region, averaging something like 2×10^4 volt/cm. Since the depth of the space charge region is not much smaller than the half-thickness of a typical photographic grain, then it is clear that these electric fields and potential differences may be of some significance to the photographic process. We will see below that the dynamics of the photocarriers, the operation of the ionic step of the Gurney-Mott cycle, and the efficiency of surface traps all are influenced by these electrical inhomogenieties.

THE ELECTRON AND HOLE

Returning now to the Gurney-Mott cycle, we must inquire into the properties of the electron and hole which were produced by the absorption of the photon. To be useful photographically, the electron must neither recombine with the hole nor become trapped in the interior of the grain. To understand how this comes about, we consider in this section such dynamic microscopic properties as the mobility (defined as the velocity per unit electric field), the lifetime against trapping in the interior of the grain, and the effective mass ; the traps themselves are then discussed in the following sections.

The transport properties of electrons and holes in silver halide crystals have recently been reviewed by Brown[19,20]. In discussing carrier mobility, one must distinguish between the microscopic mobility, μ_m, which in the silver halides is limited by scattering from longitudinal optical phonons, and the drift mobility, μ_d, which is less than μ_m if the carrier spends part of its lifetime in shallow traps. The microscopic mobility may be determined by Hall effect measurements on carriers injected either by illumination or (for holes) by contact with excess halogen gas. Pulsed photoconductivity experiments yield data on μ_d.

For electrons in large single crystals at room temperature, it is found that $\mu_m = \mu_d$ in both AgCl and AgBr, indicating that shallow trapping is negligible. The values of these mobilities are 50 cm^2/volt-sec in AgCl, and 60 cm^2/volt-sec in AgBr. For holes in AgBr, μ_m is 1.7 cm^2/volt-sec, and μ_d has the slightly lower value of 1.1 cm^2/volt-sec, indicating a small amount of temporary trapping.

Quite a different situation exists in microcrystals. Hamilton and Brady studied the displacement of carriers under the combined influence of pulsed illumination and electric fields. For their particular AgBr grains, μ_d was found to be only 0.2 cm^2/volt-sec for electrons and about 0.001 cm^2/volt-sec for holes, indicating extensive multiple shallow trapping, especially for holes. This is presumed to be a result of the effect of the surface charge on the sub-surface density of point defects. For example, the enhanced concentration of interstitial Ag$^+$ might well provide the extra shallow traps for photoelectrons ; the nature of the additional hole traps is not known.

The drift mobility of a carrier also determines its diffusion coefficient, D. By the Nernst-Einstein relation, $D = (kT/e).\mu$. Recalling that the carriers in a photographic grain experience an electric field due to the surface charge, it is instructive to estimate the relative importance of drift in this field as compared to random diffusion. In a time t, the mean displacement produced by diffusion is $2(Dt)^{1/2}$ and that produced by a field E is μEt. Thus, the ratio of the time required to drift a distance d to that for diffusion through the same distance is 4kT/eEd. Since at room temperature, kT/e is approximately 1/40 volt, this ratio equals 1/10dE. If d is about 10^{-5} cm and we take E to be the average value of 2 x 10^4 volt/cm quoted above, then we find that the drift time is one-half the diffusion time. Thus, one certainly cannot neglect the effect of the sub-surface field in comparison with random diffusion of the carriers.

Turning now to carrier lifetimes, these are found to be extremely dependent on specimen purity, perfection, and thermal history. For example, with large single crystals, electron lifetimes as long as several micro-seconds have been obtained with very pure AgCl or with crystals annealed in air (this presumably removes heavy metal electron traps by formation of inert oxides) ; with crystals annealed in chlorine or vacuum, however, the lifetime is often as low as 10^{-7} or 10^{-8} sec. A beautiful technique for studying holes in large crystals has been developed by Malinowski[21], who detects their arrival at the rear face of an illuminated slab by the destruction of a previously deposited latent image distribution. He finds that in addition to a "normal" lifetime of some microseconds, there is a surface component with values as great as several hundreds of microseconds ; this perhaps involves an equilibrium between liberated atomic halogen and the silver halide crystal.

Hamilton and Brady have measured the lifetimes of electrons and holes' in emulsion grains by the synchronous pulsed light/pulsed field technique mentioned above. By delaying the application of the field some microseconds after the light flash, one can observe the decrease in carrier displacement that results from loss of carrier during the delay time. Hamilton and Brady found that for their emulsion grains, the electrons had a lifetime of 3 μsec while the hole lifetime was 15 μsec. These lifetimes were found to be independent of intensity, thus demonstrating that they are limited by trapping, not recombination.

The effective masses of the carriers are determined by cyclotron resonance. For AgBr, these are found to be, in units of the electron mass, 0.29 for the electron at the bottom of the conduction band, and 1.1 for the hole at the top of the valence band. These masses are not the "band" masses of the bare carriers, but are polaron masses ; i.e., they include the effect of the associated lattice polarization which moves with the carriers. The larger mass for the hole is part of the reason that the hole mobility is considerably less than that of the electron ; it also makes the trap depths at point defects somewhat greater for the hole than for the electron, and perhaps partially explains why the shallow trapping observed in emulsion grains by Hamilton and Brady (described above) is so much more extensive for holes than for electrons. In AgCl, the polaron mass for electron is, in units of the electron mass, 0.43.

INTERNAL TRAPS

Fortunately for the photographic process, the large dielectric constant and small effective mass make coulombic trap depths in the interior of the crystal quite shallow. For silver bromide, the low-frequency dielectric constant is 12.5 and the polaron mass is 0.29 ; thus the hydrogen-like ionization energy is reduced from 13.6 eV by a factor of $0.29/(12.5)^2$, giving a binding energy of only about 0.03 eV. Of course, there may be traps which bind carriers more strongly than this, such as cations which have two possible stable valence states. Thus, high purity, other than controlled additives, is an essential requirement in the preparation of photographic emulsions. Also, for surface traps, the coulombic binding is likely to be somewhat greater than in the interior, because the full dielectric screening is not available ; this is fortunate, since it is usually desired that the latent image form on the surface.

Largely because the internal traps within the grain are often so shallow, there is little information available on the trap depths and cross sections of defects and impurity ions. It is known that dislocations can trap photoelectrons (and thereby become "decorated" with photolytic silver), but not holes. Since the dislocations are negatively charged, this result seems surprising, but may be understood in terms of the liklihood that it is the positive space charge surrounding the dislocation which actually does the trapping. This speculation is strengthened by the observations of C.Childs[22] that the decoration of dislocations is very sensitive to impurities, on the part-per-million level. As for the effects of point defects, the low value of μ_d observed for electrons in AgBr emulsion grains, by Hamilton and Brady, has been attributed to shallow trapping by interstitial silver ions, with trap depths estimated to be only a few hundredths of an eV. Malinowski's experiments[21] on the mobility and lifetime of holes in large crystals have indicated that silver ion vacancies provide traps for holes with a depth of approximately 0.2 eV. It is possible that an adjacent impurity ion, such as a divalent cation or an iodide ion, may also be involved, but there is no firm evidence for this. In any event, optical absorption and electron paramagnetic resonance experiments have thus far been unsuccessful in identifying a hole trapped at a vacancy, although a long-lived photoresponse in AgBr : Cd at liquid nitrogen temperature appears to be due to electrons compensating such a center[23].

There is a small amount of data on the effects of various impurity cations. For example, optical absorption, EPR, and photoconductivity investigations have demonstrated that Cu^+ is a deep hole trap, while Ni^{2+} is an electron trap. Luminescence experiments have shown that iodide is an excellent hole trap in AgBr - to such an extent, in fact, that the dominant, green, "intrinsic" emission is really from the recombination of an electron with a hole trapped at an iodide. Also, the addition of various cations introduces a red luminescence in both AgBr and AgCl, suggesting a rather deep (several tenths of an eV) trap ; Riesenfeld[24] has recently presented evidence that of the two carriers, it is the hole that is trapped.

It would seem that detailed studies of luminescence and thermoluminescent glow peaks would yield ex-

tensive information on the properties of various do-
pants as traps. Unfortunately, there are a number of
complications. In some cases, the additive simply dis-
torts and shifts the luminescent spectra, rather than
introducing discrete new emissions. A second difficulty
is the identification of a given emission as involving
a hole trap or an electron trap. Also, the assignment
of energy levels is complicated by possible Stokes
shifts and the non-equality of optical and thermal excitation energies. Lastly, there is the problem of the
association of aliovalent impurity ions with vacancies:
the cation-vacancy association energy is typically
0.2 - 0.4 eV, and it is rarely certain what sort of association (and aggregation of these complexes) has been
frozen in during the cool-down to the temperatures at
which luminescence measurements may be made. Also, whe-
ras EPR studies indicate that a great many of these
complexes energetically prefer a next-nearest-neighbor
association at low temperatures, it is not certain that
this is the dominant configuration at room temperature.
The recent work of Fatuzzo and Oggioni[25] is an example
of the techniques of multiple excitations and trap de-
pletions that are being used to approach some of these
problems.

In the pure silver halides, more definitive and
quantitative results have been forthcoming. It has been
shown that the hole in AgCl --but not in AgBr-- becomes
self-trapped. Quite unlike the self-trapped hole in the
alkali halides, this one is centered on a silver site,
forming an Ag^{2+} which is complexed with its chloride
neighbors[26]. This center is stable up to about liquid
nitrogen temperature, and is the site of the dominant,
intrinsic, green luminescence of AgCl. Detailed studies
of optical absorption and emission in the pure silver
halide crystals and their solid solutions, by Kanzaki
and co-workers[27], have characterized in some detail
the excitons, their trapping, and their radiative decay in these materials.

SURFACE TRAPPING AND SENSITIZATION

If, as is generally the case in highly sensitive
emulsions, the photoelectron does not become trapped
in the interior of the microcrystal, it ultimately reaches the surface, where it is now trapped and becomes
the precursor to the latent image speck. The nature of
the surface trapping sites is obscure. It is tempting
to think of such a site as a positively charged jog on
a surface ledge. Since such a jog has an effective charge of 1/2e, capture of an electron would convert the

charge to $-1/2e$, thereby providing a coulombic potential to aid in the subsequent ionic step (if, on the
other hand, the electron were trapped at a site originally of charge e, then after trapping there would be
no electrostatic field to attract an interstitial Ag^+).

Actually, for efficient formation of the latent
image, the emulsion grains must first be chemically
sensitized. In contrast to dye or spectral sensitization, which extends the threshold for photon absorption out to longer wavelengths, chemical sensitization
serves to increase the efficiency with which the grain
uses the photoelectrons in forming the image speck.
Three types of chemical sensitization are used : reduction sensitization, sulfur sensitization, and gold sensitization. Each type performs a different function.

Reduction sensitization, achieved by bathing the
emulsion in a solution of a mild reducing agent, increases the photographic sensitivity somewhat at all in
tensities of exposure, but especially at low intensities. It is believed that this treatment forms dispersed centers, perhaps consisting of single atoms or
pairs of silver, at sites which will not grow into latent image specks under illumination. The function of
these silver atoms is to capture holes as they arrive
at the surface, thereby forming Ag^+ and minimizing the
attack by holes on the growing latent image. There are
two types of evidence for this model. First, it is possible to observe the presence of reduction sensitization centers by treatment with gold salts, to render
them developable, followed by partial development and
then electron microscopy. One finds that exposure does
not cause such centers to become proper latent image
specks, but actually destroys them. Second, Moisar has
prepared grains on which a further overgrowth of AgBr
was added after reduction sensitization, thereby bu-
rying the centers under the new growth. It is found
that upon exposing such grains the latent image still
forms on the surface, and not in the interior, where
the reduction centers were located.

Sulfur sensitization is obtained by treatment with
compounds containing reactive sulfur (to some extent,
gelatin is included in this category). Apparently, discrete specks of Ag_2S or some related silver sulfide are
formed. These specks act as the locus of the latent i-
mage upon exposure, apparently by stabilizing either
the trapped electron, the resulting silver atom, or
both. Thus, in overgrown grains, in which the sulfide
is now in the interior, the latent image forms inside,

whereas in the usual emulsions, in which the sulfide is on the external surface, the latent image is formed on that surface.

Thus, sulfur sensitization increases the photographic sensitivity at moderate intensities of exposure, and it is especially potent at low intensities, where its stabilizing effect suppresses low-intensity reciprocity failure. At all but high intensities, there may form only one latent image speck per grain in properly sensitized emulsions, and the process whereby successive photoelectrons are all concentrated into the one most favorable latent image precursor will be discussed in more detail below. At high levels of intensity, however, electron microscopic studies show that sulfur sensitization causes a multiplicity of latent image specks per grain ; this decreases the photographic sensitivity --more photons are needed to produce the extra specks-- and thereby causes an enhanced high-intensity reciprocity failure. This disadvantage can be counteracted by treating the emulsion with gold salts.

Gold sensitization is generally used in conjunction with sulfur sensitization, so that the product is probably a mixed sulfide of silver and gold. It efficacy in decreasing the sulfur-aggravated high-intensity reciprocity failure is believed to result from the effect of gold atoms on the minimum size required for a latent image : gold is suggested to be a better catalyst for development, so that a latent image which contains one or more gold atoms mixed with the silver need not be quite so large as otherwise.

It should also be pointed out in connection with the growth of the latent image --with or without sensitization-- that there is some controversy as to the exact sequence of events. For example, Hamilton has argued that the site destined to grow into the latent image is not initially a deep trap for electrons, but only becomes so after the arrival of an interstitial Ag^+. This argument is based on his observation that, under a variety of conditions, the lifetime of the photoelectron is almost exactly equal to the ionic conductivity relaxation time. Also of interest are recent calculations of Baetzold[28] on the electronic energy levels of a growing speck of silver, which suggest that the electron affinity oscillates as successive atoms are added. These results indicate that after Ag_2 is formed, still another Ag^+ may next be added, followed then by the successive capture of two photoelectrons.

It is not clear, however, that Baerzold's neglect of d-electron interactions has not perturbed the results of his extended Hückel calculations.

THE ROLE OF THE INTERSTITIAL SILVER ION

Having trapped the photoelectron at a sensitivity center on the grain surface, the next step in the Gurney-Mott cycle, the ionic step, requires the prompt arrival of an extra silver ion. It is here that a major difference between the silver halides and many other ionic materials plays a critical role, because the availability of excess, mobile cations is a prescription for the dominance of a cationic Frenkel defect.

An early demonstration that Frenkel defects do indeed predominate in silver halides was provided by studies of the effect on the ionic conductivity of addition of divalent cation, which introduces additional cation vacancies. The conductivity versus vacancy concentration was found initially to decrease, passing through a minimum, and then eventually rising. This initial decrease proves that in the pure substance the main contributor to conductivity is not the cation vacancy, but must be another, more mobile defect, the concentration of which is supressed by addition of cation vacancies. Since radiotracer studies of self-diffusion in AgCl and AgBr show that defects in the halide sub-lattice cannot contribute more than about 0.1 % of the ionic conductivity, then the dominant, mobile defect must be the interstitial silver ion.

Detailed analysis of the ionic conductivity as a function of temperature and addition of divalent cation have given the following values for the defect parameters[29,30] :

	AgCl	AgBr
Frenkel pair formation enthalpy, H_f (in eV) :	1.47	1.06
Migration enthalpy of Ag^+ interstitial, H_m^i (eV) :	0.05	0.06
Migration enthalpy of cation vacancy, H_m^v (eV) :	0.29	0.34

Actually, the interstitial migrates by displacing a substitutional cation, and there are two possibilities: collinear or non-collinear. The values quoted here for H_m^i refer to the collinear process, which has the smaller activation energy, and is the more important of the two at all but rather high temperatures. The cor-

responding entropies of formation and migration are also known. For AgCl these are (in units of k) :
$S_f = 9.95$, $S_m^i = -2.65$ and $S_m^v = -0.50$.

At temperatures near to the melting point, the
fractional concentration of Frenkel defects is approximately 0.01 for AgCl and 0.02 for AgBr. These values
greatly exceed the concentrations of Schottky defects
found in the alkali halides. They also somewhat exceed
by more than a factor of 2, the values calculated from
$\exp -(H_f-TS_f)/2kT$, because of the effect of Debye-Hüc-
kel electrostatic screening of the charged defects as
high concentrations, and perhaps also because of the
very large thermal expansion of the lattice that sets
in within about 100 degrees of the melting point.

At room temperature, the intrinsic concentrations
of Frenkel defects --i.e., the concentrations that
would be found in an infinite, perfect, and absolutely
pure crystal-- are 1.5×10^{-10} for AgCl, and 2.5×10^{-8}
for AgBr. In real crystals, however, there is typical-
ly about 1ppm of divalent cation impurity. This would
be expected to suppress the concentration of interstitial silver ions to about 6×10^{-10}, a value so low
that a typical grain in a photographic emulsion might
be expected to have only of the order of one interstitial Ag^+. On the other hand, Hamilton and Brady have
shown that the ionic step is indeed accomplished by
the transport of interstitial Ag^+ ; for example, any
dopant or adsorbant which should decrease the concentration of interstitial Ag^+ also correspondingly increases the time required for the ionic step, as evaluated from the onset of high-intensity reciprocity
failure. The answer to this apparent dilemma is found
in the effect of the negative charge carried by the
surface of the grain : the corresponding lowering of
the electrostatic potential near the surface produces
an increase in the concentration of positively charged
defects by a factor of $\exp(e\Delta V/kT)$, where ΔV is the
change in electrostatic potential. For a potential difference of 0.15 volts, this factor is 400 at room temperature.

A high mobility of the Ag^+ insterstitial is also
critical to the photographic process. Taking the vibration frequency to be about 3×10^{12}/sec in each of
the four possible jump directions (the interstitial
sits in a site of tetrahedral symmetry), one estimates
that at room temperature the jump frequency is 10^{11}/sec.
In a time of 10 microseconds, the diffusion range will
thus be several tenths of a micron, which is compara-

ble to the thickness of a typical grain. By way of comparison, the jump frequency of the cation vacancy at room temperature is only about 10^8/sec.

THE CONCENTRATION MECHANISM

We have seen that a sensitivity speck must accumulate approximately four silver atoms before it can function as a latent image. If, as one might expect, there were very many such electron trapping sites on the grain surface, then it would require the absorption of an exceedingly large number of photons before any one of these competing specks became developable. Experimentally, however, it is found that with high-sensitivity emulsions, a substantial fraction of the grains are rendered developable upon absorption of only 4 - 6 photons, and that at least half of all the grains require no more than 10. This is an astonishing result! It tells us that on most of the grains there are only one or a few efficient trapping sites which can form atomic silver. Indeed, electron microscopic studies of partially developed grains do verify that, providing the intensity is not too high, fine grains often form no more than one latent image speck per grain, while larger microcrystals usually form no more than a few per grain.

These results cannot be understood by postulating that the one point site which happens to have the highest positive charge will attract all of the photoelectrons. The large dielectric coefficient of the silver halides causes the potential of a point charge to fall to within kT (1/40 volt at room temperature) in a distance of about 15 atomic spacings. Thus, the area of influence of a point charge covers no more than 0.01 % of the surface area of the grain.

It has been pointed out[31] that one can perhaps understand the operation of this latent image concentration mechanism in terms of the negative charge on the surface. The resulting potential difference will repel photoelectrons from the immediate surface, thus causing them to sample the entire crystal. Surface traps that are not deep compared to the 0.15 volt potential difference will be ineffective, and only an exceptionally deep trap can then produce latent image. Moreover, Fatuzzo and Coppo[32] have recently shown that Ag_2S and metallic silver take on a positive potential when in contact with AgBr --and this is also consistent with Baetzold's calculations[28]-- as required by this mechanism. Thus, the electron energy bands of the silver

428

halide bend upward at the surface everywhere except in the vicinity of an Ag_2S sensitization speck or a growing silver image speck : the sensitivity site represents a window in an otherwise repulsive wall. One could thus understand the occurence of only one image speck per grain if, during sulfur sensitization, the sulfide speck nucleated at a singular surface site, such as a positively charged jog, and then grew autocatalytically, so that a small grain would be provided with one, or at most a few, electrostatic windows through which the successive photoelectrons could emerge.

THE UNIQUENESS OF THE SILVER HALIDES

It is perhaps of interest, now, to summarize those properties of AgBr and AgCl microcrystals --some of them quite unusual-- that contribute to the high efficiency of the photographic process. (1) There is no appreciable electronic conductivity in the absence of illumination. (2) The near-coincidence of the silver 4d energy levels with the halide p-levels bends the valence band up, thus allowing photoelectron production with long-wavelength photons. (3) This transition, although indirect, is of relatively high probability because of good coupling with the plentiful phonons, and because of the conservation law relaxation in mixed crystals. (4) The quantum efficiency for photoelectron production by absorbed photons is almost unity. (5) The probability of recombination between electron and hole is very low ; this is perhaps a result of the separating effect of the sub-surface electric field, the screening of the coulombic interaction by the large dielectric constant, and the fact that the recombination involves an indirect transition. (6) The internal lifetime of the photoelectron is long, perhaps due to the effect of the large dielectric constant and the low effective mass in reducing the coulombic depth of possible permanent traps, and also to the perfection of the microcrystal. (7) The low Frenkel formation and migration energies assure the availability of mobile cations with positive effective charge. (8) Successive photoelectrons are captured at the site of the trap of the first electron ; this concentration mechanism probably depends on the existence of surface charges and the possibility of formation of deep-lying trapping levels at a sensitivity center. (9) The discrimination between the presence or absence of a latent image speck enables the development process to provide an amplification exceeding 10^8.

HIGH-ENERGY RADIATION

In the track of a high-energy particle, such as a beta ray or a secondary electron, the ionization density can be very high. For example, absorption of a 1 A photon can produce 200 Ag atoms. For a grain lying in such a track, therefore, the photographic effect is equivalent to an exposure of short duration and very high intensity. The statistics for latent image formation are now of the one-hit type ; there is no threshold on an H & D plot, and developed density rises linearly with increasing radiation dose.

Because of the instantaneous release of a number of photoelectrons in the grain, the efficiency of production of a developable latent image speck is very low : the process suffers from high-intensity reciprocity failure. There can be considerable recombination of electrons and holes (recombination probability increases as the square of intensity), and those electrons which do survive are not efficiently collected by a single site. Thus, much of the silver is dispersed in a number of small specks, instead of the more useful single speck, and there can be considerable formation of internal silver. The effects of sulfur sensitization, which works to consolidate the latent image, are therefore very strong in this situation. Also, because the image specks are often small, the required times for development are somewhat longer.

Callaby[33] has measured the ratio of numbers of latent image to sub-image specks in emulsions with various grain sizes, after exposure to Co-60 gamma rays. His analysis indicates that under these conditions, it is necessary to produce approximately 40 photoelectrons within a grain in order to render it developable. For the fast electrons produced by the gammas, this requires a path length within the grain of at least 0.3 microns. Thus, emulsions with grain sizes smaller than this are found to be much less sensitive.

It has been shown by Childs[34] that the tracks of ionizing particles can be made visible within large single crystals of AgCl. It appears that the thermal strains produced by the energy released in the track leave a highly deformed cylinder, which can then trap electrons if these are subsequently swept through the specimen. The internal print-out silver which forms at the sites of the trapped electrons provides a sharp image of the track, one which is not deformed by any development process. Although singly charged particles are not registered by this method, particles of higher charge do give visible tracks, as shown in the two figures.

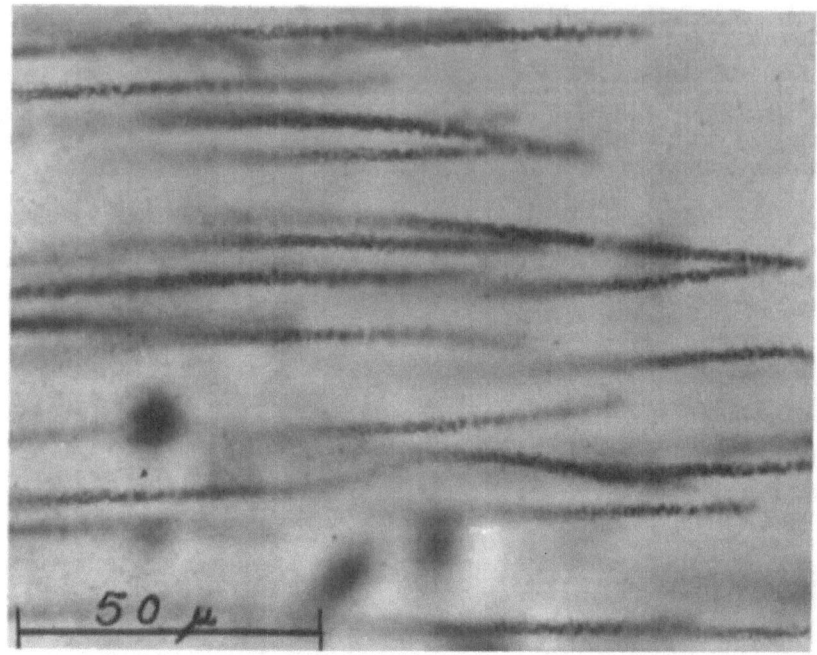

Tracks of alpha particles recorded in $AgCl:Pb^{2+}$ (Photo courtesy C.B.Childs)

The process is very sensitive to trace quantities of impurities, and appears to work best in crystals which have been annealed in air prior to exposure and which contain a few ppm Pb^{2+}. The air-anneal presumably converts potential internal image sites into inert metal oxides. The function of the Pb^{2+} is speculative. It is a large ion with a very small solubility in AgCl, and is believed to aggregate around the crystal dislocations. If it also resists oxidation during the air-anneal, it could serve as an electron trap for those electrons which reach the dislocations that were already present. Since Pb^{+} is not very unstable in some ionic systems, this trapping may thus prevent formation of photolytic silver at the back-ground dislocations, thereby preventing loss of optical contrast of the decorated tracks.

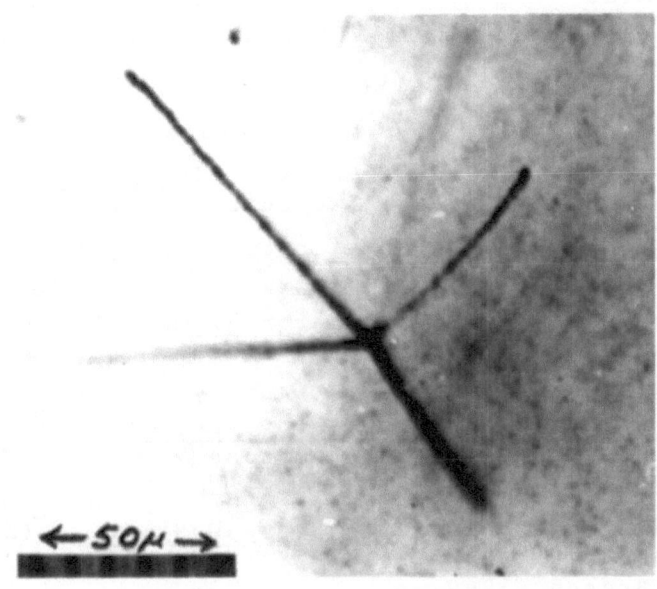

A nuclear event recorded in a crystal of $AgCl:Pb^{2+}$;
the incoming proton is not registered
(Courtesy C.B.Childs)

An interesting variation of this scheme has been
developed by Haase et al[35]. They use AgCl crystals which have been heavily doped with Cd^{2+}. It is found that
the image can be developed by subsequent exposure to
UV and that such detectors only function permanently
if they had been illuminated simultaneously with yellow
light at the time of exposure to the high-energy particles. Hence, they can be switched on and off. By varying the level of dopant, one can also control the threshold energy-loss-rate at which particles are detected.
It seems very likely that the action of such detectors
depends on the re-solution of $CdCl_2$ precipitates and
smaller aggregates in the wake of the energetic particle, thus perturbing the local concentration of vacancy (temporary hole traps) and interstitials. Also, it
is presumed that the intense ionization in the region
nucleates silver specks and perhaps halogen bubbles,
which can each later be enlarged by UV illumination during the "development" stage. The detailed mechanism,
however, is still unknown.

432

<u>REFERENCES</u>

The first eight references are to reviews of the photographic process, listed chronologically.

1. Mitchell J.W., Rep.Prog.Phys., <u>20</u>, 433, (1957)

2. Mees C.E.K. and James T.H., <u>The Theory of the Photographic Process</u>, 3rd edition (Macmillan, 1966)

3. Brown F.C., <u>The Physics of Solids</u>, p.374, (Benjamin W.A., 1967)

4. Klein E., <u>In Festkörperprobleme</u>, (Ed.by O.Madelung) Vol.8, p.74, (Frieder.Vieweg & Sohn, 1968)

5. Frieser H., Haase G. and Klein E., <u>Die Grundlagen der Photographischen Prozesse mit Silberhalogeniden</u>, Vols. 1-3, (Akademische Verlagsgesellschaft, 1968)

6. Slifkin L., in <u>Solid State Dosimetry</u>, (Ed.by Amelinchx S., Batz B. and Strumane R.), p.241 (Gordon & Breach, 1969)

7. Slifkin L., Sci.Prog. Oxf. <u>60</u>, 151 (1972)

8. Brown F.C., in <u>Solid State Chemistry</u>, (Ed.by Hannay N.B.), Chapt.10, (Plenum, in press)

9. Pontius R.B. and Willis R.G., Photogr.Sci. & Engr. <u>17</u>, 326 (1973)

10. Mueller W.E., Photogr.Sci. & Engr. <u>17</u>, 94 (1973)

11. Bayer B. & Hamilton J., J.Opt.Soc.Am. <u>55</u>, 439, 538 (1965) ; <u>56</u>, 1088 (1966) ; Hamilton J., Photogr. Sci. & Engr. <u>14</u>, 102 (1970)

12. Donecker J., Phys.Stat.Solidi <u>37</u>, 275 (1970)

13. Tani T., Photogr.Sci. & Engr. <u>17</u>, 11 (1973)

14. Levy B. and Lindsey M., Photogr.Sci. & Engr. <u>16</u>, 389 (1972) ; <u>17</u>, 135 (1973)

15. O'Brien D., Photogr.Sci. & Engr. <u>17</u>, 226 (1973)

16. Wullschleger J. and Steiger R., to be published

17. Baetzold R. and Hamilton J., Surf.Sci., <u>33</u>, 461 (1972)

18. Slifkin L., Proc.Marseille Conf.on Defects in Ionic Crystals, 1973 (to be published by J.de Physique)

19. Brown F., in <u>Point Defects in Solids</u>, (Ed.by Crawford J. & Slifkin L.) Vol.1, p.491 (Plenum, 1972)

20. Wei J. and Brown F., Photogr.Sci. & Engr. <u>17</u>, 197 (1973)

21. Malinowski J., Photogr.Sci. & Engr. <u>14</u>, 112 (1970)

22. Childs C., Private Communication ; to be published

23. Cordone L., Fornili S., Micciancio S. and Palma M., Phys.Rev.Lett. <u>26</u>, 135 (1971)

24. Riesenfeld J., Photogr.Sci. & Engr. <u>17</u>, 213 (1973)

25. Fatuzzo E. and Oggioni R., Photogr. Sci. & Engr. <u>17</u>, 319 (1973)

26. Moser F., Van Heyningen R. and Lyu S., Sol.State Comm. <u>7</u>, 1609 (1969)

27. Kanzaki H. and Sakuragi S., Photogr.Sci. & Engr. <u>17</u>, 69 (1973)

28. Baetzold R., J.Sol.State Chem. <u>6</u>, 352 (1973)

29. Corish J. and Jacobs P., J.Phys.Chem.Solids <u>33</u>, 1799 (1972)

30. Müller P., Phys.Stat.Solidi <u>12</u>, 775 (1965)

31. Slifkin L., Mc Gowan W., Fukai A. and Kim J.S., Photogr. Sci. & Engr. <u>11</u>, 79 (1967)

32. Fatuzzo E. and Coppo S., J.Appl.Phys. <u>43</u>, 1467 (1972) ; J.Photogr. Sci. <u>20</u>, 43 (1972)

33. Callaby D., J.Photogr.Sci. <u>20</u>, 157 (1972)

34. Childs C. and Slifkin L., Brit.J.Appl.Phys. <u>16</u>, 771 (1965)

35. Haase G., Zorgiebel F., Schopper E., Granzer F., Henig G., Schott J. and Wendnagel F., Photogr. Sci. & Engr. <u>17</u>, 409 (1973).

CHEMICAL ASPECTS OF RADIATION DAMAGE PROCESSES : RADIOLYSIS

K.D. ASMUS

Hahn-Meitner-Institut für Kernforschung
Berlin GmbH, Bereich Strahlen chemie

1 Berlin 39, Germany

PRIMARY SPECIES

a) - Formation of Primary Species

Interaction of high energy radiation with matter (gaseous, liquid, and solid) leads primarily to ionization and excitation, the end product of which is observable physical and chemical damage. In the present discussion, the chemical aspects will be considered.

It is well known that a high energy particle on penetrating matter loses its energy in relatively small amounts in successive stages along its track. The energy released by a Compton electron for example, amounts to an average of ca.100 eV per stage. This in general allows at least one ionization act, e.g.

$$AB \rightarrow AB^+ + e^- \qquad (1)$$

The electron will be ejected from the molecule and depending on its kinetic energy will travel a certain distance before it becomes thermalized. The positive ion owing to its larger mass and bulk generally stays behind without much translational motion taking place.

Both the electron and the positive ion may, of course, undergo chemical reactions with an appropriate scavenger. Most of these reactions are essentially diffusion controlled, i.e. they occur with maximum bimolecular rate constants of ca.$10^{10} - 10^{11}$ M^{-1} sec^{-1} (in liquids). At typical scavenger concentrations of $10^{-5} - 10^{-1}$M the half-lives for these reactions can be calcu-

lated as ca.10^{-10} - 10^{-5} sec. For a direct reaction
of AB^+ and e^- formed in the ionization process via
eq.1 a life-time of at least 10^{-10} sec is therefore
required.

Quite often the primarily produced species undergo physical or chemical changes in times less than
10^{-10} sec, and different species are present on the
time scale which is interesting for chemical reactions
with scavengers. The ionization of water, for example,

$$H_2O \rightarrow H_2O^+ + e^- \qquad (2)$$

leads to electrons and H_2O^+ ions. Within less than
10^{-11} sec the electron will be solvated

$$e^- \xrightarrow{\quad H_2O \quad} e^-_{aq} \qquad (3)$$

forming a species of completely different properties.
The H_2O^+ ion is also unstable and undergoes a fast
proton transfer reaction

$$H_2O^+ + H_2O \rightarrow H_3O^+ + OH. \qquad (4)$$

The chemically important "primary" species in the
radiolysis of water are therefore hydrated electrons
and hydroxyl radicals rather than H_2O^+ and the "dry"
electron.

Similar solvation and ion-molecule reactions occur in a variety of other systems during the early
stages of radiolysis. In addition dissociation of a
positive ion may occur within the period of a molecular vibration (ca.10^{-13} sec)

$$AB^+ \rightarrow A^+ + B \qquad (5)$$

The primary positive ion from neopentane, for
example, immediately splits off a methyl radical

$$neo\text{-}C_5H_{12}^+ \rightarrow CH_3 + C_4H_9^+ \qquad (6)$$

In the absence of suitable scavengers positive
ions and electrons may recombine

$$AB^+ + e^- \rightarrow AB^{\ddot{}} \qquad (7)$$

to produce an excited species. This process is only
possible, however, if at least part of the recombination energy (=ionization energy) can be transferred

to a third body or be dissipated as vibrational excitation over the molecule.

Excited states may also be formed by direct excitation

$$AB \rightsquigarrow AB^{*} \qquad (8)$$

The fate of these excited states is either collisional deactivation or, chemically more important, dissociation into radicals :

$$H_2O^{*} \rightarrow H\cdot + OH\cdot \qquad (9)$$

$$c\text{-}C_6H_{12}^{*} \rightarrow H\cdot + c\text{-}C_6H_{11}\cdot \qquad (10)$$

$$R_1 \rightsquigarrow R_2^{*} \rightarrow R_1 \rightsquigarrow \cdot + \cdot \rightsquigarrow R_2 \qquad (11)$$

In addition to radicals and ions molecular species are formed as primary products in many cases. In γ-irradiated water, for example, molecular H_2 and H_2O_2 are produced which cannot be scavenged at normal scavenger concentrations. Similarly in cyclohexane H_2 and dicyclohexyl is produced the yield of which cannot be suppressed by H atom and cyclohexyl radical scavengers, respectively. These observations are explained by either molecular elimination from an excited molecule such as

$$c\text{-}C_6H_{12}^{*} \rightarrow H_2 + c\text{-}C_6H_{10} \qquad (12)$$

or fast combination of radicals produced in the immediate vicinity of each other (spur reaction)

$$OH\cdot + OH\cdot \rightarrow H_2O_2 \qquad (13)$$

$$C_6H_{11}\cdot + C_6H_{11}\cdot \rightarrow C_6H_{11}\text{-}C_6H_{11} \qquad (14)$$

b) - Radiation Chemical Yields

Radiation chemical yields are usually measured in G-units rather than concentrations. G is defined as the number of molecules produced or destroyed per 100 eV absorbed energy, i.e.

$$G = \frac{N_L \cdot c \cdot 10^2}{D \cdot 10^3}$$

($N_L = 6.023 \cdot 10^{23}$ molecules/mole ; c = concentration in mole/liter ; D = dose in eV/g).

An average 30 eV are necessary for one ionization process (and simultaneous excitation) which gives a mean of G (ion pair) $\approx$ 3.

Typical G-values for primary species may in fact be anything up to ca.6, as can be seen from the following table :

Some Primary yields in γ-irradiated

Water	Cyclohexane
$G(e^-_{aq}) = 2.7$	$G(e^-) = 4.0$
$G(OH\cdot) = 2.8$	$G(H\cdot) = 1.5$
$G(H.) = 0.6$	$G(c\text{-}C_6H_{11}\cdot) = 5.2$
$G(H_2)_{mol} = 0.45$	$G(H_2)_{mol} = 4.2$
$G(H^+_{aq}) = 2.7$	$G(C_6H_{11}\text{-}C_6H_{11})_{mol} = 0.5$

CHEMICAL SCAVENGING OF PRIMARY SPECIES

a) - Scavenging Kinetics

The yield of radicals and ions produced during irradiation may be determined by chemical scavenging and analysis for the reaction products. Organic radicals, for example, readily react with iodine via

$$R\cdot + I_2 \rightarrow RI + I\cdot \qquad\qquad (15)$$

The yield of RI as a function of I_2 concentration is shown in fig.1[1] (γ-irradiated-iso-octane).

At low iodine concentrations G(RI) is seen to increase with the scavenger concentration. The curve then reaches a plateau clearly indicating that all radicals available are scavenged. The plateau value of G(RI), therefore, can be regarded as a true measure of the free radical yield. Since iodine is a very efficient radical scavenger concentrations as low as about 10^{-6}M are sufficient for complete scavenging. At very high I_2 concentrations (> 10^{-2}M) the curve slightly increases again. This is explained in terms of increasing interference with radical processes which would otherwise have led to the formation of molecular products.

Quite a similar picture is observed for the scavenging of hydrated electrons in water (and other ions in the gas phase or polar liquids). Relatively small scavenger concentrations (10^{-4} - 10^{-5}M) are sufficient to scavenge all of the free, i.e. homogeneously distributed ions.

A different situation with respect to the scavenging of ions and electrons exists for an irradiated

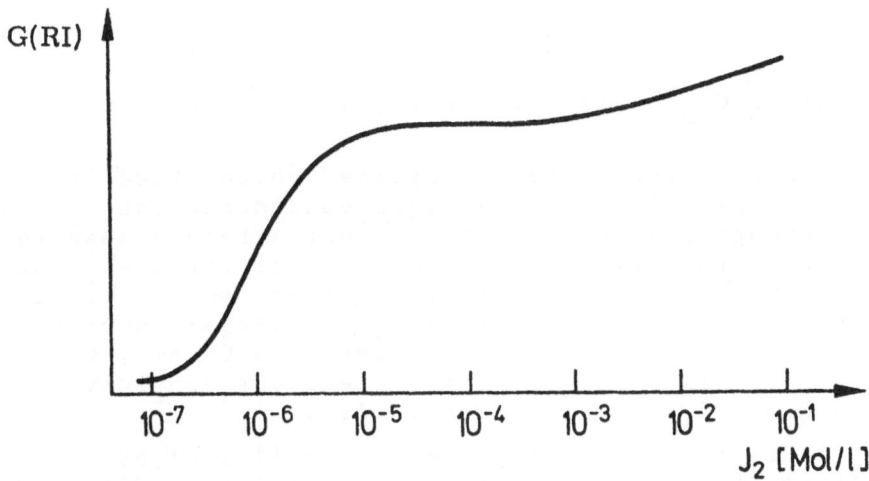

Fig.1 : Yield of radicals scavenged by iodine in γ-irradiated iso-octane.

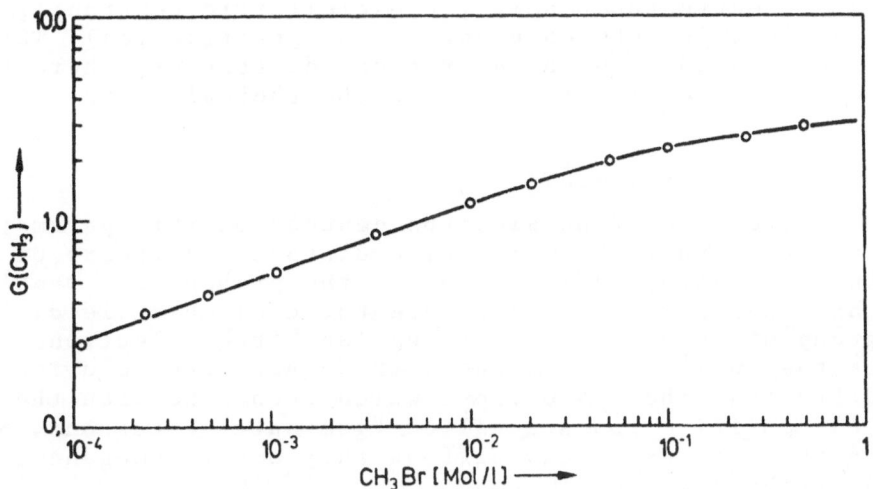

Fig.2 : Yield of methyl radicals from the reaction of methylbromide with electrons in γ-irradiated cyclohexane[2].

solvent of low dielectric constant. In a hydrocarbon solution the yield of electrons can be determined[2] via

$$CH_3Br + e^- \rightarrow CH_3 + Br^- \qquad (16)$$

A plot of $G(CH_3)$ versus scavenger concentration is shown in fig.2[2].

An S-shaped curve is obtained which steadily increases. Even at high scavenger concentrations complete scavenging i.e. a plateau is not attained despite the fact that methyl bromide is one of the most powerful known electron acceptors. Furthermore the yield of scavenged electrons at ca. $10^{-4}M$ scavenger concentration amounts to only a few tenths of a G-unit, whereas in a polar solvent like water electrons are scavenged with G = 2.7 at this concentration.

This finding is explained as follows : as a result of the ionization process electrons are ejected from the molecules and depending on their kinetic energy they can travel a certain distance before being slowed down to thermal energies (kT). Opposing this process however the electron will of course be attracted by the positive ion due to the coulombic potential given by

$$\frac{e^2}{\varepsilon.r}$$

(e = electrostatic unit ; ε = dielectric constant ; r = distance between electron and positive ion). This allows calculation of a critical distance r_c where the coulomb potential is equal to the thermal energy

$$r_c = \frac{e^2}{\varepsilon.kT}$$

The fate of an electron depends on this parameter and the actual thermalization distance. A theory of Onsager[3] allows calculation of the probability $\phi = e^{-r_c/r}$ that the electron will escape the coulombic field. The group of such electrons is called "free" electrons since they will be distributed throughout the solution by diffusion. Those electrons which recombine with the parent positive ion are called "geminate" electrons. Since they cannot freely diffuse they are inhomogeneously distributed.

The range of electron energies, i.e. the range of thermalization distances r is very similar for all irradiated solvents. Differences in the escape probabilities ϕ are, therefore, mainly determined by the parameter r_c which is dependent on the dielectric constant and the temperature. At room temperature (ca.20°C) r_c is calculated as ca. 7 Å and 300 Å for water ($\varepsilon = 80$) and cyclohexane ($\varepsilon = 2-3$), respectively. The overall probability of an electron's escaping the coulombic field is, therefore, much higher in water ($G = 2.7$) than in cyclohexane ($G = 0.1$).

The different yields of "free" and "geminate" electrons and also of positive ions have a direct bearing on the chemistry of an irradiated system. Free, homogeneouly distributed electrons and ions are quantitatively scavenged by suitable compounds at low scavenger concentration. Geminate electrons and ions are much more difficult to scavenge because the coulombic forces do not allow a free diffusion and their overall distribution is inhomogeneous[2,4].

For practical purposes a semiempirical equation[2] allows one to estimate the yield of scavenged ions as a function of scavenger concentration in a non-polar liquid, i.e. calculation of the curve shown in fig.2

$$G_{si} = G_{fi} + G_{gi} \frac{\{\alpha\,[s]\}^{1/2}}{1 + \{\alpha\,[s]\}^{1/2}}$$

where α is a reactivity parameter proportional to the reaction rate constant, $[s]$ is the scavenger concentration, and the indices si, fi and gi refer to "scavenged ions", "free ions", and "geminate ions", respectively. This equation for the scavenging of inhomogeneously distributed ions becomes important for all non-polar solvents where G_{gi} exceeds G_{fi}.

Scavenging of neutral radicals is, of course, not affected by the dielectric properties of the solvent. These species diffuse randomly and are distributed homogeneously throughout the solution.

b) - Competition reactions

The kinetics of competition reactions also depend on whether the reactive species are distributed homogeneously or inhomogeneously.

If, for example, the homogenously distributed hydrated electron in water can react via two reactions paths

$$A + e_{aq}^- \xrightarrow{\ k_A\ } A^- \qquad\qquad (17)$$

$$B + e_{aq}^- \xrightarrow{\quad k_B \quad} B^- \qquad (18)$$

(k_A, k_B = bimolecular reaction rate constants) the
yields of A^- and B^- are given by

$$G(A^-) = G(e_{aq}^-) \frac{k_A [A]}{k_A [A]+k_B [B]}$$

and

$$G(B^-) = G(e_{aq}^-) \frac{k_B [B]}{k_A [A]+k_B [B]}$$

respectively. Where more than two competitors are present this gives the general form

$$G(A^-) = G(e_{aq}^-) \frac{k_A [A]}{\sum_i k_i c_i}$$

(c_i = concentration of the i-th competitor), i.e. the
fraction of electrons reacting with the various scavengers depends on the respective products of rate
constant times scavenger concentration.

These equations in general apply to all radical
and free ion reactions. The quantitative treatment of
the competition of geminate electrons and ions has to
be based on the inhomogeneous scavenging kinetics[2].
The yield of A^- from the reaction of A with electrons
in cyclohexane, for example, is then expressed by

$$G(A^-) = \left[G_{fi}+G_{gi} \frac{\{\sum_i \alpha_i c_i\}^{1/2}}{1+\{\sum_i \alpha_i c_i\}^{1/2}} \right] \frac{\alpha_A c_A}{\sum_i \alpha_i c_i}$$

The important factor in this case is that the total yield of electrons (or ions) scavenged increases
with the total scavenger concentration (given by the
expression in the square brackets), whereas in the homogeneous system the total yield of scavengable species is independent of the scavenger concentration.

The relative distribution of the electrons or ions
in the inhomogeneous system among the various scavengers also depends on the individual reactivities and
concentrations.

c) - Brief outlining of some experimental methods

It is evident from the discussion so far that the chemistry of the primary species produced upon irradiation plays a key role for the overall radiolysis of a system. Since the lifetime and consequently the actual concentration of these highly reactive species is generally rather small very sensitive and fast analytical methods are required for investigation of their properties. Two non-conventional analytical methods, therefore, shall be discussed briefly.

Let us, for example, consider the reaction

$$CH_3Br+e^- \rightarrow CH_3+Br^- \tag{16}$$

which was used to determine the yield of scavenged electrons in an irradiated hydrocarbon. In order to determine the curve given in fig.2, particularly at low methylbromide concentration, extremely small yields of $CH_3^\cdot$ radicals have to be analysed. This became possible by using radioactive iodine as a scavenger for methyl radicals[2,5].

$$CH_3^\cdot + {}^{131}I_2 \rightarrow CH_3{}^{131}I + {}^{131}I \cdot \tag{19}$$

$CH_3{}^{131}I$ can be separated by gaschromatography and measured by a scintillation counter. This method allows quantitative analysis of as little as 10^{-10} moles in 0.5 cc with an accuracy of ± 1 %. ($t_{1/2}$ of ${}^{131}I_2$ is 8.05 days).

Short lived species and their reactions in general can be investigated by several methods :

1. by characteristic chemical reactions which yield a stable product

2. by increasing their lifetime by working at lower temperatures and in the solid, particularly glassy state

3. by direct observation with a fast detection method. Among these techniques pulse radiolysis has proven to be a most useful one, and the most essential features will now be discussed.

Short electron pulses of highly energetic electrons ranging from pico- to microsecond duration from a van de Graaff or linear accelerator are directed into a solution or solid, usually contained in a quartz cell. This produces the primary, highly reactive species at concentrations up to ca.10^{-5}M. Their physical and che-

444

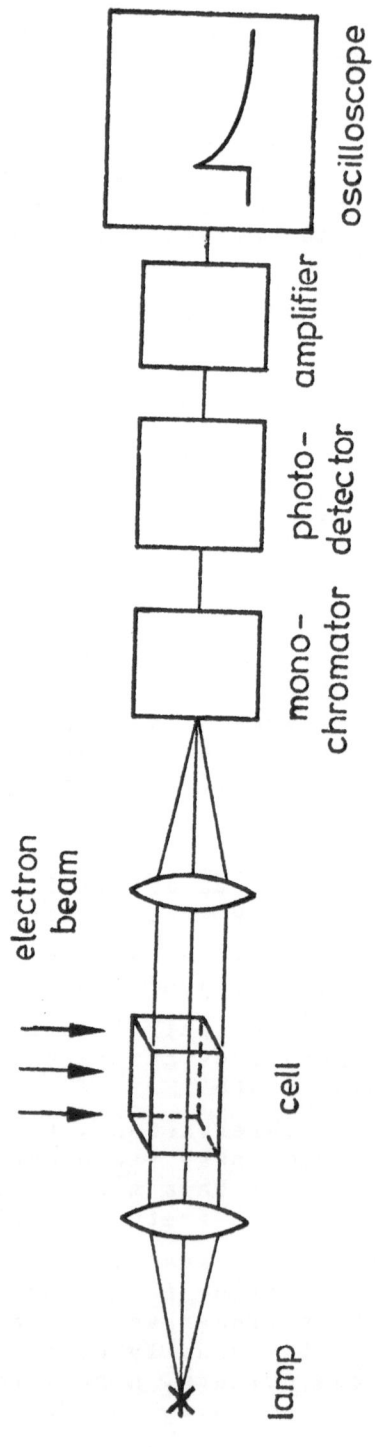

lamp   electron beam   cell   mono-chromator   photo-detector   amplifier   oscilloscope

Fig.3a : Optical method for the observation of primary species produced upon irradiation.

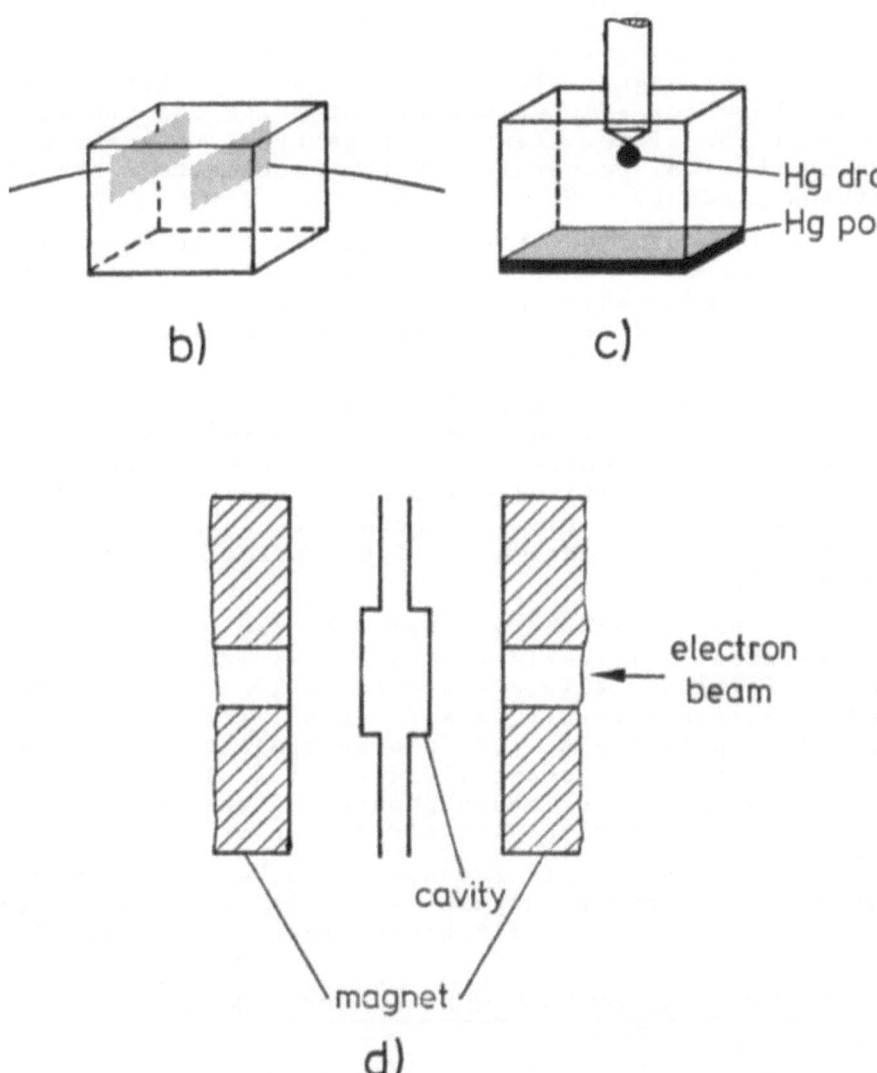

Fig.3 : Experimental methods for the study of primary species
produced upon irradiation.

mical properties can be observed by various direct methods.

(a) Changes in optical absorption as a function of time at various wavelengths in the visible and UV can be detected by the arrangement shown in fig.3a. Absorbing species produced during the pulse will subtract from the intensity of a light beam (e.g. from a xenon lamp). A typical absorption-time signal is shown on the oscilloscope. It indicates the fast formation of an absorbing species during the pulse and subsequent decay of the signal (taken at a particular wavelength) due to chemical conversion of the species to a non absorbing product.

Such absorption-time curves permit the determination of reaction kinetics, and absorption spectra of transient and stable species (and extinction coefficients), and the direct observation of chemical reactions with added reactants.

(b) Location of two electrodes in the cell (fig. 3b) with an applied voltage of ca. 10-100 V allows measurement of conductivity changes as a function of time in the pulse irradiated solutions[6]. Such experiments are useful for the identification of charged species, particularly if no or only slight optical absorption is observed[7].

(c) Polarographic measurements can be conducted for species of life-times of a few microseconds[8] (fig. 3c). This method allows the observation of reduction and oxidation currents as a function of time at various electrode potentials, i.e. the determination of half-wave potentials of short lived species. In addition to the investigation of redox properties this technique is also of use for the identification of species.

(d) Electron spin resonance can be applied for identification and structural analysis of transient radicals (fig. 3d) with lifetimes greater than ca. 10 μsec[9].

RADIATION CHEMISTRY OF AQUEOUS SOLUTIONS AS AN EXAMPLE FOR POLAR SOLVENTS

Irradiation of water leads to the following primary species on a timescale of interest for chemical reactions :

$$H_2O \rightarrow e_{aq}^-, OH\cdot, H\cdot, H_2, H_2O_2 \qquad (20)$$

The most important (most highly reactive) species are the hydrated electron, the hydroxyl radical, and the hydrogen atom. In a neutral solution they are formed with the yields of $G(e_{aq}^-) = 2.7$, $G(OH\cdot) = 2.8$, and $G(H\cdot) = 0.6$.

a) - <u>Reduction Reactions by Hydrated Electrons</u>

The hydrated electron is the most powerful reducing agent in aqueous solutions (reduction potential = -2.6V), and as such reacts readily with a large number of compounds. The kinetics of such reactions can easily be investigated by means of pulse radiolysis since the hydrated electron has a strong optical absorption in the visible and near IR range (λ_{max} = 720 nm with ε_{720nm} = 1.8 x $10^4 M^{-1} cm^{-1}$). Several types of processes can be distinguished :

(I) Associative electron capture, as for example the reduction of metal ions

$$Fe^{3+} + e_{aq}^- \rightarrow Fe^{2+} \tag{21}$$

$$Cu^{2+} + e_{aq}^- \rightarrow Cu^+ \tag{22}$$

$$Cd^{2+} + e_{aq}^- \rightarrow Cd^+ \tag{23}$$

$$Zn^{2+} + e_{aq}^- \rightarrow Zn^+ \tag{24}$$

or other molecules as

$$O_2 + e_{aq}^- \rightarrow O_2^-. \tag{25}$$

$$NO_3^- + e_{aq}^- \rightarrow NO_3^{2-}. \tag{26}$$

$$(CH_3)_2CO + e_{aq}^- \rightarrow (CH_3)_2\dot{C}O^- \tag{27}$$

thereby forming radical ions.

Such reactions are possible to a large extent since in water, i.e. in the condensed phase, the excess energy (= electron affinity) can rapidly be transferred to surrounding molecules. Most of the reactions of the hydrated electron are essentially diffusion controlled processes.

(II) Another group of high electron affinity compounds undergoes a dissociative electron capture process.

$$N_2O + e^-_{aq} \xrightarrow{H_2O} N_2 + OH\cdot + OH^- \qquad (28)$$

$$CCl_4 + e^-_{aq} \rightarrow CCl_3\cdot + Cl^- \qquad (29)$$

$$HgCl_2 + e^-_{aq} \rightarrow HgCl + Cl^- \qquad (30)$$

$$C(NO_2)_4 + e^-_{aq} \rightarrow C(NO_2)_3^- + NO_2 \qquad (31)$$

(III) Both I and II may happen simultaneously. The reduction of halogenated aromatics such as fluorobenzene[10] occurs to 20 % via dissociative capture

$$C_6H_5F + e^-_{aq} \rightarrow C_6H_5\cdot + F^- \qquad (32)$$

and to 80 % via interaction of the electron with the π-system of the aromatic ring and subsequent neutralization of the anion thus formed to a cyclohexadienyl radical

(33)

A very important factor for the nature of the electron reaction product is the pH of the solution. Most of the radicals or radical ions formed in the reduction process exist in an acid-base equilibrium. The O_2^- ion, for example, is the anion of the HO_2 radical

$$HO_2\cdot \rightleftharpoons O_2^- + H^+_{aq} \qquad (34)$$

and the pK of this equilibrium is 4.8[11].

The reduced form of acetone similarly exists in an equilibrium

$$(CH_3)_2\dot{C}OH \rightleftharpoons (CH_3)_2\dot{C}O^- + H^+_{aq} \qquad (35)$$

with a pK of 12.2[12].

Almost all radicals with functional groups containing oxygen, nitrogen, sulfur and to some extent carbon-hydrogen bonds

$$- O - H \qquad - S - H$$
$$- N - H \qquad - C - H$$

can dissociate to form anions. The respective pK values are strongly influenced by the nature of additional functional groups. Electron withdrawing groups ($-NO_2$, $-CN$, $> CO$, halogen) favour the dissociation, i.e. decrease the pK. Electron releasing groups such as $-NH_2$, $-CH_3$, $-OCH_3$ etc. on the other hand decrease the probability of dissociation thereby increasing the pK.

Knowledge of the acid-base properties of radicals is of great importance for the understanding of the radiation chemistry of a system. Very often the anionic and the neutralized form exhibit different chemical reactivities. In general the basic form has a longer lifetime and is a much more powerful reducing agent.

b) - <u>Oxidation Reactions of Hydroxyl Radicals</u>

The hydroxyl radical exists in an acid base equilibrium

$$OH. \; \rightleftarrows \; O.^- + H^+_{aq} \qquad (36)$$

The pK is 11.9. In very basic solutions, therefore, O^- is the predominant form, and it shows a quite different behaviour as will be discussed later.

In general, certain types of hydroxyl radical reactions have to be considered.

(I) Abstraction reactions :

$$RH + OH^{\cdot} \; \rightarrow \; R^{\cdot} + H_2O \qquad (37)$$

Abstraction of a hydrogen atom is particularly prevalent in aliphatic hydrocarbons (and also substituted compounds). Cleavage of a tertiary H atom is favoured to a secondary H atom, and the highest activation energy is required for the abstraction of a primary H atom. If several types of hydrogen atoms are available abstraction may occur at various positions[13]. The reaction

$$OH^{\cdot} + CH_3CH_2OH \; \rightarrow \; H_2O + \overset{\bullet}{C}H_2CH_2OH \qquad (38a)$$
$$CH_3\overset{\bullet}{C}HOH \qquad (38b)$$
$$CH_3CH_2O. \qquad (38c)$$

yields three different species with 13.2 % (38a), 84.3 % (38b), and 2.5 % (38c) efficiency. The relative yields are important for the overall chemistry ; the α-radical $CH_3\overset{\bullet}{C}HOH$ has strong reducing properties, the oxyradical $CH_3CH_2O.$ good oxidizing properties.

(II) Addition reactions of the hydroxyl radical occur at centers of high electron density, particularly in aromatic systems, and those containing olefinic and other double bonds :

$$CH_2 = NO_2^- + OH \cdot \rightarrow CH_2 - \overset{\cdot}{N}O_2^- \qquad (39)$$

$$\underset{/}{\overset{\backslash}{C}} = \underset{\backslash}{\overset{/}{C}} + OH\cdot \rightarrow \cdot\underset{|}{\overset{|}{C}} - \underset{|}{\overset{|}{C}} - OH \overset{OH}{} \qquad (40)$$

These reactions are based on the electrophilic nature of the hydroxyl radical. At this point it is interesting to note that the basic form $O^{\overline{\cdot}}$ abstracts a hydrogen atom rather than undergoing addition. This, however, is easy to understand, since its negative charge reduces the electrophilic character.

(III) The hydroxyl radical can also act as a simple oxidizing agent by taking over an electron as in the following reactions

$$Fe^{2+} + OH\cdot \rightarrow Fe^{3+} + OH^- \qquad (41)$$

$$NO_2^- + OH\cdot \rightarrow NO_2\cdot + OH^- \qquad (42)$$

Depending on the nature of the compounds, two or all three types of reactions may, of course, occur simultaneously. A thioether, for example, reacts with hydroxyl radicals forming 3 different products[14]

$$R-S-R + OH\cdot \rightarrow R-S-R(-H)\cdot + H_2O \qquad (43a)$$
$$\rightarrow (R-SOH-R)\cdot \qquad (43b)$$
$$\rightarrow R-S-R^+ + OH^- \qquad (43c)$$

The rate constants of addition and electron transfer reactions are very often only controlled by the diffusion of the reactants, whereas abstraction reactions generally require an activation energy to break the C-H bond.

c) - Reactions of hydrogen atoms

In neutral water hydrogen atoms are produced with only a relatively small yield of $G(H\cdot) = 0.6$. In acid solutions, however, they are formed with $G = 3.3$ since all hydrated electrons are converted to H atoms via

$$H_{aq}^+ + e_{aq}^- \rightarrow H\cdot \qquad (44)$$

In radiation chemistry as a whole hydrogen atoms play a role almost as important to that of electrons, since they are to be found in almost every system of hydrogen containing molecules. Even in irradiated solids highly mobile hydrogen atoms have been observed.

Hydrogen atoms undergo the same type of reactions as hydroxyl radicals such as abstraction

$$RH + H \cdot \rightarrow R \cdot + H_2 \qquad (45)$$

and addition

$$\text{(benzene ring)} + H \cdot \rightarrow \text{(cyclohexadienyl radical, } C \cdot \text{ with two H)} \qquad (46)$$

In addition hydrogen atoms are also involved in electron transfer processes. In contrast to the hydroxyl radical, however, they act as a reducing agent (reduction potential $\approx -2.0V$) :

$$H \cdot + O_2 \rightarrow HO_2^{\cdot} \rightleftarrows O_2^{-} + H_{aq}^{+} \qquad (47)$$

$$H \cdot + CCl_4 \rightarrow CCl_3^{\cdot} + Cl^{-} + H_{aq}^{+} \qquad (48)$$

$$H \cdot + Hg^{2+} \rightarrow 1/2 Hg_2^{2+} + H_{aq}^{+} \qquad (49)$$

d) - Electron Transfer Reactions

A most important type of reaction in radiation chemistry is represented by electron transfer processes They have to be considered in almost all systems containing several scavengers of reasonably high electron affinity. This will be illustrated by two examples.

The O_2^{-} ion which is formed by the reaction of oxygen and hydrated electrons, for example, can easily transfer an electron to another compound of high electron affinity

$$O_2^{-} + C(NO_2)_4 \rightarrow O_2 + C(NO_2)_3^{-} + NO_2 \qquad (50)$$

On the other hand oxygen itself may accept an electron from another radical

$$CO_2^{-} + O_2 \rightarrow CO_2 + O_2^{-} \qquad (51)$$

α-Alcohol radicals (which are formed in the reaction of alcohols with hydroxyl radicals and hydrogen atoms, or by reduction of ketones and aldehydes) very often can also undergo electron transfer processes

$$\begin{array}{c} R \\ R \end{array}\!\!\dot{C}\text{-OH} + X \rightarrow X^{-} + \begin{array}{c} R \\ R \end{array}\!\!C = O + H_{aq}^{+} \qquad (52)$$

where X is a molecule which has a higher electron affinity than the radical.

In fact, all types of radicals may undergo similar processes. The extent to which these transfer reactions occur primarily depends on the relative electron affinities of the donor and acceptor molecules, i.e. their structure and the nature of their functional groups.

These reactions certainly complicate the overall radiation chemistry of a complex system, since the final stable products of radiolysis cannot necessarily be explained directly in terms of the primary reactions.

e) - General Considerations

It is clear that most of the aspects discussed in this "aqueous" section are of general relevance to all kinds of radiolysis processes. This is particularly true for the various types of electron and hydrogen atom reactions, the acid-base equilibria of radicals, and the electron transfer reactions. In going from liquid water to another system the effect of dielectric properties, structure etc. on the various thermodynamic and kinetic parameters has, of course, to be taken into account.

RADIATION CHEMISTRY OF CYCLOHEXANE AS AN EXAMPLE OF NON-POLAR LIQUIDS

From the chemical point of view the same kind of reactions occur in the main in non-polar liquids. Owing to the large fraction of geminate electrons and ions, i.e. inhomogeneously distributed species, however, the scavenging kinetics are quite different compared with those in a highly polar system. It has been indicated already that it is virtually impossible to scavenge all geminate electrons and ions. Therefore, the chemical fate of the unscavenged part becomes equally important.

a) - Electron Reactions

In general electrons which are formed via

$$c-C_6H_{12} \rightarrow c-C_6H_{12}^+ + e^- \qquad (53)$$

are found to react with the same compounds in the hydrocarbon as they do in an aqueous solution ; i.e. they are easily scavenged by all kinds of halogen containing compounds such as CCl_4, $c-C_6F_{12}$, CH_3Br, SF_6 etc., and other molecules with strong electronegative functional groups as $C(NO_2)_4$, $C_6H_5NO_2$ etc. Nitrous oxide, molecular oxygen, and carbon dioxide are similarly good electron scavengers. However, a high rate constant for the reaction of a solute with electrons in water does not necessarily mean a high reactivity in a hydrocarbon as well. Iodine, for example, in water is one of the most efficient electron scavengers $(k(I_2+e_{aq}^-) = 5 \times 10^{10}$ $M^{-1}sec^{-1})$ whereas in cyclohexane it does not seem to react at all[2]. (This may be due to I_3^- formation in water).

The main difference between electron reactions in a polar and a non-polar liquid is the efficiency of scavenging[2]. We have seen already (fig.2) that it is virtually impossible to scavenge all electrons even at high scavenger concentrations. Furthermore the yield is generally dependent on the solute concentration. For a CH_3Br solution in cyclohexane, for example, the yield of scavenged electrons can be calculated via the equation for inhomogeneous scavenging kinetics

$$G(e^-)_{scav} = G_{fi} + G_{gi} \frac{\{\alpha \, [CH_3Br]\}^{1/2}}{1 + \{\alpha \, [CH_3Br]\}^{1/2}}$$

with G_{fi} = 0.12, G_{gi} = 3.8 and α = 16.2 $M^{-1})^2$

At 10^{-2}M CH_3Br $G(e^-)_{scav}$ = 1.21, and at 10^{-1}M CH_3Br $G(e^-)_{scav}$ = 2.25. Considering that a total of $G(e^-) \approx 4.0$ is available only a fraction is seen to be scavenged while the rest of the electrons will undergo different reactions.

b) - Positive Ion Reactions

The scavenging kinetics of electrons can also be applied to positive ions, since they are trapped in the field of mutual coulombic attraction to the same extent[2,4]. In other words, the positive ions are also distributed inhomogeneously in irradiated cyclohexane and the yield which can be scavenged by suitable compounds is a function of the scavenger concentration. The reactivity parameter α, however, is generally lower by a factor of ca. 20 compared with that for electron scavengers. This is due to the much smaller mobilities of the large positive ions compared with the highly mobile electrons.

Typical scavengers for positive ions are ammonia, amines, water, olefines like ethylene, aromatics like benzene, naphthaline, pyrene etc. and small ring compounds like cyclopropane. The type of reaction undergone by a positive ion can be compared to some extent with those of the hydroxyl radical, i.e. an RH^+ formed via

$$RH \rightarrow RH^+ + e^- \tag{54}$$

has oxidizing properties and can thus remove an electron from aromatics, for example,

454

$$RH^+ + Ar \rightarrow RH + Ar^+ \qquad (55)$$

In addition they transfer protons as in the following reaction[15,16]

$$RH^+ + NH_3 \rightarrow NH_4^+ + R. \qquad (56)$$

The reaction of positive cyclohexane ions with cyclopropane finally results in ring opening and H_2-transfer to form propane, C_3H_8, as a stable reaction product[17,18].

c) - Ion Recombination Processes

It is generally known that ions are much less stable in non-polar than in polar liquids due to the effect of solvent dielectric properties on dissociation constants. Furthermore, in an irradiated hydrocarbon the majority of electrons and ions including their primary reaction products remain within the sphere of mutual coulombic attraction, and consequently fast ion recombination will occur.

Let us consider, for example, an irradiated solution of positive ion scavengers (NH_3) and electron scavengers (CH_3Br) in cyclohexane. Depending on the scavenger concentration RH^+ will partially react with NH_3 to form NH_4^+ ions (eq.56) and part of the electrons will react with methyl bromide to methyl radicals and bromide ions. Ion recombination then occurs via the processes

$$RH^+ + e^- \rightarrow RH^* \qquad (57)$$
$$RH^+ + Br^- \rightarrow R. + HBr \qquad (58)$$
$$NH_4^+ + e^- \rightarrow NH_3 + H. \qquad (59)$$
$$NH_4^+ + Br^- \rightarrow NH_4Br \qquad (60)$$

(the excited RH^* may then dissociate into radicals, or is deactivated in collisional processes). In general the products of all possible ion recombination have to be taken into account for the overall chemical process.

A further important feature of solutions of both electron and positive ion scavengers must now be discussed. The positive ion, RH^+, has a certain lifetime with respect to its neutralization by electrons (eq.57). If the electron is now converted into a large and much less mobile anion the rate of neutralization of RH^+

(e.g. by eq.58) will be slowed down[19]. Consequently the lifetime of RH^+ will increase and so will the probability for a reaction of a positive ion with a suitable scavenger. In other words, the reactivity parameter α for a positive ion scavenger and thus the product yield from positive ion reactions increases with electron scavenger concentration.

d) - Radical Reactions

The one-electron reduction and oxidation reactions, the addition and abstraction processes, and the dissociation of excited states in most cases yield short lived radicals or radical ions. The chemical fate of these species is, of course, of direct relevance for the final, stable radiolysis products. Although they are discussed in the "cyclohexane" section the radical reactions are of general importance independent of the matrix, i.e. the principles apply to polar liquids, gas phase, and solid phase as well as for the non-polar liquids.

I) Some of the reaction types have already been presented in the previous chapters in terms of hydroxyl radicals and hydrogen atoms. Thus radicals are very likely to undergo hydrogen abstraction reactions from C-H bonds thereby forming new radicals :

$$CH_3^{\cdot} + c\text{-}C_6H_{12} \rightarrow CH_4 + c\text{-}C_6H_{11}\cdot \qquad (61)$$

$$R_1^{\cdot} + R_2H \rightarrow R_1H + R_2\cdot \qquad (62)$$

II) Very good radical traps are aromatics, olefins, and other compounds with double bonds (addition reactions)

$$R. + ^{\diagdown}C=C^{\diagup} \rightarrow ^{|}_{|}C-C^{|}_{|}-R \qquad (63)$$

III) These reactions play an important role for polymerisation processes where one radical may start a chain of reactions if the radical formed in reaction (63) adds to another monomer double bond compound.

Another chain reaction which can be observed for example in a mixture of cyclohexane and carbon tetrachloride,

$$CCl_3^{\cdot} + c\text{-}C_6H_{12} \rightarrow CCl_3H + c\text{-}C_6H_{11}\cdot \qquad (64)$$

$$c\text{-}C_6H_{11}^{\cdot} + CCl_4 \rightarrow c\text{-}C_6H_{11}Cl + CCl_3^{\cdot} \qquad (65)$$

is responsible for high yields of chloroform and cyclohexylchloride. The initiating $CCl_3^\bullet$ radicals are formed by reaction of CCl_4 with electrons and hydrogen atoms, the c-$C_6H_{11}\cdot$ radicals by reaction of c-C_6H_{12} with hydrogen atoms or decay of excited cyclohexane.

IV) Termination of a chain reaction is normally achieved by a radical -radical reaction. Such processes, which also occur between radicals not involved in chain reactions, are found to take place either via addition

$$OH\cdot + OH\cdot \rightarrow H_2O_2 \qquad (66)$$

or disproportionation

$$2\ C_6H_5NO_2H \rightarrow C_6H_5NO_2 + C_6H_5NO + H_2O \quad (67)$$

Steric effects and redox properties are important factors for these processes. Highly structured radicals, for example, disproportionate rather than combine. This is quite well demonstrated in the following example of the deactivation of alcohol radicals. In this case both disproportionation and addition occur simultaneously

$$2\ R_1R_2\overset{\bullet}{C}OH \left\{ \begin{array}{ll} R_1R_2CHOH + R_1R_2CO & (68a) \\ (R_1R_2COH)_2 & (68b) \end{array} \right.$$

The ratio of disproportionation (D) to combination (C) has been found to be $D/C = 0.57$, $1.9.$, and 6.6 for methanol, ethanol and isopropanol radicals, respectively[20].

V) Electron transfer reactions from radicals to other molecules of high electron affinity are also of general importance. They have already been discussed in detail in the "aqueous" section.

VI) Reactions of radicals with molecular oxygen occur via fast reactions if even traces of O_2 are present in an irradiated system. Usually peroxy radicals are formed

$$CH_3^\bullet + O_2 \rightarrow CH_3O_2^\bullet \qquad (69)$$

$$H\cdot + O_2 \rightarrow HO_2^\bullet \qquad (70)$$

$$R\cdot + O_2 \rightarrow RO_2^\bullet \qquad (71)$$

Such radicals often combine to tetroxides[21]. Unfortunately there is still a lack of information on the chemistry of the peroxy radicals which very often precludes the establishment of complete reaction mechanisms for irradiated oxygen containing systems.

VII) Radicals which result from the dissociation of excited molecules may carry some excess kinetic energy. These socalled "hot" radicals mainly exist in the gas phase. To a certain, though smaller extent (higher collision frequency) they are also observed in the condensed phase. They essentially undergo abstraction reactions since owing to the excess kinetic energy of these radicals only little if any extra activation energy is required. If abstractions may occur from the solvent this means that the probability for other reactions will be almost zero.

EXCITED STATES AND ENERGY TRANSFER

The importance of excited states in the radiation chemistry of a system has been indicated already in terms of the possible dissociation of an excited molecule into radicals.

For chemical processes the relatively long lived triplet states are more important, since singlet excited states which are formed in comparable yields are generally deactivated by physical processes (collision, fluorescence).

In addition to dissociation into radicals or molecules an excited species may also transfer its energy

$$A^* + B \rightarrow A + B^* \tag{72}$$

$$\text{e.g.} \quad \text{[naphthalene]} + (CH_3)_2CO_T^* \rightarrow [\text{naphthalene}]_T^* + (CH_3)_2CO \tag{73}$$

which has been observed in irradiated solutions of small concentrations of naphthalene in acetone[22]. Particularly in the solid phase excited states are probably involved in energy transfer processes.

Excitation energy transfer may even lead to ionization of the acceptor as has been observed in the well known example

$$He^* + Ar \rightarrow He + Ar^+ + e^- \tag{74}$$

(The lowest excitation level of Helium is higher than the ionization potential of argon ; He^* : 19.8 eV, IP(Ar) = 15.6 eV).

Owing to the presence of two unpaired electron-spins triplet states can be regarded in some respects

as biradicals. Thus it is not surprising that triplet
excited molecules sometimes undergo typical radical
reactions as is shown in the following example :

$$(C_6H_5)_2CO\overset{..}{H} + (CH_3)_2CHOH \rightarrow (C_6H_5)_2\overset{\bullet}{C}OH + (CH_3)_2COH \quad (75)$$

where benzophenone triplet abstracts a hydrogen atom
from isopropanol.

REACTIONS IN THE SOLID STATE

From the chemical point of view the same type of
reactions that have been discussed for the liquid phase can also occur in the solid state, where irradiation also leads to ionization, i.e. formation of electrons and positive holes (positive ions., electron defects). In addition, neutral radicals can be formed
via excitation and ion recombination processes.

The formation of trapped electrons is a well-do-
cumented phenomenon in the solid state ; in irradiated
alkali halide crystals, for example, the formation of
"colour centers" (F, V, M, etc.) is attributed to electrons trapped at various sites within the lattice. In
this discussion, however, we shall not deal further
with this kind of essentially physical trapping of e-
lectrons and ions but concentrate on chemical trapping
in the presence of suitable scavenger molecules.

For example in an irradiated glass of 3-methyl-
pentane containing diphenyl (DP) and tetramethylphenyl-
enediamine (TMPD), the following reactions have been
observed

$$DP + e^- \rightarrow DP^- \qquad\qquad (76)$$
$$TMPD + \boxed{+} \rightarrow TMPD^+ \qquad\qquad (77)$$

($\oplus$ =positive hole) via the characteric absorption
bands of the DP$^-$ and TMPD$^+$ ions. The rate controlling
factor for these and similar reactions is the transport
of electrons or positive holes to the acceptor molecules. The latter cannot, of course, freely diffuse as in
a liquid. Electrons, however, even if they are trapped
initially may be rendered mobile, and statistically
(Maxwell-Boltzmann) a certain fraction of electrons is
always "detrapped", even at low temperatures. Higher
yields of mobile electrons can be obtained by heating
or illumination of the sample. These detrapped, mobile
electrons can travel very fast through material particularly if promoted into conduction bands.

Another mechanism which has to be considered for

electron migration is quantum-mechanical tunnelling. This in particular applies to solids at low temperatures where insufficient energy is available for complete liberation of an electron from a physical trap ; nevertheless, at high electron acceptor concentrations the formation of negative ions -via electron tunneling- can be observed.

Positive holes can also be transported very fast through crystalline and glassy material even if the positive ion (RH^+) and the acceptor (TMPD) are large, immobile molecules separated by a certain distance :

$$RH^+ \xleftarrow{e^-} RH \xleftarrow{e^-} RH \xleftarrow{e^-} RH \xleftarrow{e^-} TMPD$$

$$\textcircled{+} \quad \rightarrow$$

The positive charge is carried from the RH^+ to the TMPD by consecutive electron "hopping" (especially where H-bonding is present) eventually being trapped at the TMPD molecule.

Ion recombination processes, i.e. trapping of an electron by a positive hole leads to excited states which may dissociate into radicals. Direct excitation is, of course, also a potential source of radicals. Hydrogen atoms, formed in almost all hydrogen containing matrices on irradiation can be regarded as the most important radicals in the solid state since they are highly mobile species. Thus they can be identified directly, e.g. by ESR, only at low temperatures (e.g. in pure water < 20K). At higher temperature they immediately combine to H_2 or undergo reactions with other molecules.

The fast diffusion of electrons, positive holes and radicals are also responsible for many of the surface reactions observed. Oxygen which is adsorbed at the surface of irradiated silica gel, for example, can trap electrons which were formed within the silica lattice and transported to the surface :

$$e^- + O_{2,ads} \rightarrow O_{2,ads}^- . \qquad (78)$$

The adsorbed O_2^- ions formed in this reaction can be determined quantitatively by treating the silica gel with acidified water and analysis for hydrogen peroxide[23] :

$$H_{aq}^+ + O_{2,ads}^- \rightarrow HO_2^{\cdot} \qquad (79)$$

$$2 \ HO_2 \ \rightarrow \ H_2O_2 + O_2 \qquad\qquad (80)$$

The chemistry of large and immobile radicals in a solid is, of course, limited to either unimolecular decay or reaction with neighbouring molecules. In the latter case favourable orientation of the reactive centers is most often required.

LET-EFFECTS

The discussion so far has been devoted to the chemical effects in various liquids and solid states which have been subjected to γ-irradiation or fast electrons from an accelerating device. Another parameter of importance when considering overall chemical processes is the nature of the radiation used. All the ion and radical yields mentioned in the previous chapters refer to low LET (linear energy transfer) radiation of the order of 0.02 - 0.05 eV $\overset{\circ}{A}^{-1}$ which means that an average 6-15 ion pairs are formed along a distance of 1 μm , i.e. approximately one ion pair every 1000$\overset{\circ}{A}$. Much higher local concentrations of irradiation products are formed along the track of more energetic and consequently higher LET radiation. The following table gives a survey on LET and average ion pair formation for various kinds of radiation.

Radiation	LET [eV $\overset{\bullet}{A}^{-1}$]	Ion pairs per 1 μm
γ, acc.electr.	0.02 - 0.05	ca. 10
tritium-β	0.4	120
Po-α	9	2 700
$^7Li/^4He$ from $^{10}B(n,\alpha)^7Li$	24	7 000
U-fission products	370	110 000

At low LET the concentration of ions, electrons and radicals formed is relatively small and individual species have a fair chance of diffusing randomly. High LET radiation, however, produces ions and radicals at local concentrations of several moles per liter which has certain, quite obvious consequences :

1. higher yields of ion recombination
2. more radical-radical reactions
3. more "molecular" products

4. more secondary reactions of primary reaction products
5. less diffusion of individual reactive species out of the track of the ionizing radiation
6. consequently less scavenging of the reactive species by molecules which are present in the matrix at comparatively small concentrations.

This can be illustrated by the following example on the yield of primary reactive species (e_{aq}^-, OH., H.) and molecular products (H_2, H_2O_2) of the water radiolysis by irradiation with $^{60}Co-\gamma$ (low LET) and Po-α (high LET) rays :

	$^{60}Co-\gamma$	Po-α
$G(e_{aq}^-+H.)$	3.3	0.45
$G(OH.)$	2.7	0.6
$G(H_2)$	0.45	1.7
$G(H_2O_2)$	0.7	1.4

The yield of scavengable, i.e. freely diffusing e_{aq}^-, H. and OH. is seen to decrease with increasing LET, whereas the yield of products from radical-radical reactions in the spur (H_2 and H_2O_2) increases.

One of the final consequences is of course, that the range of products and in particular of product yields depends on the kind of irradiation. For example, no molecular oxygen is formed in the low LET radiolysis of water. Considerable yields of O_2 are observed, however, at high LET owing to a much higher local $HO_2^{\cdot}$ radical concentration which favours the reaction

$$2\ HO_2^{\cdot}\ \rightarrow\ H_2O_2 + O_2 \tag{80}$$

REFERENCES

The lectures are mostly based on the following books on radiation chemistry and the references cited therein :

I. Henglein A., Schnabel W. and Wendenburg J., "Einführung in die Strahlenchemie", Verlag Chemie, Weinheim (Germany) 1969

II. Swallow A.J., "Radiation Chemistry", Longman, London 1973

III. Gäumann T. and Hoigné J. "Aspects of Hydrocarbon Radiolysis", Academic Press, London and New-York 1968

IV. Matheson M.S. and Dorfman L.M., "Pulse Radiolysis" The MIT Press, Cambridge, Mass. 1969

V. Hart E.J. and Anbar M., "The Hydrated Electron", Wiley-Interscience, New-York, 1970

Some of the more recent developments in radiation chemistry are indicated in the text by reference to the following publications :

1. Schuler R.H., J.Phys.Chem. 61, 1472, 1957

2. a. Warman J.M., Asmus K.D. and Schuler R.H., Advances in Chemistry Series n° 82, American Chemical Society, Washington DC, 1968, p.25-57

   b. Warman J.M., Asmus K.D. and Schuler R.H., J.Phys. Chem. 73, 931, 1969

   c. Asmus K.D., Warmann J.M. and Schuler R.H., J. Phys.Chem., 74, 246, 1970

3. Onsager L., Phys.Rev. 54, 554, 1938

4. a. Rzad S.J., Infelta P.P., Warman J.M. and Schuler R.H., J.Chem.Phys., 50, 5034, 1969

   b. Warman J.M. and Rzad S.J., J.Chem.Phys., 52, 485, 1970

   c. Rzad S.J., Infelta P.P., Warman J.M. and Schuler R.H., J.Chem.Phys., 52, 3971, 1970

   d. Schuler R.H. and Infelta P.P., J.Phys.Chem., 76, 3812, 1972

5. Schuler R.H. and Kuntz R.R., J.Phys.Chem., 67, 1004 1963

6. Beck G., Int.J.Radiat.Phys.Chem. $\underline{1}$, 361, 1969

7. Asmus K.D., Int.J.Radiat.Phys.Chem. $\underline{4}$, 417, 1972

8. a. Lilie J., Beck G. and Henglein A., Ber.Bunsenges. Phys.Chem., $\underline{75}$, 458, 1971

   b. Grätzel M., Henglein A., Lilie J. and Scheffler M., Ber.Bunsenges.Phys.Chem., $\underline{76}$, 67, 1972

9. Fessenden R.W. and Schuler R.H., in "Advances in Radiation Chemistry", ed.Burton M. and Magee J.L., Vol.2, Wiley-Interscience, New-York 1970

10. Köster R. and Asmus K.D., J.Phys.Chem., $\underline{77}$, 749 1973

11. a. Rabani J., Nielsen S.O., J.Phys.Chem. $\underline{73}$, 3736 1969

    b. Behar D., Czapski G., Dorfman L.M., Rabani J. and Schwarz H.A., J.Phys.Chem. $\underline{74}$, 3209, 1970

12. a. Asmus K.D., Henglein A., Wigger A. and Beck G. Ber.Bunsenges.phys.Chem.,$\underline{70}$, 756, 1966

    b. Simic M., Neta P. and Hayon E., J.Phys.Chem., $\underline{73}$, 3794, 1969

13. Asmus K.D., Mockel H. and Henglein A., J.Phys.Chem. $\underline{77}$, 1218, 1973

14. Meißner G., Henglein A. and Beck A., Z.Naturforsch. $\underline{22}$ $\underline{b}$, 13, 1967

15. a. Busler W.R., Martin D.H. and Williams F., Disc. Faraday Soc., $\underline{36}$, 102, 1963

    b. Williams F., J.Am.Chem.Soc., $\underline{86}$, 3954, 1964

16. Asmus K.D., Int.J.Radiat.Phys.Chem. $\underline{3}$, 419, 1971

17. Scala A.A., Lias S.G. and Ausloos P., J.Am.Chem. Soc., $\underline{88}$, 5701, 1966

18. Rzad S.J. and Schuler R.H., J.Phys.Chem., $\underline{72}$, 228, 1968

19. Rzad S.J., Schuler R.H. and Hummel A., J.Chem.Phys. $\underline{51}$, 1369, 1969

20. Sonntag C.V., "Fortschritte der chemischen Forschung", Band $\underline{13}$, Vol. 2, Springer-Verlag Berlin 1969 and references cited therein

21. Stockhausen K., Henglein A. and Beck G., Ber.Bunsenges.phys.Chem., $\underline{73}$, 567, 1969

464

22. Arai S. and Dorfman L.M., J.Phys.Chem., 69, 2239, 1965

23. Rabe J.G., Int.J.Radiat.Phys.Chem. 5, 301, 1973

Part 3

Applications

ION IMPLANTATION

J.C. PFISTER

C.E.N. Grenoble
Département de Recherche Fondamentale
BP 85
38041 GRENOBLE Cedex /france

GENERAL FEATURES AND APPLICATION OF ION IMPLANTATION

The study of ion trajectories in solids immediately yields the basic geometric properties of ion implantation at convenient energies, i.e. $\lesssim$ 1 MeV.

The penetration of the implanted ions is usually a few 10^3 Å, at most a few microns, depending on ion mass and energy and nature of the substrate. For the vast majority of effects, whether favourable or unwanted, this defines the thickness of the surface region affected by implantation.

The lateral definition corresponding to the unavoidable spreading of the beam inside the sample, is generally of the order of 100 Å, which is normally much better than can be achieved for any of the techniques used to define the implanted region at the surface of the sample. Ion implantation is thus potentially a very accurate surface treatment.

The main effects of an ion beam inside a solid belong to two very distinct classes although the experimental separation is often difficult :

1. Foreign atoms are introduced at low temperatures where ordinary thermal diffusion is·completely negligible. Any ion can be introduced into any solid, regardless of phase equilibria or solubility problems.

2. Fast moving ions displace matrix atoms and excite electrons, leading to the creation of high concentra-

468

tions of various lattice defects which do not necessarily involve the implanted impurities themselves. This effect can be reduced, for instance by implantation in a channelling direction, but it can never be completely suppressed.

Practical applications of ion implantation make use of at least one of the distinctive features of the technique.

a) Creation of a localized region of very high defect density. Examples are : confinement of the laser beam in Ga Al As lasers by local modifications of both conductivity and refraction index through H^+ implantations ; surface hardening of metals due to the introduction of both impurity precipitates and dislocation tangles ; irradiation of SiO_2 to make it more sensitive to a subsequent chemical etch.

b) Low temperature introduction of doping impurities. Examples are numerous in semiconductor technology and may become more and more important with the increased use of compound semiconductors where chemical decomposition at high temperature is a serious problem. Another interesting field is the introduction of probes for measuring hyperfine fields in magnetic materials.

c) Geometrical properties of the implantation experiment are important in all applications. Specific applications occur in microelectronic devices, the most important single one being the self-alignment of MOS transistors obtained by implanting source and drain electrodes after depositing the grid which then serves as a mask.

STRUCTURE OF THE IMPLANTATION DAMAGE

Damage structure in a 0° K implantation

A widely used and very convenient approximation distinguishes two distinct parts in the trajectory of a high energy ion :

At high energy, and for light ions electronic stopping predominates and most of the energy loss goes into electronic excitation, whereas direct atomic displacement is negligible as far as stopping power is concerned. In this part of the trajectory, there is a difference in the number of defects produced in metals or semiconductors, where only nuclear collisions are effective, and insulating materials where pure ionization can lead to defect production. The nuclear process in this range of energy can be considered, with reaso-

nable accuracy, to be Rutherford scattering with the
result that most of the collisions will lead to small
energy transfer and thus to the production of relatively simple defects (single vacancies and interstitials
or low-order complexes).

At low energies and for heavy ions, the dominant
energy loss process is the elastic scattering on atoms,
involving defect production by direct atomic displacements. The scattering cross sections involved are very
high with no or little enhancement of low energy transfers because the scattering becomes more and more isotropic ("hard-sphere collisions") as energy decreases.
The result is a massive creation of lattice defects in
a very small region near the end of the ion path, the
detailed structure of which is poorly known and probably different from one material to another (e.g. amorphous for valence semiconductors and still crystalline
for metals).

The boundary between high energy and low energy
ranges is usually taken as A keV, A being the mass
number of the incident ion. This is obviously a very
crude assumption, neglecting in particular all effects
of chemical bonds and electronic shell structure, as
well as effects like channeling which may change the
ratio between electronic and nuclear stopping power.

To evaluate the number of lattice defects produced
in an implantation experiment, one has to take into
account the complete path of the incoming ion and the
spatial variation of ion energy and energy straggling
inside the sample. The defect structure thus obtained
is very complex and difficult to describe accurately.
It is possible to gain some physical insight by using
the simple theory of Kinchin and Pease relating the
number of defects produced to the energy lost in nuclear
collisions. The basic assumption is that it is possible
to define a displacement threshold E_d such that the
probability of an atom being displaced is unity if the
energy transferred T is above E_d, zero if $T < E_d$. The
result is that the total number of displacements, taking into account displacement cascades, is simply given by

$$n_d = \frac{E_{nucl}}{2E_d}$$

E_{nucl} is the total energy transferred into nuclear collisions. In a typical ion implantation experiment,
most of that energy is given up in the last part (low

energy range) of the ion trajectory, so that the defect production rate shows a peak at a depth slightly lower than the mean penetration depth, with a width that is larger than the range straggling of the ions, due to the finite thickness of the individual damage zones.

This model may be considered as qualitatively correct, although it is grossly oversimplified for three main reasons :

1. The Kinchin-Pease model is already highly simplified and usually overestimates even low-temperature damage measurements by a factor of the order of two.

2. The very high defect densities expected at the end of an ion path will lead to immediate athermal anneal of part of the damage. The extent of this effect is completely unpredictable.

3. Interaction of an ion path with defects already present (due to earlier ion tracks) is entirely neglected. This is an important consideration for practical applications where the doses needed are often very high. The results of these interactions are strongly temperature dependent and will be considered in the next section.

Implantation at finite temperatures in silicon

At sufficiently low temperatures (most experiments being made at 77° K), the elementary ion damage as described earlier is retained because all defects are immobile. The resulting structure is thus, for low doses, a combination of "simple" defects near the surface (corresponding to the high energy range) and small zones of highly distorted material at the end of the individual ion paths (in the case of low energy, heavy ions , the first part may be missing). As irradiation proceeds, the highly distorted zones will overlap and eventually merge into a continuous layer of amorphous material (at least if the individual zones can be considered amorphous).

If the implantation is continued, the amorphous layer thickness then increases slowly in both directions, ultimately leading to an amorphous layer reaching from the surface to slightly more than the average range of the ions in the material.

At medium temperatures (around room temperature in the case of Si), the elementary defects are mobile during the implantation and this leads to the creation

of secondary defect structures. The details may be
expected to differ greatly from one crystal to another
and the following description is valid only for silicon, which is the best known material. In this temperature range, the larger defect structures are still
unable to anneal and two types of processes will occur
due to the migration of simple defects :

1. Some annealing of the larger defects (ion tracks)
will occur due to reaction with simple defects (e.g.
one can expect a vacancy aggregate to attract interstitials more strongly than it does vacancies, leading
to a gradual decrease in size). The relative importance of these processes will depend critically on the
ratio between the densities of "simple" and "complex"
defects, and thus on the mass and energy of the ions,
with lower mass and higher energy enhancing this "radiation anneal". The extreme case is the recent experimental evidence showing that high energy electron
irradiation in a high voltage microscope anneals out
a previously created amorphous region. This "radiation
anneal" is a very important process in determining the
critical amorphization dose, which in the medium temperature range will be strongly dependent on ion mass
(because of the ratio of simple to complex defects),
flux and temperature (defining the time scales for
creation and annealing of defects).

2. The migrating simple defects will interact and
this will lead to some agglomeration and the creation
of larger defects such as vacancy loops (the interstitials being in most cases highly mobile and escaping
to surfaces), and ultimately to the creation of dislocation tangles which will not anneal until the sample
is annealed at very high temperatures, in the self-
diffusion range of the material, e.g. 1100°C for Si :

At high temperatures (above 400°C for Si), all
simple defects are highly mobile and their steady state concentrations remain negligible, so that no nucleation of aggregates is possible, whereas thermal evaporation of vacancies from small aggregates is possible. No extensive defect formation is observed in pure materials and no amorphous region is created for
any dose. The extended defects observed are related
less to direct radiation damage than to metallurgical
problems such as strains due to the gradient in implanted impurity concentration or precipitation of
impurities, either implanted or already present. This
precipitation is made possible by the enhanced diffusion due to the implantation, and maximum solubilities

472

are generally much lower than they are at the temperatures where the crystal is originally grown, so that it is an important problem in some practical applications.

Post-implantation anneals

Three temperature ranges can again be distinguished :

- From 200 to 400° C, the "simple defects" anneal and, although the larger defects are still present and have an influence on most physical properties, the effect (e.g. electrical activity) of the implanted impurity are manifest if no amorphous layer has been formed. Unfortunately, many impurities are still not in their normal lattice sites, and this depends on the chemical nature of the impurity. The most spectacular example is the case of silicon where implantation of group V donor impurities leads to high electrical activity and channeling measurements show that typically 80-90 % of the impurities are substitutional, whereas an implantation of group III acceptor impurities leads to low electrical activity and channeling experiments show that equal amounts of impurity atoms are in substitutional and interstitial positions, suggesting a pairing between these two sites.

- Around 600° C, the amorphous layer recrystallizes epitaxially from the substrate, yielding a monocrystalline layer with small stacking faults and vacancy loops.

- Above 800°C, the dislocation loops and stacking faults start annealing and the perfect lattice is restored. This is the self-diffusion range and some modifications in the impurity profile may be expected due to diffusion if annealing is too extensive.

Defect Structure in Other Materials (mainly compounds semiconductors)

The following section is an attempt at generalization of the known situation in silicon, relying on various experiments mainly on Ga As. Some general features may a priori be expected to be found in any material. These are mainly :

1. The general behaviour of defect concentration vs temperature with of course some difference in the temperature scale, as well as the defect vs dose curve at medium temperatures (i.e. a critical dose for nucleation of large defects and subsequent supralinear increase with dose up to a saturation value).

2. A large difference in the temperature necessary to restore a perfect crystal during or after the implantation.

Some extra problems are easily foreseeable due to the fact that different kinds of atoms are involved in a compound :

1. The variety of elementary defects is larger (two types of vacancies and interstitials, possibility of replacements...)

2. Implantation of only one type of ion will perturb the lattice site equilibrium because the implanted atoms will generally have a preference for one type of site. An example is the implantation of Zn in Ga As : Zn is expected to be stable on a Ga site where it acts as an acceptor. The complete reaction taking site equilibrium into account will, however, involve creation of one extra site on each sublattice and may be written

$$Zn_{beam} \rightarrow Zn_{Ga} + V_{As}$$

The net effect of introduction of the Zn atom with an As vacancy is to make the crystal more n-type, unless action is taken to remove the As vacancies, e.g. by increasing the As vapor pressure or simultaneous As implantation.

3. Stoichiometry will be affected by any thermal treatment either during or after implantation, and the same mechanisms leading to self-compensation in large gap materials will operate, although with widely different initial conditions and efficiency of thermal treatments.

Experimental results show that the general behaviour is considerably more complicated than in Si :

- Even in a 400°C implantation, the lattice is highly perturbed and implanted ions do not occupy their regular lattice sites unless the crystal is annealed at least to 900°C, i.e. in the self-diffusion range.

- Post implantation anneal shows complex structure including reverse anneals in the 600-700°C range.

- Implanted p-n junctions often show a p-i-n behaviour of unknown origin. The formation of an intrinsic layer is not simply due to implantation defects because the thickness of the i layer can be much larger than the range of implanted ions (up to 100 μ in some cases). Diffusion of intrinsic defects is certainly involved in in this effect, although the detailed mechanism is not yet understood

DIFFUSION AND IMPURITY PRECIPITATION

Both phenomena are obviously related and occur only during relatively high temperature treatments when lattice defects are mobile. Such a treatment is always included in the process leading to any practical application, either during implantation ("hot implants") or during post-implantation anneal.

In the simplest case of pure vacancy diffusion with no interaction between defects and impurities, the impurity diffusion is identical to self-diffusion and the number of jumps per atom (lattice or impurity) is just the ratio of the total number of atomic (i.e. vacancy) jumps to the density of lattice sites. If L_v is the diffusion length of vacancies before annihilation, the diffusion length of impurities during a complete experiment can be obtained easily as

$$L_i = L_v \sqrt{\frac{N_v}{N_o}}$$

where N_v is the total number of vacancies migrating during the experiment and N_o the number of lattice sites (note that in the case of a high temperature implantation N_v may be larger than N_o).

In most cases L_v is an unknown quantity and may vary during the course of the treatment depending in particular on the dislocation density. Direct measurements have been attempted in Si and Ge by various groups with essentially two types of results :

- for high temperature experiments with low instantaneous defect concentrations and consequently no loop formation, L_v is a few microns.

- for low temperature experiments presumably involving extensive loop formation during the early stages, the measured diffusion lengths are very short, down to 100 Å.

This result may be expected to be generalizable L_v is large if the defect concentration <u>during vacancy migration</u> is small.

Practical effects of enhanced diffusion

1. If L_i is greater than the average distance between any type of impurities, precipitation will occur if the solubility limit is exceeded. This is an important

limiting factor to the use of ion implantation to obtain thermodynamically unstable phases or very high impurity concentrations.

2. If L_i is comparable to the range straggling of the implanted ions, the concentration profile will be modified. Two limiting cases may be of interest :

- for large L_v, the shape of the profile is modified almost as in the case of thermal diffusion, with only a modification in the absolute value of the diffusion coefficient

- for short L_v the modification of the impurity profile is strongly different from that obtained by thermal diffusion and very abrupt profiles are expected in favorable cases, which would be very interesting for electronic devices. Unfortunately, short L_v implies a high local defect density in the active region, which is of course detrimental to device properties.

DAMAGE TO NUCLEAR MATERIALS

R.S. NELSON

AERE Harwell

Didcot, Bershire, U.K.

INTRODUCTION

Perhaps the most important area where radiation damage effects have a direct technological significance is in the field of Nuclear Energy, and this alone has been the major motivation for its study. In this brief review we can only outline some of the major problem areas, and in doing so we will concentrate on the underlying physical phenomena rather than discussing in detail their implications. Essentially, radiation effects are manifest in two areas ; the nuclear fuel and its cladding and although we will discuss these separately there are many common aspects.

NUCLEAR FUELS

As a consequence of nuclear fission, fission fragments of many hundreds of MeV are released into the fuel (e.g. U or UO_2). During their slowing down process these loose energy which is manifest as heat and atomic displacement. Whilst the heat is required to produce electricity, the radiation damage can result in deleterius effects. However, it is the chemical nature of the fission fragments themselves which can cause the most serious problems. Figure 1 shows a typical atomic mass distribution of fission fragments where it is readily seen that about 10 % of all fragments are the inert gases Kr and Xe. Let us now consider the fate of these gas atoms within a solid.

It is well known that the inert gases are highly insoluble in solids, and provided their mobility is

478

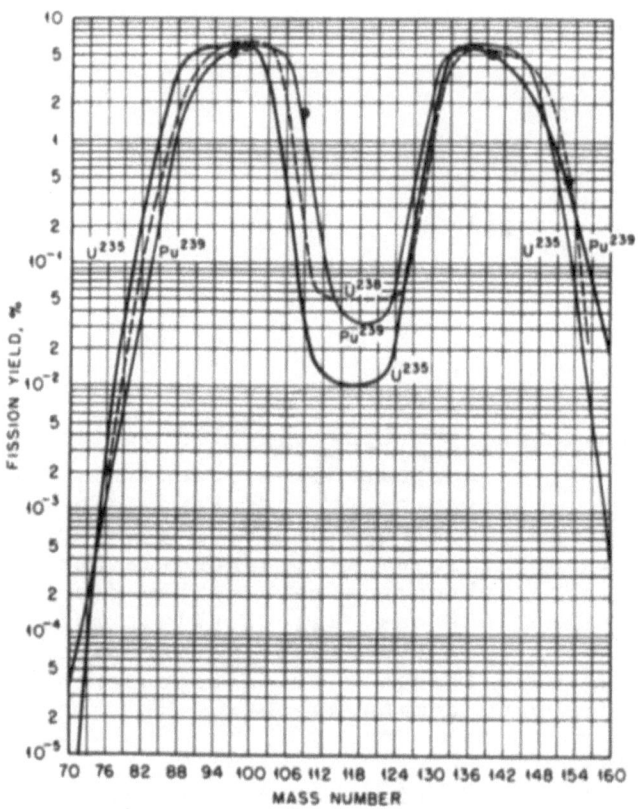

<u>Fig.1</u> : Fission fragment distribution.

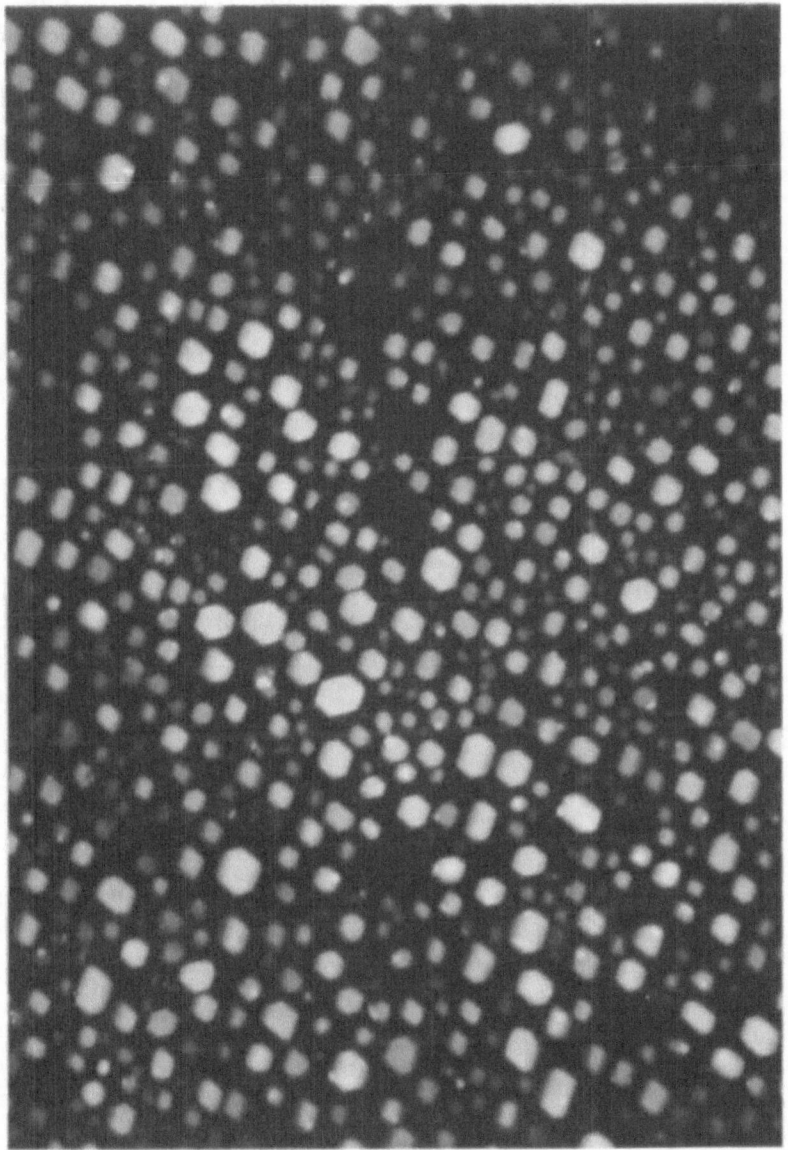

Fig.2 : Argon gas bubbles in copper -showing facets.

sufficient, precipitation will occur for even the smallest concentrations. In this particular case because of the inability of the inert gases to form compounds, precipitation takes the form of an agglomerate of inert gas atoms which can best be described as a bubble. It is the behavior of such bubbles in nuclear fuels which have given the engineers most concern. Let us therefore briefly discuss the physics of gas bubbles in solids prior to outlining their technological significance in nuclear fuels.

The Physics of Gas Bubbles

Consider a gas bubble in a solid which has a surface free-energy γ. The total free energy of the system is

$$F_T = F + F_s + A\gamma + \phi$$

where F is the free energy of the gas within the bubble F_s is the free energy of the solid, A is the surface energy of the bubble surface, and ϕ is the elastic strain energy in the solid surrounding the bubble. If the temperature is sufficiently high such that thermal vacancies are generated in copeous quantities at, for instance, the free surfaces, the bubble will change its size to lower the total free energy of the system, then in equilibrium

$$\frac{\partial F_T}{\partial V} = \frac{\partial F}{\partial V} + \frac{\gamma \partial A}{\partial V} + \frac{\partial \phi}{\partial V} = 0$$

However, for bubbles in excess of about 10 $\overset{\bullet}{A}$ radius $\frac{\partial \phi}{\partial V}$ can be neglected. Furthermore $(\frac{\partial F}{\partial V})_T = -p$ and so our equilibrium conditions is defined by

$$p = \gamma \frac{\partial A}{\partial V}$$

For a spherical bubble of radius r, this simply reduces to

$$p = \frac{2\gamma}{r}$$

However, due to the crystallographic nature of solids, the surface free-energy varies with orientation being lowest for certain low index planes. In practice, therefore, equilibrium bubbles in solids often take on a polyhedral form, see for example figure 2.

Provided the gas bubble is above about 100 $\overset{\bullet}{A}$ radius, the gas it contains will obey the perfect gas laws i.e. pV = mkT where m is the number of gas atoms in the bubble. Inserting expressions for p and V we then obtain

$$m = \frac{8\pi \gamma r^2}{3kT}$$

Clearly the vacancies which constitute bubbles result in the transfer of material to the free surface. This in turn results in an overall volume change of the material which is given by the expression

$$\frac{\Delta V}{V} = \frac{4}{3} \pi r^3 N$$

where N is the number of bubbles per unit volume of average radius r.

About ten years ago it was generally realised that gas bubbles could actually move within solids. At sufficiently high temperatures any atom on the surface of a bubble can be transferred to another position by either surface diffusion, volume diffusion or evaporation. In the absence of any chemical potential gradient or driving force such a bubble will migrate randomly within the solid by Brownian motion. We can therefore asign a diffusion coefficient to the bubble as a whole of D_B. If then a chemical force, F, acts on the bubble it will have a net velocity in the direction of the force

$$v = \left(\frac{D_B}{kT}\right) F$$

Before discussing in detail the actual diffusion mechanisms let us consider the nature and magnitude of the forces which might occur in nuclear fuels. The forces which have been identified are, temperature gradient, dislocation motion grain boundary motion, and stress. The relevant expressions have been derived previously (Barnes and Nelson[3]) viz : -

temperature gradient $\quad \frac{F_{dt}}{dx} \sim \frac{4\pi r^3}{\Omega} k\left(\frac{dT}{dx}\right)$

dislocation $\quad F_{dis} \sim \mu b^2$

grain boundary $\quad F_{gb} \sim \pi \gamma_{gb} r$

$$\text{stress gradient} \qquad F_{stress} \sim 8\pi r^3 \left[\frac{d\sigma}{dx}\right]$$

where r is the bubble radius, Ω is the atomic volume, μ is the shear modulus, b is the burgess vector, γ_{gb} is the grain boundary energy and σ is the stress. These forces can conveniently be represented graphically as shown in figure 3 which illustrates their variation with bubble radius.

Similarly, the equations governing the diffusion mechanisms have all been derived previously and it will simply be sufficient to quote the results

$$\text{surface diffusion} \qquad v_{s.d} = \frac{9}{\pi^3} \frac{\Omega^{4/3}}{r^4} \frac{D_s}{kT} F$$

$$\text{volume diffusion} \qquad v_{v.d} = \frac{\Omega}{\pi r^3} \frac{D_v}{kT} F$$

$$\text{evaporation} \qquad v_{e.v} = \frac{\Omega}{2\pi r^3} \frac{D_g p_T \emptyset}{k^3 T^3} F$$

where D_s and D_r are the surface and volume diffusion coefficients respectively, D_g is the gaseous diffusion coefficient of evaporated atoms through the gas in a b bubble, p_T is the vapour pressure of the solid, and $\emptyset$ is the activation energy for evaporation.

Let us now apply these ideas to the nuclear fuel itself and in particular.to UO_2. In a typical fuel rod large temperature gradients up to 10^3 °C/cm, are likely and as such are expected to provide the major driving force. In accordance with the above expression, bubbles are expected to migrate up the temperature gradient, but in so doing they will become pinned on dislocation lines and grain boundaries. Furthermore, bubbles will collide with one another and coalesce, especially on grain boundaries. As the bubbles become larger the migration mechanism will change from that of diffusion controlled to that of evaporation and they will rapidly migrate to the hot central region of the fuel elements. Apart from the swelling which ensues from the coalescence and growth of gas bubbles, another feature is the phenomenon of gas release. For instance as a consequence of power changes in the reactor, the oxide fuel is continually subjected to thermal cycling which results in the cracking. Due to the weakening of grain boundaries by gas bubbles, the fuel is most likely to crack

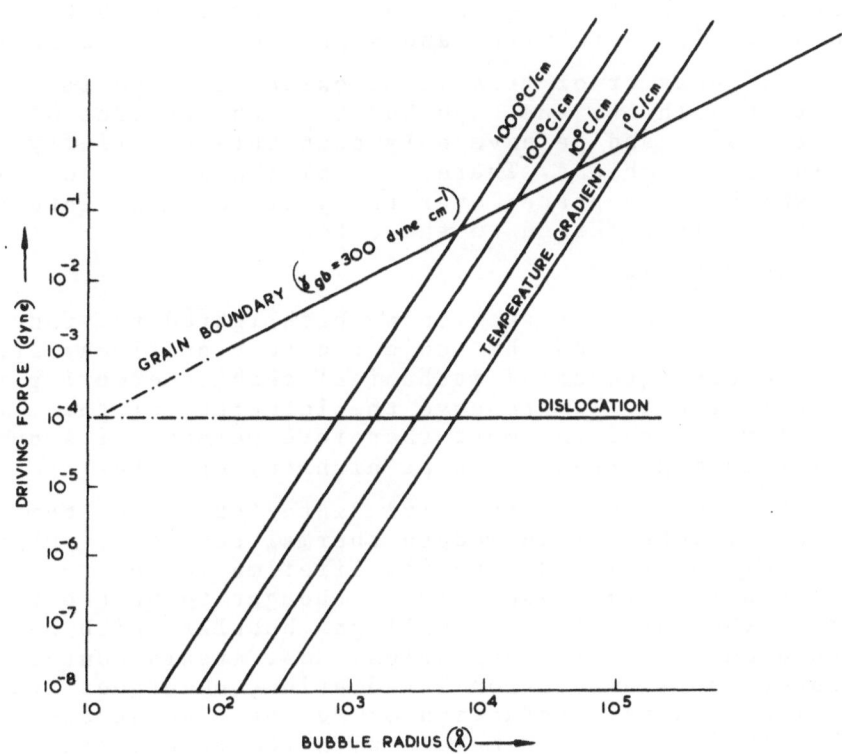

<u>Fig.3</u> : Comparison of driving forces on gas bubbles as a function
bubble radius.

along such boundaries, with the consequent release of gas trapped. Unless special care is taken this released gas will become trapped within the fuel-can, where it can eventually build up to very high pressure and can cause deformation of the can. The fuel can itself must therefore be able to withstand, not only direct deformation as a consequence of fuel swelling but must also be designed to withstand significant gas pressures.

The behavior of irradiated oxide fuel and its interaction with the fuel can has been the subject of much research and we have only been able to briefly outline the problems. There is a plethora of information which has appeared over the years and the interested reader is referred to these texts.

CLADDING MATERIALS

In the previous section we highlighted the fact that the fuel can may be subjected to significant stresses as a consequence of mechanical changes occuring in the fuel itself. To preserve the integrety of the total fuel element, the can must therefore maintain its ductitlity during irradiation at high temperature.

It is well known that irradiated steel, at temperatures of interest in modern thermal reactors, suffers a loss of ductility during its lifetime in the reactor. In this particular case this is thought to be the result of the formation of small gas bubbles which become trapped on the grain boundaries. Most steels contain B which during slow neutron irradiation, undergoes transmutation with the production of He. He like Xe and Kr, is highly insoluble in metals with the result that gas bubbles readily form at temperatures where vacancies are created by thermal activation. Such a situation exists in stainless steel cladding in thermal reactors with the consequence that gas bubbles form and grow and become trapped on the grain boundaries so preventing their movement. This clearly results in a loss of ductility, and if the fuel swelling is particularly severe, can lead to fracture.

However, perhaps the most important radiation damage phenomenon which occurs in cladding materials, is that of void formation which has recently been highlighted in the context of fast reactor development. Void formation is essentially a manifestation of radiation damage created by fast neutrons at high dose. It was therefore only with the advent of the fast reactor that sufficiently high neutron doses at elevated temperatures could be achieved. In 1966, C.Cawthrone and E.J.

Fulton[5] working at the Dounreay Experimental Reactor
Establishment of the UKAEA, discovered that austenitic
stainless steel fuel cladding, exposed to high doses of
fast neutrons, developed an internal porosity in the
form of small $\sim$ 100 Å cavities. Such cavities were
shown to be essentially empty and were therefore called
"voids". Subsequent work has shown that void formation
in irradiated metals can result in overall volume changes of several per cent. The phenomenon of void formation is a topic of immense interest to both theoretical
and experimental physicists concerned with irradiation
effects in metals ; see, for example, the proceedings
of the two international conferences on void formation
(Pugh et al[9], Corbett and Ianniello[6]). However, its e-
xistence is also of interest to engineers concerned
with the design and operation of nuclear power stations,
especially the new generation of Fast Breeder Reactors,
in which materials will experience fast neutron doses
far greater than those hitherto achieved. We must therefore ask the question ; is void formation solely a
scientific curiosity or has it important implications
in the technology of Fast Reactor design ?

In order to answer this question, we must briefly
consider some aspects of Fast Breeder Reactor design.
Clearly the ultimate aim is the production of economic
power, and we must therefore assess the implication of
void swelling in relation to the attainment of high
fuel burn-up, station reliability and availability. A
consideration of these implications will constitute the
final section of this review, and for the purposes of
this introduction we will simply put the phenomenon in
perspective in the context of the above question.

We will assume the reactor coolant to be sodium,
the fuel to be (U, Pu)O_2, the fuel cladding to be austenitic steel and other structural components within
the vicinity of the core also to be steel. In this environment the maximum volume changes likely to be experienced in the reactor core structural components can
be of the order of 10 %. However, because the neutron
flux and sodium temperature are not uniform within the
core, some components will only experience this maximum
swelling over part of their length. This non-uniformity
can therefore give rise to differential swelling and
distortions, which in some circumstances, could influence the mechanical properties of components, or lead
to servicing difficulties. Void swelling is therefore
certainly more than a scientific curiosity and must be
taken into account in fast reactor technology. However,

486

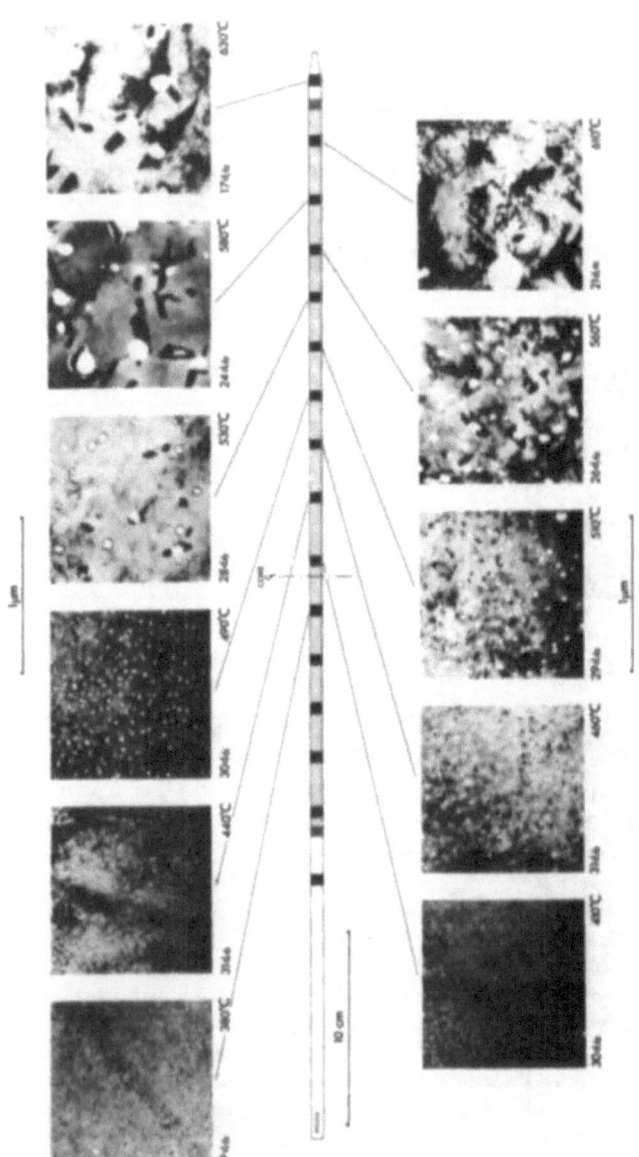

Fig.4 : Void and defect structures produced in D.F.R. fuel pin cladding. The pin was 20 % cold-worked M316.

the reactor designer is quite familiar with distortion
and mechanical property changes, which are produced purely from raising the sodium temperature up to 600°C ;
for instance volume changes of ∿ 3 % due to thermal expansion are quite common in steel components and thus
the eventual contribution of the physicist is to provide the design engineer with reliable data in order
that he may properly account for such dimensional changes in his design.

We have already discussed the basic principles behind void formation, so it remains for us to outline
some of the more important experimental results together with attempts to alliviate the problem. The most
comprehensive accounts of reactor results on void swelling, are to be found in the proceedings of the two major conferences at Reading (Pugh et al[9]) and Albany
(Corbett and Ianniello[6]). Results are of two basic types, those from fuel cladding (which may therefore be
stressed because of excessive fuel swelling) and those
from unstressed components or test specimens. Most published data has originated from irradiation either in
the Dounreay Fast Reactor in Northern Scotland (DFR)
or from the American Experimental Breeder Reactor II
(EBRII). However, the French have published some data
obtained from their Rapsodie Fast Neutron Breeder Reactor. Clearly it is impossible to refer to all the published work in this section, so we will select just a
few results from the plethora of available data to illustrate the main trends.

Figure 4 illustrates a schematic diagram of a DFR
fuel pin together with transmission electron micrographs
taken from selected regions, (Bagley et al[2]). Due to
the variation in fast neutron flux within a reactor,
not all sections experience an identical damage dose ;
furthermore, as the sodium temperature varies from top
to bottom of the core, there is a steady temperature
gradient along the pin. The damage doses and temperatures are therefore recorded for each micrograph as shown.
Fuel pin data is of two basic kinds ; direct diameter
changes measured from the outside of the pin, and immersion density changes obtained from selected sections
of the pin. In general, diameter changes and density
changes correspond remarkably well, suggesting that essentially the whole of the diameter increase can be ascribed to void swelling, though this itself may be influenced by stress. However, we must remember that the
fuel itself, within the cladding can swell due to fission gas bubble growth, and for this reason diameter

488

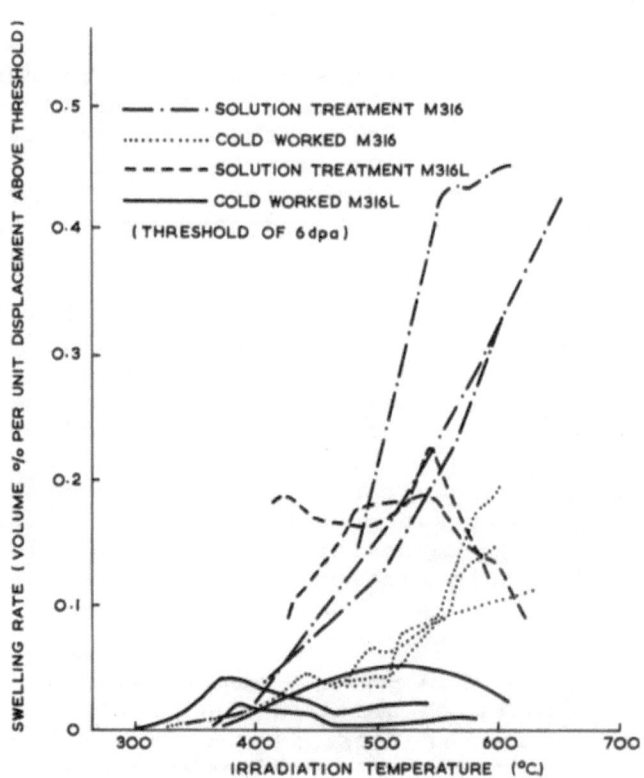

Fig.5 : Swelling rates for M316 and M316L

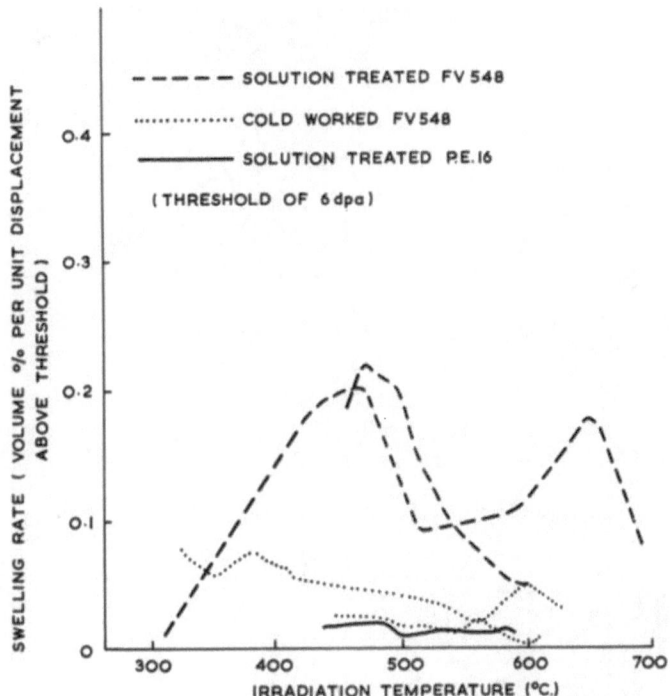

<u>Fig.6</u> : Swelling rates for FV548 and Nimonic PE16 cladding.

490

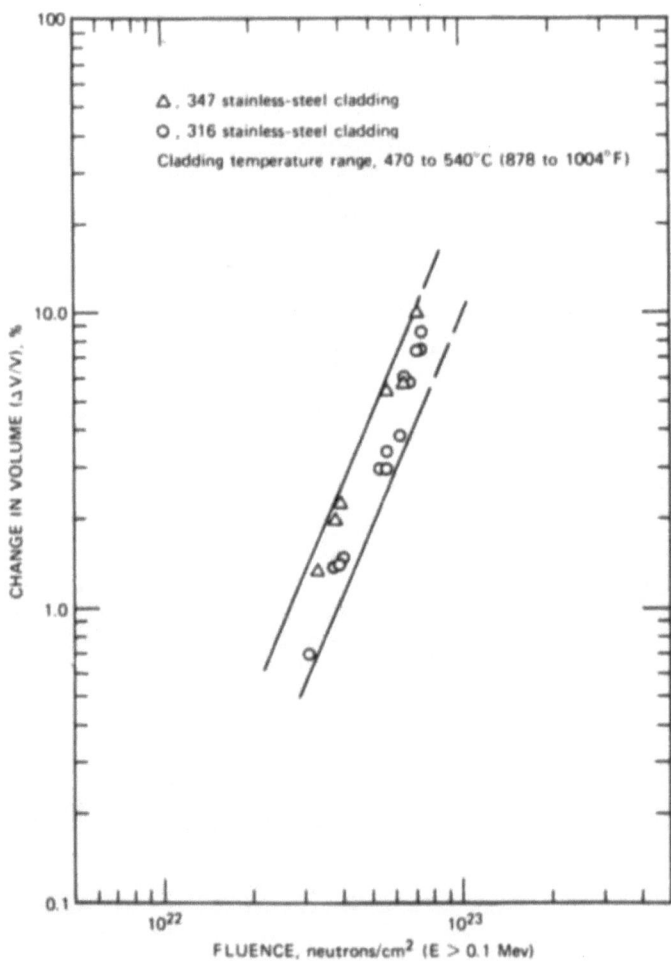

__Fig.7__ : Cladding volume increase v.s. neutron fluence in 347 and 316 stainless steel cladding.

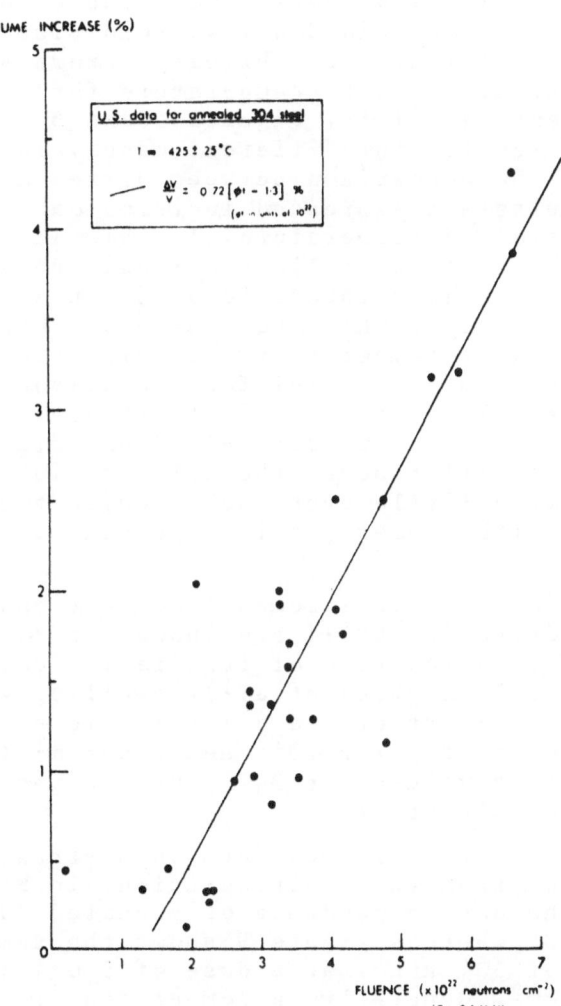

Fig.8 : Void swelling as a function of dose in 304 steel at 425°C.

changes will not necessarily always give unambiguous data on void swelling.

A glance at figure 4 clearly indicates a marked effect of temperature on void density, void size and total swelling ; so let us first briefly report variations of these parameters with temperature for a selection of different materials. Figures 5 and 6 illustrate a selection of results for different materials. In order to help in the comparison between different materials, the results are expressed in terms of a swelling rate as a function of temperature. In this it has been assumed that the swelling follows a linear relationship with dose above a common threshold of 6 d.p.a., it is the slope of this curve which has therefore been plotted as a function of temperature. Clearly this is only an approximation, but is useful for comparison purposes and is consistent with the simple theoretical expressions. The results for M316 and M316L (figure 5) show significant differences, the M316 showing a peak swelling rate at a little over 600° C while the lower carbon steel, M316L, shows profiles peaking at about 500°C.

Immersion density measurements on sections cut from pins irradiate in EBR II are shown for 347 and 316 steel cladding as a function of irradiation dose (470 540°C) in figure 7 (Appleby et al[1]). Swelling values of up to 10 % in 347 steel and 8.7 % in 316 steel were measured for doses of $\sim$ 7 x 10^{22} neutrons/cm^2 (E>0.1 MeV). The swelling values for 347 pins are generally greater than for 316 steel.

The majority of published data on unstressed components has come from the US irradiations in EBRII. Figure 8 shows the dose dependence of annealed 304 steel at 425° ± 25° C, whereas figure 9 shows the temperature dependence of 304 steel at a dose of 5.0 ± 0.5 x 10^{22} neutrons/cm^2. Generally, a comparison of results between fuel pins and unstressed components provides a reasonable agreement. However, some difference suggest that stress may play an important role in ultimately determining the final swelling, although this is not as yet clearly understood theoretically.

It should be pointed out that the use of fast reactors to accumulate radiation damage necessitates the requirement of rather long irradiations -of the order of 1 year or more- so it is therefore rather difficult to collect the necessary data in a short time. Some guidance as to the long term effects can be provided

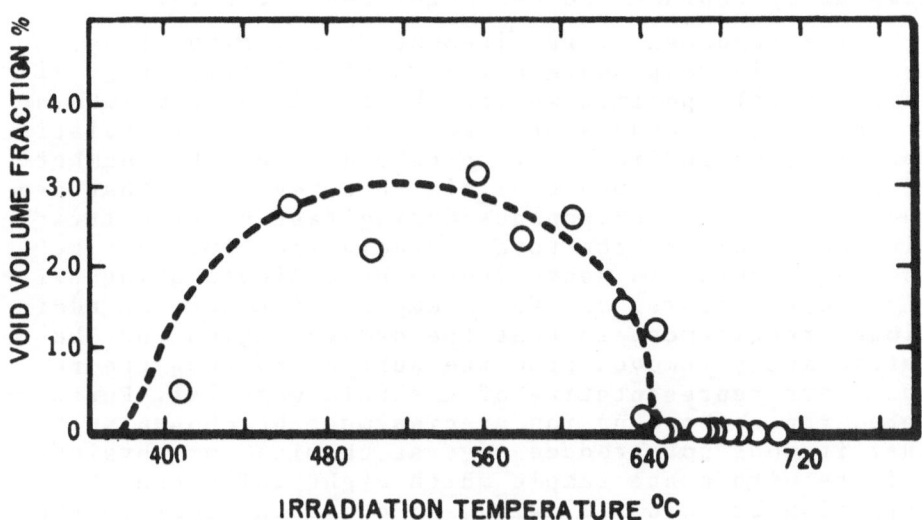

<u>Fig.9</u> : Temperature dependence of 304 steel at 5.10^{22} n/cm^2

by simulation experiments using charges particles from
powerful accelerators, such as the Harwell Variable
Energy Cyclotron (VEC) and 1 MeV electrons in the High
Voltage Electron Microscope (HVEM). In both cases the
damage rate can be many orders of magnitude greater
than even the most powerful fast reactor. However, the
effects of such increases in damage rate are inevitably
rather complex and, in practice, whilst the swelling
data obtained by simulation techniques can certainly
give a useful guide to expected material behaviour and
thereby directly assist in material selection it cannot
completely replace the essential reactor data.

The fundamental requirement of any simulation
technique is to produce a region of uniform damage within a metal specimen whilst the specimen is maintained
at an elevate temperature. Furthermore, the irradiation
dose must be sufficient to create a comparable number
of atomic displacements within this region to that produced in reactor components during fast neutron irradiation. However, the total irradiation time must be
no longer than can conveniently be achieved using existing accelerators, say for exemple, 24 hours. An additional requirement is that the damage region must be
sufficiently removed from the surface so that the results are representative of the bulk material. Furthermore, the bombarding ion species must be chosen such
that it does not produce adverse chemical or physical
effects within the sample which might influence the
formation of voids. Finally, in order to simulate the
helium which is produced within the material as a result of (n,α) reactions, trace quantities of helium
($\sim 10^{-5}$ atom/atom) must be uniformly implanted throughout the sample prior to irradiation with the damaging
ions. These requirements have been discussed in detail
previously (Nelson et al[8], Pugh et al[9] and Corbett and
Ianniello[6]), and in this instance it will suffice to
simply outline some typical results.

The acquisition of quantitative data depends on
relating the total damage produced during ion bombardment to that produced during neutron irradiation. In
other words, we must check to see that irradiations to
the same calculated damage dose in the reactor and accelerator do in fact correspond. We must, of course, be
careful to choose a system that will permit any dose-
rate effects to be neglected or at least accounted for.
In this context, it is possible to choose corresponding
temperatures for nickel, which at the peak swelling
temperatures show little or no effect of dose rate on

495

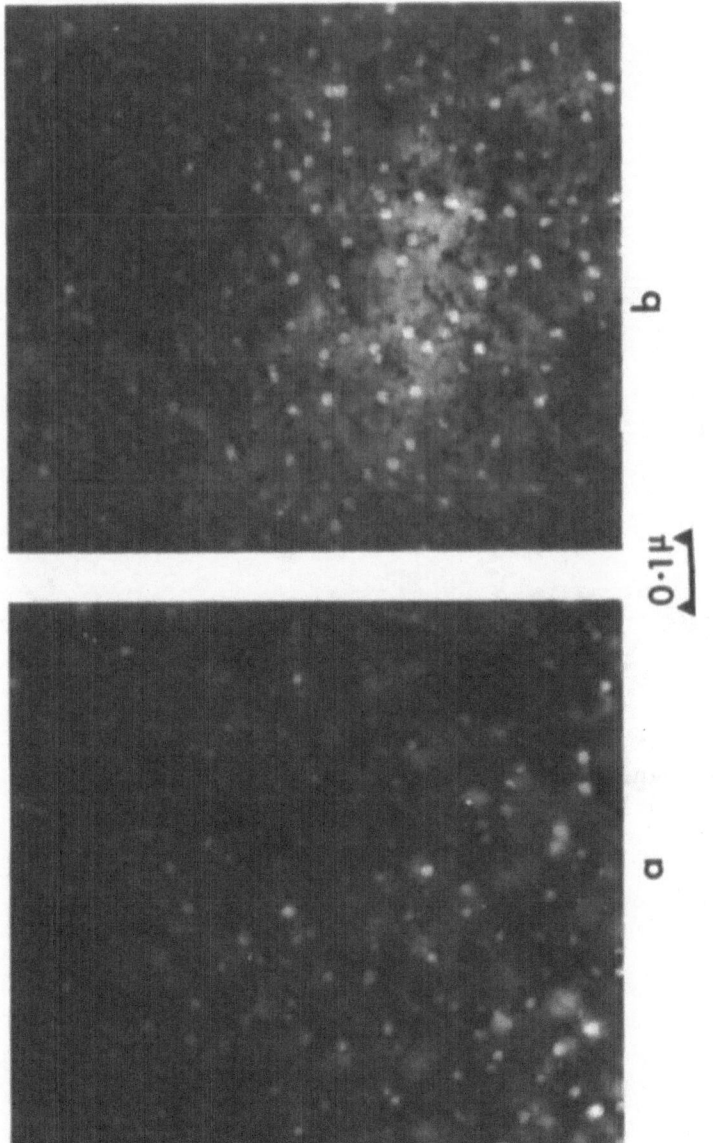

0·1μ

Fig.10 : Comparison between DFR and VEC irradiated Ni containing 10^{-5} He. Both samples were irradiated to the same calculated dose of 5 dpa.

496

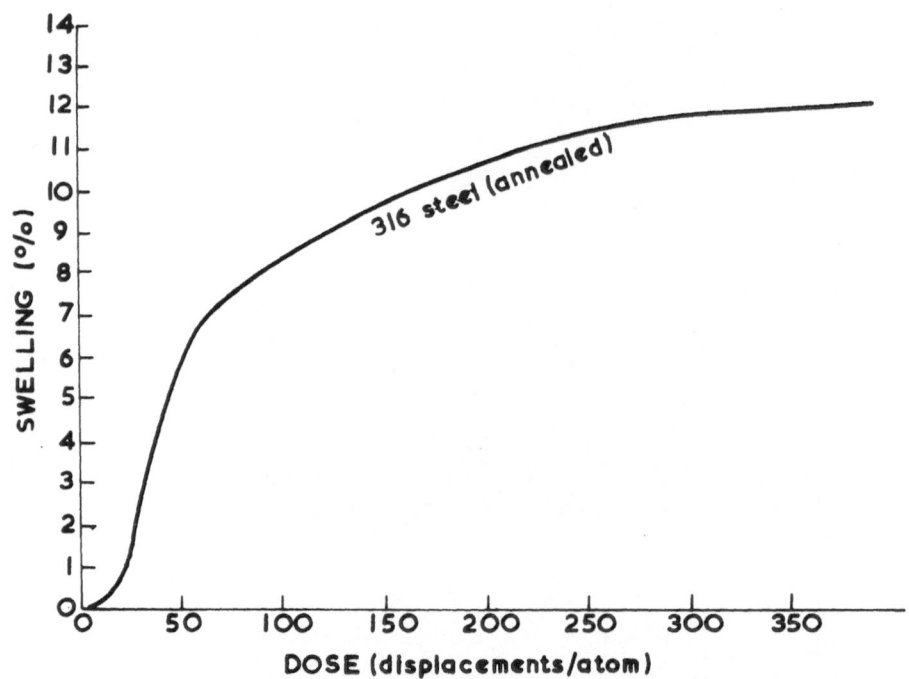

Fig.11 : Dose dependence of swelling in 316 steel at 525°C
(10^{-5} He)

the magnitude of the swelling, $\Delta V/V$. To eliminate the variability that might result from different helium concentrations, it was decided to perform these check irradiations with nickel samples taken from the same parent material, each sample having been previously annealed to the same temperature and implanted uniformly with identical concentrations of helium (10^{-5} atom/atom). A selection of samples was then irradiated in either DFR or the VEC at corresponding temperatures to equal total damage models. Figure 10 shows two electron micrographs of voids which illustrate the results of this experiment. In both cases, a computation of swelling from a knowledge of the void density, the void size and the foil thickness gives answers that, within the expected experimental error, agree remarkably well. This result therefore gives us confidence both in the simulation technique and in the model calculating damage.

As our understanding of void formation has grown, it has become increasingly evident that one of the most important parameters that influences void growth is the radiation-produced dislocation density. In relating the results of simulation experiments to those obtained from reactor irradiation, it is therefore essential to ensure that the dislocation structures are similar in nature and of comparable density. Owing to the limited reactor data, especially at high doses, it is not always possible to check this point. However, experiments on 316 stainless steel, and the above mentioned nickel experiment do, in fact, show good agreement. On the other hand, the results from low energy ion irradiations, where the penetration depth is only a few thousand angstrom units, show a somewhat reduced dislocation density relative to either the reactor irradiations or the higher-energy VEC irradiations.

The dose dependence of swelling in 316 stainless steel at 525°C is shown in figure 11 whereas the temperature dependence at one dose, 40 dpa, is shown in figure 12 for both solution treated and 20 % cold-worked steel. Figure 13 shows the dose dependence of the nimonic alloy PE 16 at 525°C for comparison.

In the above we have outlined the physics behind the important technological phenomenon of void swelling in structural components of fast reactors. It only remains for us to briefly discuss the implications of void swelling in relation to the design and economics of fast reactor design. Two important articles discussing these particular points have been published as

498

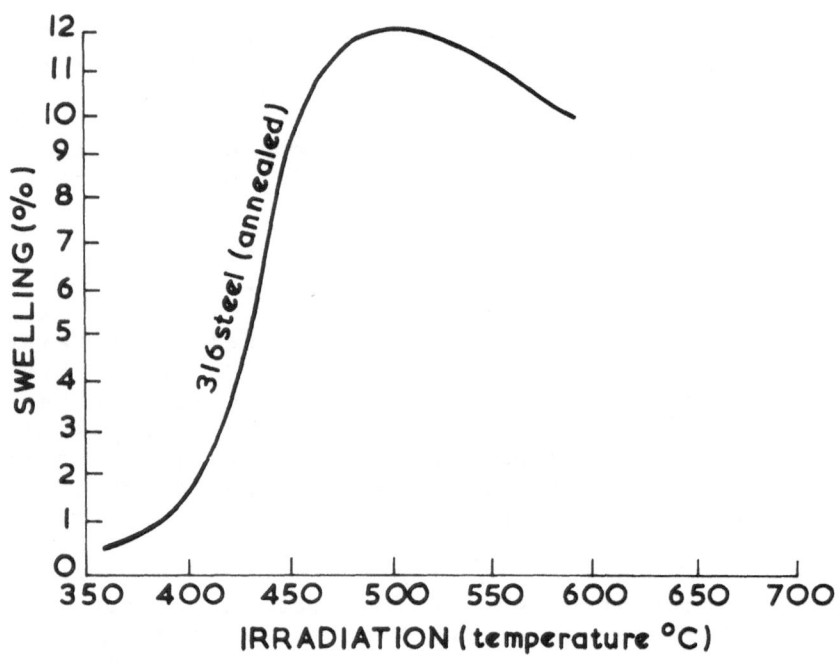

Fig. 12 : Temperature dependence of swelling in 316 steel at 40 dpa $(10^{-5}$ He)

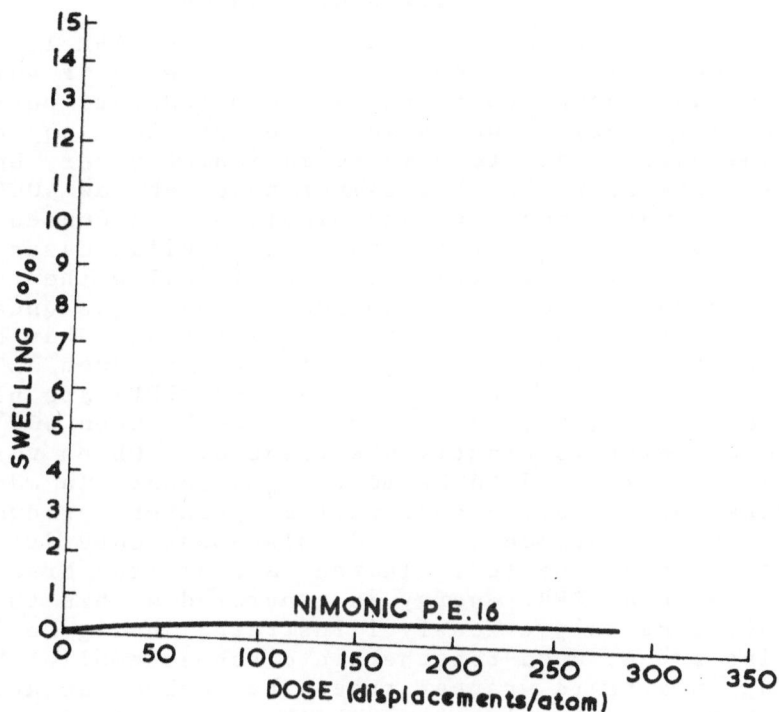

Fig.13 : Dose dependence of swelling in PE16 at 525°C (10^{-5} He)

part of the two major international conferences, e.g.
Bishop[4] and Huebotter and Bump[7]. In this review we
simply outline the salient features highlighted by
these authors; the interested reader is referred to the
original articles for detailed discussion.

Existing fast reactors such as DFR and EBR II were
designed and built long before the phenomenon of void
formation was known. Furthermore, both reactors perform
quite satisfactorily and as yet have not shown any de-
leterious effects due to void formation. However, both
reactors work at rather low temperatures -below 400°C-
where we do not expect serious problems as a consequence of voidage. On the other hand, in specific cases
the coolant flow is restricted so as to allow the temperature of operation to reach 700°C, and it is under
such a situation that most of the reactor data has been
obtained. More advanced designs of reactor, such as
the proto-type fast reactor at Dounreay (PRF) are plan-
ned to operate in the temperature range between 400°C
and 700°C, which is exactly the range over which void
formation is expected to be most significant. In addition, the total neutron dose will be greater, producing
for instance about double the displacement dose during
the life-time of the fuel. In the case of fast breeder
reactors such as PFR, energy is generated within the
long fuel pins as previously illustrated in figure 4.
Such pins are grouped together into sub-assemblies by
mounting in a rigid wrapper tube. The sodium coolant
is then required to flow through the system in a uniform and controlable manner.

The main requirement of cladding is to preserve
the overall integrity of the fuel pin in respect of
sodium leaks etc. Void swelling, as such, is not thought
to be a major problem in this context, except for possible restrictions to sodium flow. However, the greatest concern is that in some circumstances, voids could
give rise to a loss in ductility.

The major concern is the ability of sub-assembly
wrappers to withstand swelling. As we have seen void
swelling is a complex function of both damage dose and
temperature, and such wrappers will elongate by different amounts depending upon their exact position within
the core. They will also dilate, so that the design must
allow for sufficiently large gaps between sub-assemblies. Potentially the most serious effect is that differential swelling across a single sub-assembly due to
neutron flux variation away from the centre line, will
result in its bowling along its length. For instance,

it is estimated that a peak swelling of only 5% will
result in the displacement of PFR wrappers by up to
∿ 2.5 cm, if such wrappers are simply fixed at the
bottom of the core. A possible solution is the periodic rotation of sub-assemblies, however, this implies
both operational and economic penalties. Alternatively
the whole core can be restrained by a clamping ring,
but this in turn may introduce very large stresses which
are themselves undesirable. The choice of suitable materials which show resistance to void swelling is therefore of great importance.

In fact, there are many possible designs and operational solutions to accommodate void swellings in
fast reactors. In general, all such solutions involve
some economic penalty ; however, the economic penalty
is reduced if the magnitude of the swelling is itself
reduced. For this reason research into those materials
which show void resistance is a major priority. To quote the paper by Heubotter and Bump[7] ; "The cost-benefit
analysis for the development of low-swelling alloys
indicated that having to accommodate only 5 % swelling
in breeder reactors over the period 1970 to 2020 has a
1970 present worth of from $ 864,000,000 to $ 5,600,
000,000 relative to the case of 15 % swelling". Even
should these estimates prove to have been somewhat on
the high side, the potential benefit would seem to justify the research which has been carried out in this
field.

REFERENCES

1. Appleby W.K., Sandusky B.W. and Wolf W.E, in "Radiation Induced Voids in Metals" Corbett J.W. and Ianniello L.C., Eds, 1972, 156

2. Bagley K.Q., Bramman J.I. and Cawthrone C., in " "Voids Formed by Irradiation of Reactor Materials", Pugh S.F., Loretto M.H. and Norris D.I.R., Eds., 1971, 1

3. Barnes R.S. and Nelson R.S., Proc.Conf.Rad.Effects in Solids, Asheville U.S.A., 1965

4. Bishop J.F.W, in "Voids Formed by Irradiation of Reactor Materials", Pugh S.F., Loretto M.H. and Norris D.I.R., Eds., 1971, 301

5. Cawthorne C. and Fulton E.J., Nature 216, 575, 1966

6. Corbett J.W. and Ianniello L.C., Eds. "Radiation Induced Voids in Metals" Proc.Int.Conf. Albany, U.S.A., 1972

7. Huebotter P.R and Bump J.R., in "Radiation Induced Voids in Metals", Corbett J.W. and Ianniello L.C. Eds, 1972, 84

8. Nelson R.S., Mazey D.J. and Hudson J.A., J.Nucl. Mat. 37, 1, 1970

9. Pugh S.F., Loretto M.H. and Norris D.I.R., Eds., "Voids Formed by Irradiation of Reactor Materials" Proc. B.N.E.S. European Conference, Reading, 24th-25th March, 1971.

APPLICATION OF RADIATION DAMAGE EFFECTS IN DOSIMETRY

D.POOLEY

Materials Physics Division
AERE, Harwell
Great Britain

INTRODUCTION

One of the growth industries of our time is the industrial and medical use of radiation, particularly of high energy x-rays. In medecine they are used to diagnose bone fractures and discover tumours, as well as for the treatment of certain diseases and the sterilization of medical supplies. In industry x-rays are used for checking that structures are free from cracks and electrons are beginning to be used for welding and paint curing. Perhaps the most obvious sources of high energy radiation are the nuclear reactors now widely used for electrical power generation, although in practice they are so carefully controlled that they emit almost no radiation. Nevertheless a large number of people are in potential contact with radiation which can hazard their health, an it is essential that their contact with it should not be excessive. The law therefore requires that the radiation dose received by everyone likely to be in contact with radiation should be regularly and continuously monitored, hence the need for cheap and effective radiation dosimeters.

GENERAL ASPECTS OF RADIATION DOSIMETRY

Strictly speaking radiation doses to human tissue should be measured in rem (roentgen equivalent man) in which the biological effectiveness of a particular kind of radiation is taken into account as well as the energy deposited in the tissue. Thus for fast neutron irradiation a given absorbed energy will cause about ten

times as much biological damge as for x-ray irradiation. However for x-rays and fast electrons the biological effectiveness is such that a biological dose of 1
rem is achieved for a physical dose of about 1 rad, the
rad corresponding to an energy of 10 mJ. per kilogram
of absorber.

The effects of large radiation doses are well documented and very unpleasant (table 1), but the effect
of smaller doses is less well understood. Doses due to
natural background are usually around 1.5 rems per
year and in Britain the maximum permissible artifical
doses are as given in table 2. It is fairly clear therefore that radiation monitors should be able to measure radiation doses down to at least 100 m rem, with
accuracies of about 10-20 m rem.

Most types of radiation lose most of their energy
to the electrons of the atoms with which they interact,
particularly x-rays and electrons. It is therefore not
surprising that all sensitive radiation detectors depend primarily on the excitation of electrons. In semiconductor detectors transient electron hole pairs are
created by the irradiation, but they can cause an electric current to flow and the charge is related to radiation dose. In integrating solid state personnel dosimeters, on the other hand, we are trying to dind a
material where permanent effects are caused with high
efficiency by the electron excitation, in order that
cumulative doses received by the dosimeter can be measured at convenient intervals.

Photographic film is still the most widely used
dosimeter material, in the silver halide crystals dispersed in a gelatine matrix the electron-hole pairs
cause latent darkening of the film which can be developed chemically later. Photographic film dosimeters
are certainly cheap and sufficiently sensitive, but
they have two disadvantages. Firstly film is rather
variable in sensitivity and requires complex chemical
processing, which must be very carefully controlled if
charges in the film blackening are to relate only to
the radiation dose received and not to variations in
the development. Secondly silver bromide has a mean a-
tomic number of about 40 ; very different from human
tissue which is dominated by the oxygen in water and is
therefore about 8. The absorbed dose in rads in a particular material is proportional to the absorption
cross section per atom divided by the atomic weight of
the atom, that is to

Table 1. - The Biological Effects in Man of
very large whole radiation doses

Dose	Approximate time to death	Cause of death
>100 k rem	a few minutes	) damage to brain and
10-100 k rem	up to 1-2 days	) central nervous) system
1-10 k rem	3-5 days	damage to gastro-intestinal epithe-lium
200-1,000 rem	10-30 days	Damage to bone marrow cell genera-tion
<200 rem	Some life-shortening	"Premature ageing"

Table 2. - Maximum Permissible Radiation Doses
in Great Britain (to adults over 18 years)

Part of body exposed	Maximum allowed doses	
	Designated person	Non-designated person
Whole body, blood forming organs, gonads	5 rem/year average 12 rem in one year 3 rem/quarter	1.5 rem/year
Skin and bone, except hands and feet ect.	30 rem/year 8 rem/quarter	3 rem/year
Hands, arms, feet and ankles	75 rem/year 20 rem/quarter	7.5 rem/year

$$\sigma(Z)/A$$

For gamma rays $\sigma(Z)$ can vary with Z^4, Z or Z^2 depending on whether photoelectric absorption, Compton scattering or pair production is dominant, and of course even Z/A varies slowly (it is 0.50 for oxygen and 0.43 for silver). This means that the dose received by the dosimeter material will be the same as that received by human tissue only if the atomic number of its components are all around 8. In photographic film dosimetry this problem is circumvented by covering the film with a variety of filters, which in principle allows the type of radiation to be determined and the dose which would have been received by human tissue to be inferred. This technique is not altogether satisfactory however, and other solid state dosimeters have therefore been developed.

RADIOPHOTOLUMINESCENCE

In this kind of dosimetry materials are chosen in which defects are created by the radiation which are suitable for photon excited luminescence. The materials most widely used are still the silver doped phosphate glasses proposed by Schulman et al[1] who pioneered the use of these dosimeters for US Navy purposes. Material improvements have now been made (for example Yokota and Nakajima[2]) which allow dose measurements to be made down to 10 m Rad. A typical dosimeter glass is made from a base containing 47 % $LiPO_3$ and 53 % $Al(PO_3)_3$ to which 12 % $AgPO_3$ is added. Irradiation creates colour centres which have an absorption band in the near UV at about 320 nm (figure 1), and excitation at 320 nm causes efficient luminescence to occur at 640 nm. The emission intensity is a well defined and fairly linear function of dose (figure 2) and the intensity does not change on standing at room temperature after the first few hours.

Absorption in the 320 nm band can also be used for dosimetry but photoluminescence is always much more sensitive at low doses. Suppose, for example, we have an absorbed dose of 100 m Rad (6 10^{12} eV g^{-1}) in a sample about 10 mm x 10 mm x 1 mm. This dose will create at most about 10^{10} defects, even by an efficient electronic or photochemical process. If the defects are strong absorbers like F centres in alkali halides the sample will still absorb only about 10^{-5} of the incident light at the absorption band wavelength, and such a small absorption is very difficult to detect. On the other hand if the defects created luminesce efficiently

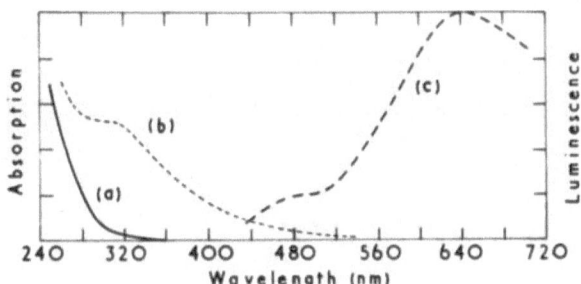

Fig.1 : Absorption and Luminescence of a silver activated phosphate glass.

      a) absorption of unirradiated glass
      b) absorption of irradiated glass
      c) luminescence of irradiated glass caused by excitation
         at 320 nm (after Schulman, 1967)

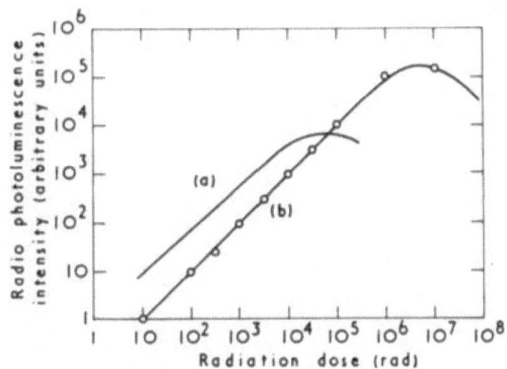

Fig.2 : The intensity of photoluminescence in
      a) silver activated phosphate glass and
      b) "pure" LiF as a function of radiation dose, glass data
        is from Schulman, 1967, LiF data from regulla, 1972.

508

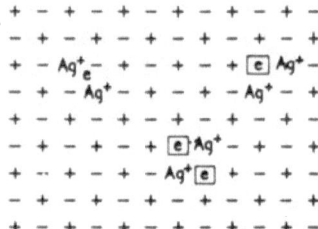

<u>Fig.3</u> : Possible atomic structures of the photoluminescence (C) centers in Ag⁺ doped KCl

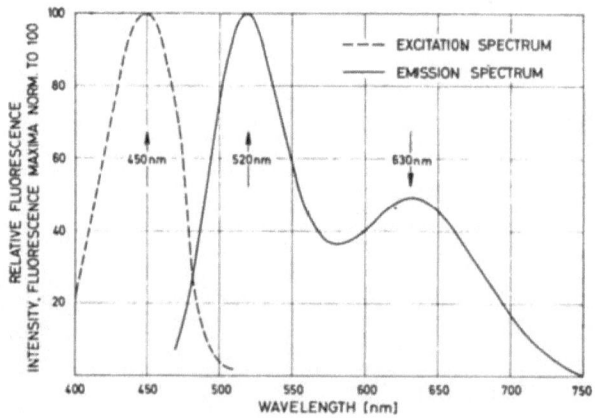

<u>Fig.4</u> : Corrected excitation and emission spectra of radiophoto-luminescence of LiF crystals (exposure : 100 kR).

then an exciting beam of only 10^{14} photons S^{-1} ($\sim$100µW) will cause 10^9 (10^{14} x 10^{-5}) defect excitations S^{-1} and up to 10^9 photons S^{-1} will be emitted. The luminescence photons have a different wavelength from those used for excitation and the two can be separated very easily with a monochromator. In photoluminescence then, distinguishing a dose of 100 m Rad from zero dose involves distinguishing between a luminescence intensity of 10^8 photons S^{-1} and zero intensity, whereas in absorption we would need to distinguish between 1.00000 10^{14} photons S^{-1} and 0.99999 10^{14} S^{-1}. At higher doses absorption becomes a relatively better technique, and eventually is more favourable than luminescence.

The way in which silver activated phosphate glass dosimeters function has been deduced from work on silver activated alkali halides (Etzel and Schulman[3]). These materials are less suitable for dosimetry because their Z is further from 8, and they are harder to prepare consistently, but they do seem to behave in the same way as the glasses and the structure of the important defects is much easier to determine in the cubic alkali halides than in the structureless glasses. Using KCl doped with Ag^+ ions it has been shown that Ag^+ ions act as traps for holes, forming Ag^{2+} ions during irradiation. Pairs of Ag^+ ions, to which chlorine ion vacancies have probably been attached, act as electron traps and during irradiation form complex defects of the type shown in figure 3. These latter centres absorb at 310 nm in KCl and emit at 556 nm, so that they seem to be analogous to the photoluminescence and centres used in the glasses.

Photoluminescence dosimetry is also possible using nominally pure alkali halides (Claffy et al[4]). In both cases M centres in LiF have been used, which absorb at 450 nm and emit at 520 nm and 630 nm (figure 4). The response given by Regulla is remarkably linear (figure 5) and he was able to measure doses as small as 10 Rad, although he used samples which were very large in achieved this (8 mm x 8 mm x 4.7 mm). The reason for the linearity is not entirely without doubt, since $M(F_2)$ centres are responsible for the emission and these are created in proportion to the square of the number of primary defects, the F centres (Sonder and Sibley[5]). However, in LiF the F centre concentration grows with the square root of the dose (Durand et al[6]) so that a linear photoluminescence dose relationship is not really surprising.

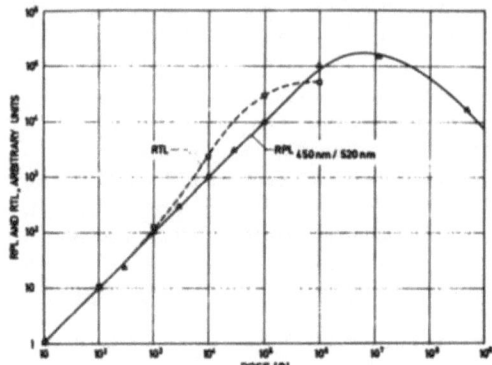

Fig.5 : RPL$_{450}$ intensity and RTL yiedl (total light) of thermoluminescent LiF (TLD-100) at a function of exposure.

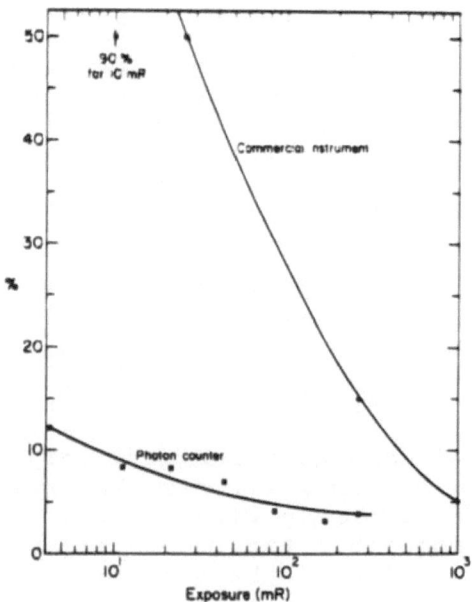

Fig.6 : Percentage standard deviation (20 TLD-100 1/8x1/8x035
ribbons).

RADIO THERMOLUMINESCENCE DOSIMETRY

Radiation induced thermoluminescence is the hot favourite to replace photographic film for routine personnel dosimetry. In this case when the dosimeter material is irradiated defects are created which are completely stable at room temperature, but which recombine when the crystal is heated and emit photons in the process. The sensitivity is potentially even higher than for photoluminescence, particularly if only simple equipment is available. No exciting light is necessary and the photons emitted can easily be counted if necessary (Schesinger et al[7]), which allows doses below 10 m Rad to be measured (figure 6).

One of the major reasons why thermoluminescence dosimetry is the favourite technique for personnel use is that LiF is a good material and has a Z very close to human tissue. $Li_2B_2O_7$ is another good material which also has a response function very close to tissue, and with these materials the range of doses which can be measured is truly remarkable (figure 7).

Although thermoluminescence in LiF has been used in dosimetry for a long time, since Boyd and Daniels[8] made the suggestion in 1949, the microscopic process involved is still very unclear. Magnesium impurities are known to be essential for proper behaviour and it is very likely that aggregates of Mg^{2+} ions and positive ion vacancies act as hole traps (Claffy[9]). It is also probable that the different thermoluminescence peaks (figure 8) correspond to different aggregates of the Mg^{2+}, positive ion vacancy dipoles (Dryden and Shuter[10]). Klick et al[11] proposed that heating caused these traps to release their holes, which then recombined with electrons in F centres to give luminescence. However, although the F centre may well act as a recombination centre in LiF containing only Mg as a deliberate impurity, it is less effective than the centres formed by small additions of a second impurity such as Ti (Zimmerman and Jones[12] ; Rossiter et al[13]). In spite of considerable research effort in this area the thermoluminescence mechanism in LiF : Mg : Ti is far from being properly defined, and a more complicated mechanism has been suggested by Mayhugh[14].

Nevertheless,thermoluminescence in LiF enjoys increasingly widespread use for dosimetry (Attox[15]) and is a good example of the usefulness of coulour centre generation in insulators.

512

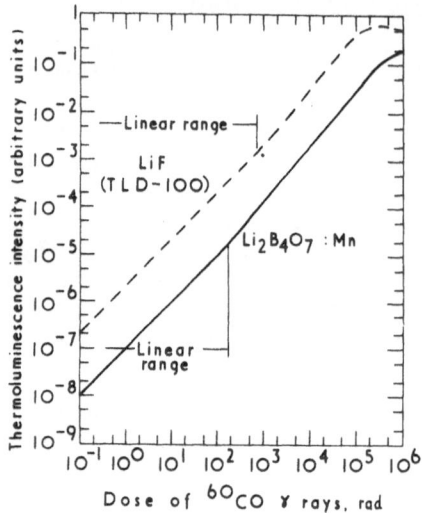

Fig.7 : Thermoluminescence output as a function of radiation dose for LiF and $Li_2B_4O_7$ (after Schulman et al, 1967)

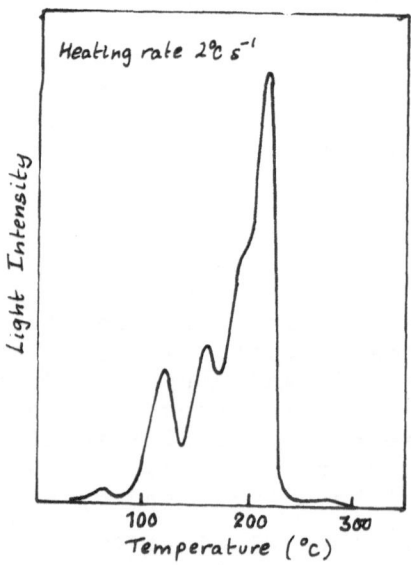

Fig.8 : Thermoluminescence in LiF : Mg

EXO-ELECTRON EMISSION

This is not a widely used dosimetry technique, to my knowledge no organisation uses it for routine personnel dosimetry. Nevertheless it is an interesting phenomena and has potential in certain areas. The dosimeter material stores energy as in thermoluminescence, in defects created by the irradiation, but in this case heating causes electrons to be emitted from the crystal surface. The electrons emitted are usually counted in a fairly conventional gas flow counter device (see for example Becker, U.S. Patent 3 484 610).

Lithium fluoride is also a reasonable exo-electron emitter but BeO is more sensitive and reproducible and the defects are more stable on standing at room temperature (Becker et al[16]). The electrons are emitted from BeO at around 300°C and the sensitivity of the surface can be destroyed by removing SiO_2 using an HF etch (figure 9). This suggests (Gammage et al[17]) that the exo-electron emission activity is associated with silicon impurities, and even that Si^{4+} can replace Be^{2+} (their respective ionic radii are 0.31 Å and 0.41 Å) and act as an electron trap.

It seems rather unlikely that exo-electron emission will seriously challenge the luminescence techniques as an x-ray personnel dosimeter, but it may eventually find use as a fast neutron dosimeter. The basis of this idea is to use a dosimeter material which is not damaged by ionisation (such as MgO), and is therefore insensitive to x-rays but sensitive to fast neutrons. The use of a hydrogenous radiator (figure 10) can give enhanced damage rates near the surface and the defects can be detected there using exo-electron emission. This method has been examined by Ritz and Attix[18], who can measure doses down to a few rads and achieve a reasonably high ratio of neutron to gamma sensitivity (figure 11).

The trend in personnel dosimetry is such that thermoluminescence will steadily replace film and photoluminescence techniques over the next decade, and that more unusual techniques, such as exo-electron emission, will make inroads only for special purposes.

514

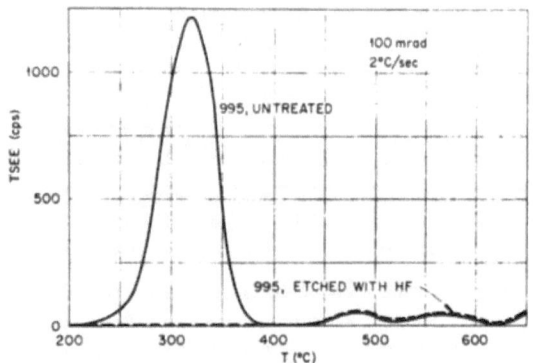

Fig.9 : Radiation induced TSEE in Thermalox 995 before and after removal of SiO_2 by etching with concentrated HF.

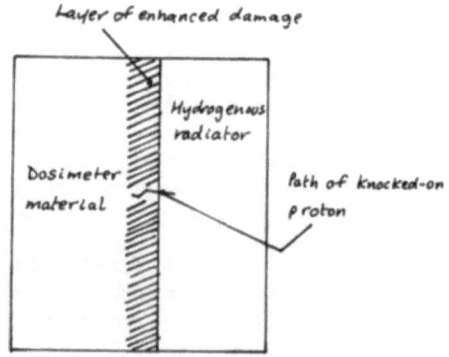

Fig.10 : Hydrogenous radiators in fast neutron dosimetry.

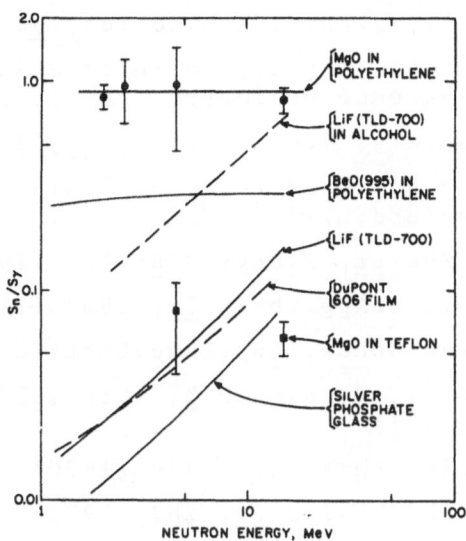

Fig.11 : Relative sensitivity to neutrons and gamma rays.

REFERENCES

1. Schulman et al., J.App.Phys. 22, 1479 (1951)

2. Yokota and Nakajima, Health Physics 11, 241 (1965)

3. Etzel and Schulman, J.Chem.Phys. 22, 1549 (1954)

4. Claffy et al., Proc. 3rd.Inst.Conf. on Lumin.Fos. Risø, 1971 ; Regulla, Health Physics, 22, 491, (1972)

5. Sonder and Sibley, Phys.Stat.Sol. 10, 99 (1965)

6. Durand et al., J.Phys.Chem.Solids, 30, 1353 (1969)

7. Schesinger et al, Proc. 3rd International Conference on Luminescence Dosimetry - Risø, 1971

8. Boyd and Daniels, J.Chem.Phys., 17, 1221 (1949)

9. Claffy, Proc.Int.Conference on Luminescence Dosimetry - Stanford, 1965

10. Dryden and Shuler, J.Phys.D 6, 123 (1973)

11. Klick et al., J.App.Phys. 38, 3867 (1967)

12. Zimmerman and Jones, App.Phys.Lett.10, 82 (1967)

13. Rossiter et al, J.Phys. D 3, 1816 (1970) ; J.Phys. D 4, 1245 (1971)

14. Mayhugh, J.App.Phys. 41, 4776 (1970)

15. Attox, Health Physics 22, 287 (1972)

16. Becker et al., Health Physics 19, 391 (1970)

17. Gammage et al., Health Physics, 22, 57 (1972)

18. Attix, App.Phys.Letters Aug.1, 1973

APPLICATION OF RADIATION DAMAGE IN INFORMATION STORAGE

D.POOLEY

Materials Physics Division
A.E.R.E. - Harwell
Great Britain

INTRODUCTION

Except where radiation damage is being discussed specifically it is probably rare for most of us to remember that photographic film, our most widely used information storage medium, relies for its operation on a ra iation damage process, but as Slifkin has described it does. As a medium for the more exacting requirements of information storage it suffers from the two major disadvantages of being irrerversible and of needing chemical development and fixing, though on the credit side it is certainly very cheap and widely used. The increasing use of electronic computers and other electronic devices has, moreover, generated a great need for cheap methods of information storage, and in aiming to satisfy it those methods which, like film, rely on radiation effects offer exciting possibilities. I will start by talking rather generally about information storage requirements and memory devices based on radiation damage, and then look in more detail at four quite different kinds of memory planes which have been developed or are being developed. Reviews which are somewhat similar are available (Towsend and Kelly[1]); Pooley and Sibley[2]).

INFORMATION STORAGE REQUIREMENTS

Information for computers is usually measured in bits (binary digits) with each bit having the value 0 or 1. Many of the most reliable electronic methods of storing and processing information, such as "flip-flop" circuits, have only two possible states and therefore make natural stores of binary information, with each "flip-flop" storing one bit. However, even a small book will contain about 1M bit of information (say 100 pages x 300 words per page x 6 letters per word x 5 bits/letter assuming a simple code such as A = 00001, B=00010, C = 00011... Z = 11010) and we therefore require a million flip-flops for its storage. Moreover, modern computers can process information at about 100M bits per second which means that the computer will need many millions of flip-flops in its memory to operate for even a fraction of a second.

On the othe hand a single TV picture also contains more than 1M bit of information (about 600 x 600 picture points with any one of about 3-50 brightness levels (5-6 bits) at each picture point. It is perhaps surprising to find that a single TV picture contains as much information as a whole book, the comparison certainly illustr tes the power and potentiel cheapness of information sotrage using a memory plane which is basically like photographic film but rather more flexible . The fundamental reason why film is so cheap and "flip-flops" are relatively expensive is that storage in film is "geographical" in character whereas "flip-flops are digital. By geographical I mean that any given bit corresponds to a particular location on an otherwise homogenous memory plane, unlike digital storage where each bit corresponds to a discrete piece of hardware, which must be manufactured individually.

Computers, then are fast and process large quantities of information. The ideal memory is therefore one whose speed can match the computers, but whose price is low enough to allow a large capacity to be obtained at a reasonable cost. Traditionally compromises have had to be made ; ferrite core memories have been fast, digital in character but rather expensive and magnetic discs and tape have been cheap, geographical in character but rather slow. Two different approaches to better memories for computers are therefore apparent, and are being taken. On the one hand"flip-flop" circuits, which are basically very fast, are being made more cheaply (figure 1) by means of putting ever more circuits on to a given area of silicon. They have already ousted ferrite cores from the market place, since they cannot be made more cheaply in this way. On the other hand geographical memories, which are basically very cheap, are being sought which can be accessed with a light or

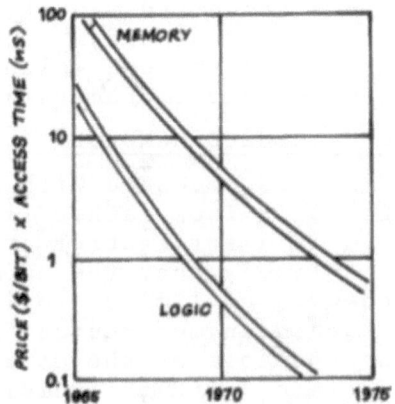

Fig.1 : Fall in semiconductor prices with time.

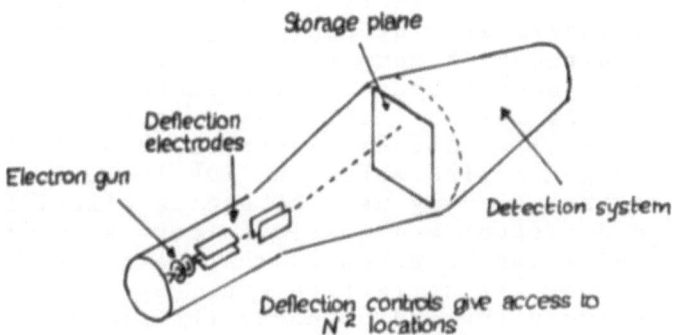

Fig.2 : An electron beam accessed memory.

electron beam rather than mechanically and can therefore be much faster. It is in this latter area that memory planes based on radiation and radiation induced defects are being so actively studied, we are seeking a new kind of photographic film, one which can be reversed and does not need chemical processing.

ELECTRON AND PHOTON ACCESSED MEMORIES

If a memory plane is accessed with an electron beam the memory device will look rather like a cathode ray tube (figure 2.) If a straightforward deflection system is used it is likely that the number of bits which can be stored will be limited to 10^6 to 10^7 per tube. A number of tubes can, of course, be connected in parallel so that the cost of the most difficult electronic task associated with the memory, which is providing very reproducible deflection voltages, can be shared between many tubes. With this kind of sharing it seems likely that the cost of such a memory would be of the order of £ 300 for 10^7 bits, that is 3mp/bit. The access time to any bit in the memory is likely to be less than 10 μs, and the data rate limit set by the electron beam device is unlikely to be less than 10MHz. This cost and access time have been added to those for more conventional memories in figure 3, which shows how attractive these memories can be. Two stage electron beam deflection tubes promise 10^9 bits for £ 1000 which is even more attractive.

Light accessed memory planes have different features. Only lasers are intense enough to achieve the speeds required by a computer and reliable lasers of fairly high power (about 1W) are expensive. Moreover the laser beam must be deflected by a fairly complex electro-optical or accousto-optical device, which can rarely manage to give more than 10^6 locations at the same time as achieving μs like access times. Since the laser and deflector are likely to cost about £ 5000, if only 10^6 bits can be accessed the overall cost (0.5p/bit) is not at all attractive. One way out of this problem is to move the medium under the field of view of the laser, but that implies mechanical movement and therefore long access times. The solution most likely to be satisfactory is to make each bit accessed by the laser into a halogram, which images the coherent laser light onto another array, which would be of photodetectors for reading (figure 4). If the laser itself can access say $10^5 - 10^6$ locations and each hologram contains 10^4 bits then a prodigious $10^9 - 10^{10}$ bits can be stored. The various hardware necessary for laser holo-

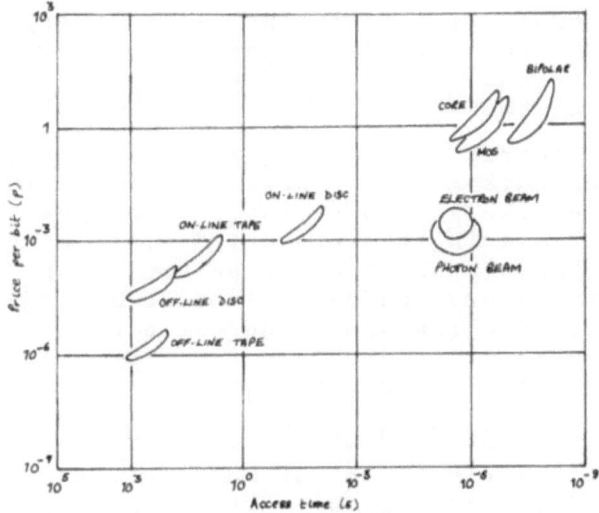

Fig.3 : Price/performance relationship of storage media.

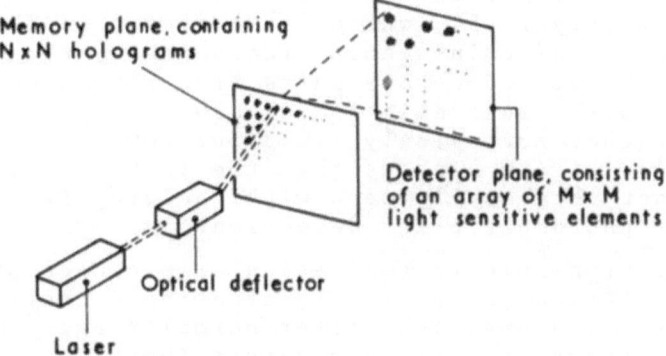

Fig.4 : A schematic diagram of readout from an optical holographic memory. The laser beam is deflected on to any one of N^2 holograms each hologram giving a reconstructed image of M^2 bits on the detector plane.

graphic storage is complex and seems likely to cost £ 20-30,000 even when developed, but still it gives a per bit price of only 0.3-3mp and the access time is likely to be about 1 μs.

The electron beam memory is word oriented whereas photographic memories are page oriented ; in the first case we are likely to have input or output 10-50 bits in parallel and in the second 10^4 bits. Nevertheless a successful memory of either type could capture the bulk of the main memory and magnetic disc market, the combined value of which now stands at about £ 2000 M per year. The major obstacle in exploiting this market is in finding a good memory material.

INHIBITABLE CATHODOLUMINESCENT PHOSPHORS

One attractive class of materials for electron beam accessed memory planes has been examined by Harwell (Bishop et al[3]). The basic phenomenology is fairly simple and can best be explained by using as an example "inhibitable cathodoluminescence" in KI:Tl.

When high energy electrons, and for device operation this means 10-30 kV, interact with insulators they create electron-hole pairs, and in KI:Tl the electrons can be quickly trapped by Tl^+ ions to form $Tl^{\bullet}$ atoms. It turns out that about half the holes created with the electrons find these $Tl^{\bullet}$ atoms very quickly (within 200 ηs) and reconvert them to Tl^+. In this recombination process light is emitted in a broad band centred at 420 nm (figure 5), which is in fact a kind of perturbed exciton luminescence. Incidentally the other half of the electron-hole pairs also recombine at Tl^+ ions and also cause emission of 420 nm light, but this occurs rather more slowly (in about 200 μs at room temperature). Figure 5 shows that the light emitted, though not very easily seen with the eye, is almost ideal for photomultiplier detection.

Electron-hole recombination also causes photochemical radiation damage in KI, creating F centres and interstitial atoms. The latter normally aggregate into large clusters at room temperature (Hobbs et al[4]) but when the crystal contains large numbers of impurity atoms the interstitials combine with these rather than with each other. It turns out that the complexes of Tl^+ ions and interstitial atoms behave very differently from the bare Tl^+ ions. Electron hole recombination does still take place at the complexes, more or less as it does ar the bare Tl^+ ions, but no light is emitted

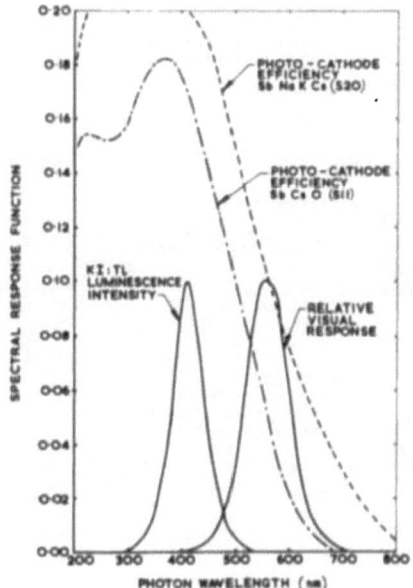

Fig.5 : Detection of luminescence from KI : Tl

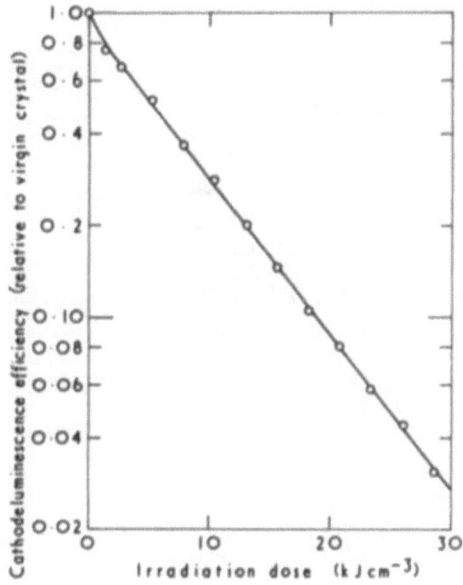

Fig.6 : The fall in cathodo-luminescence efficiency of KI : Tl
with electron irradiation dose.

1 Electron beam inhibition

2 Electron beam restoration

Fig.7 :

3 Storage density

4 Material uniformity

Fig.8 :

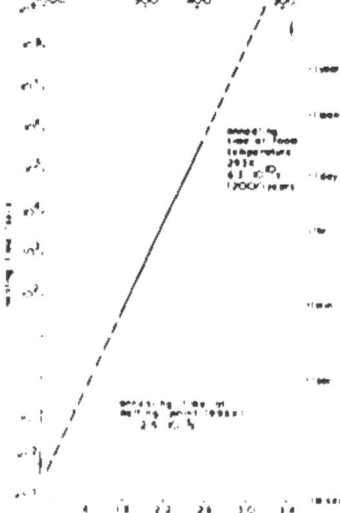

in the process.

As a result KI:Tl, which initially luminesces very brightly under electron irradiation slowly loses its efficiency if continuously excited, because the electrons create interstitial atoms which are trapped at the Tl^+ activator ions and prevent their luminescing. This fall in light output is very large, being exponential with irradiation dose and covering more than two orders of magnitude as shown in figure 6. A part of a crystal which has been inhibited in this way can easily be distinguished from a part which has not by virtue of this large difference in luminescence efficiency, see figure 7. In a memory device a fairly intense electron beam can be used to write a pattern of information which can then be read with the same electron beam at much lower current. Defects are created in the reading process just as during writing, so that reading a location a large number of times will eventually convert it into the inhibited or dim state. However, the electron dose required for reading is so small (<5pJ) that each location can in fact be read a large number of times ($10^5 - 10^8$) before this happens.

So far we have no erase process. As you might guess the defects created by electron irradiation can be destroyed by heating, when luminescence is restored (figure 8). However, what makes the KI:Tl memory plane particularly attractive is that if the electron beam used for writing and reading is made very intense the Tl^+ ion/interstitial atom complexes are destroyed by the electron beam, and the crystal appears to return to the virgin state. Although this is almost certainly due in fact to the defect annealing caused by the heating effect of the intense electron beam, the speed of the process is increased by many orders of magnitude over purely thermal annealing. Annealing with a 15 kV 3 μA beam, which heats the crystal to 400°C and should therefore take $\sim$ 1s, takes only 1 μs. The likely reason for this is that the ionisation caused by the electron beam excites defects into states where they break up and move around much more readily than in the unexcited state. Figure 7 shows how this defect annealing process can be effective, and figure 9 shows a pattern of more than 10^6 bits, stored on a crystal about 10 mm square.

In many ways the Harwell memory plane is a particularly attractive one for an electron beam accessed memory. The plane itself is easy to fabricate and made from fairly cheap materials. The electron beam determi-

526

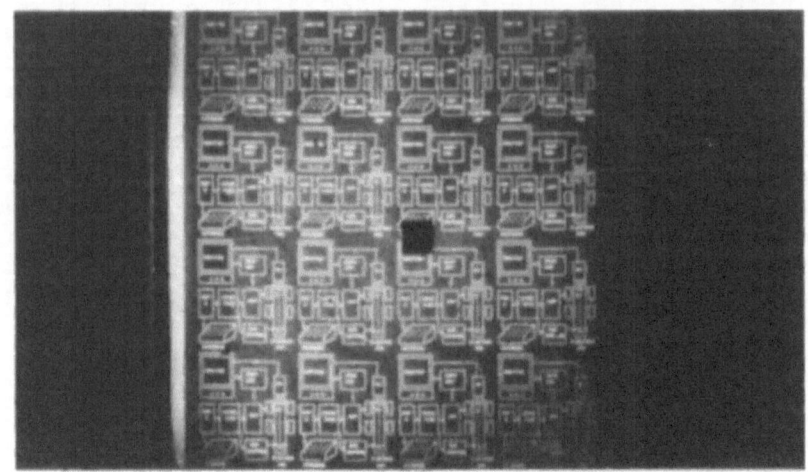

<u>Fig. 9</u> : Pattern of 10^6 bits, stored on a 10 mm square crystal

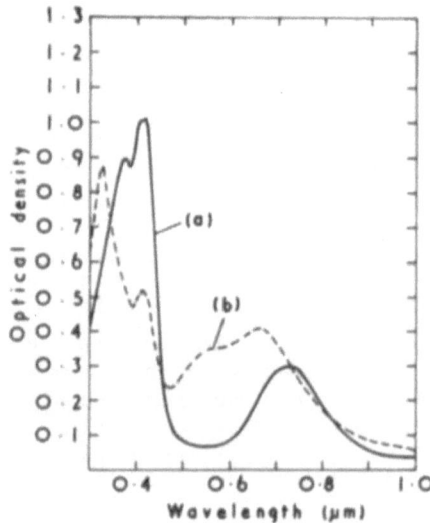

<u>Fig.10</u> : The optical absorption spectra of photochromic CaF_2:La:
<u>Na.</u>   a) stable state
       b) after irradiation in the range 380-460 nm

nes its own set of memory locations, which are automatically the same for read, write and arase processes because an electron beam of the same energy is used for each. It does not much matter if the array of locations is slightly distorted from a perfect square lattice, nor does the array have to be positioned very precisely with respect to the memory plane itself, and these tolerances make construction easier and cheaper. There are other important advantages, such as the very large number of bits which can be reached through a single input-output channel.

One obstacle in the path of the widespread application of this memory is the degredation of a particular storage element which occurs when it is cycled through the dim and bright states many times. The materials available at present have a life limited to only 10^4 reversals, but this is still potentially useful in many applications, because of the large number of times that each location can be read without refreshing and because the degradation can, in any case, be removed by thermal annealing. In other words the Harwell memory as it now stands provides a good read-mostly memory, but better materials will be needed for fully reversible operation.

PHOTOCHROMIC AND PHOTODICHROIC MEMORIES

A very large number of materials have been examined for possible use as laser beam accessed planes, but I will discuss only two, and them only briefly.

The first are based on F_{RE} centres in CaF_2 (Staebler and Kiss[5]) and have been reviewed by Duncan[6]. I have called them F_{RE} in analogy with F_A centres in alkali halides ; these are F centres with one near neighbour foreign alkali ion, and the F_{RE} centres in CaF_2 are F centres with one near neighbour rare earth ion in replacing a Ca^{2+} ion. Figure 10 shows the absorption spectrum of a crystal containing F_{RE} centres where the rare earth ions are lanthanum. There are two strong absorption bands due to the F_{RE} centres, labelled A and B in figure 10, at 400 nm and 720 nm. If a centre absorbs a photon in the 400 nm band it is ionised with very high probability, whereas absorption at 720 nm does not cause ionisation. If the crystal also contains rare earth ions which can trap the electrons which leave the F_{RE} centres, in the lanthanum case they are substitutional La^{3+} ions, then irradiation in the A band causes the following transformation to occur.

$$F_{RE} + R_E \xrightarrow{h\nu_A} F_{RE} + RE^-$$

Although the ionised state is unfortunately not very stable, it does have a lifetime of about a week when RE is Ce and of about 1 day for La, so that information can be written into the crystal with light in the A band and stored there for some time. If necessary it can be read non-destructively using light in the B band. The RE⁻ centre, which for the case of Lanthanum is La^{2+}, also has a strong absorption band which also causes ionisation, and is shown as C onfigure 10. With the aid of this light the ionised centres can be quickly returned to the unionised state as follows,

$$F_{RE}^+ + RE^- \xrightarrow{h\nu_C} F_{RE} + RE$$

giving the system its erase process.

The material therefore has write and arase processes, which are almost as efficient as in theoretically possible, and a read process which is effectively non-destructive. Moreover, no measurable degradation occurs when the material is cycled repeatedly from the switched to the unswitched state and back again. However, there are three disadvantages to the use of CaF_2 at present. First the switched state is not very stable, as we have already said. Secondly it is difficult to achieve high concentrations of centres in either the switched or unswitched states, which means that relatively thick storage planes must be used to give reasonable optical density changes. A thick plane means that fairly large areas must be used for each bit of information ; with the present materials an area of about 200 mm x 200 mm would have to be used for 1 million bits and this is rather large.

A third serious obstacle to the use of this photochromic material as an optical memory plane arises from its seemingly attractive feature of three distinct optical absorption bands for write, erase and read operations. For all optical memories lasers are the only sources of light which are sufficiently intense to carry out the read or write operations quickly enough for computer use. Laser beams can be deflected, with electro-optic or accousto-optic devices, but as yet it is impossibly difficult to arrange to have two or three different lasers scanned over the same array of locations, and halography under these conditions is impossible.

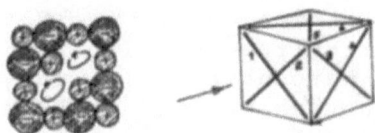

Fig.11 : The structure and possible orientations of M.centers in alkali halides

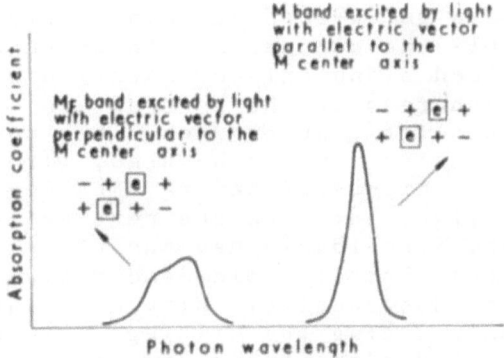

Fig.12 : The transitions of the M center in alkali halides

Photodichroic optical memories, such as were proposed by Bron et al[7] and Schneider[8], do overcome some of these problems. Schneider has concentrated on M centres in alkali halides, particularly in NaF (Collins and Schneider[9]), and I will use them as an example. The M centre consists of two neighbouring F centres and can therefore adopt 6 different orientations (figure 11). Its optical properties are illustrated in figure 12, showing that it has two strong absorption bands. The so called M band, which is at 500 nm in NaF, is excited when the electric vector of the light is parallel to the axis of the centre, while the M_F band, which is at 350 nm in NaF, requires the electric vector of the light to be polarised perpendicular to the centre. M band irradiation does not change the orientation of the centres but M_F band orientation can do.

If light in the M_F bans is incident on a crystal as shown in figure 11 and polarised parallel to orientation 1 only centres with orientation 1 are unexcited, so that eventually all the centres adopt this orientation. This is the writing process ; M_F light polarised along orientation 1 eventually causes all the centres to adopt this orientation, which we might arbitarily call a binary zero, and if the light is polarised along orientation 2 all the centres move to that orientation, a binary one. This means that write and erase processes use light at the same wavelength which is a big step forward over CaF_2 : La.

Reading can be carried out non-destructively with M light : if all the centres are in orientation 1 then M light polarised along this orientation will be strongly absorbed, whereas if the centres are in orientation 2 there will be little absorption. This does mean, however, that wo lasers have to be used, with the attendent problems we discussed before. Reading is also possible with M_F light but then the centres are reoriented during reading. Schneider[10] has therefore invented a way in which the photodichroic memory plane can be read effectively non-destructively with M_F light. This method relies on reducing the writing efficiency of the beam of M_F light during the reading operation by simultaneously irradiating the whole crystal with infra-red light. The effect of this irradiation on the writing efficiency is shown in figure 13, although it is not yet clear if a comparable effect can be readily achieved at laser intensities. Schneider supposes the effect to be due to the suppression of the M^+ centre concentration by bleaching in bands due to defects which have

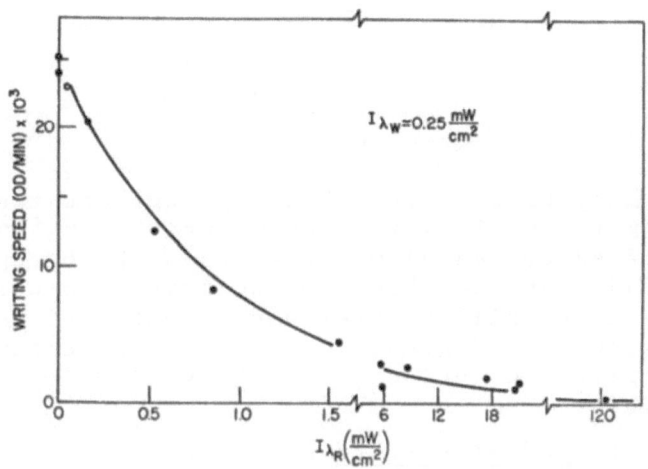

Fig.13 : Suppression of M-centre reorientation by infra-red irradiation (Schneider, 1973)

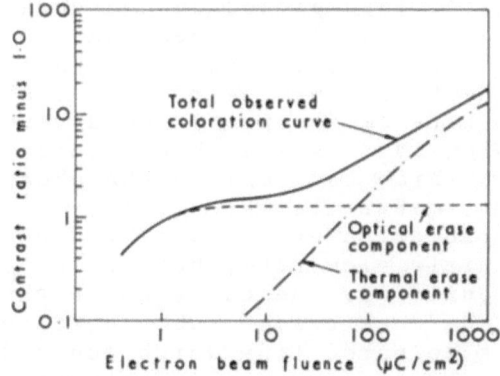

Fig.14 : The production of F centers in photosensitive sodalite by 20 kV electron irradiation. The measurement is of diffuse reflectance contrast ratio which gives a good indication of F center concentration (after Faughnan and Shidlovsky, 1972).

trapped the M centre electron, He supposes the processes involved to be as follows.

$$M + D \underset{\lambda_{IR}}{\overset{\lambda_{MF}}{\rightleftharpoons}} M^+ + D^-$$

$$M^+ \quad \overset{\lambda_{MF}}{\longrightarrow} \quad \text{reorientated } M^+$$

In many ways photodichroic planes are the most
promising for optically accessed memory devices and a
number of experimental memories using holographic techniques and a photodichroic plane have been built. As
yet they have still to overcome a number of remaining
problems such as degredation or an inconvenient operating temperature, but none of these problems appear
insuperable.

DARK TRACE DISPLAY DEVICES

Display devices represent a rather different information storage requirement. The commonest visual
display device is a TV set, but it is, in fact, rather
uncommon in that there is no need for the picture to
be stored on the tube for more than a small fraction
of a second. In most other applications, particularly
in radar and in displays driven by computers it would
be a great advantage if the display tube itself could
"store" the picture for at least several seconds. For
example if the computer had to notify the display only
when it needed to make a change in the displayed information, then the load placed on the computer by the
display would be greatly reduced. In radar also it is
convenient for the display to store information for at
least the time taken for the radar aerial to rotate,
and long persistence phosphors are a very unsatisfacto-
ry way of doing this. A search has therefore been made
for materials which can replace the phosphor in the
display tube ; materials which will colour efficiently
under the irradiation by the electron beam and will
decay at a fairly slow rate, but not so slowly that too
many images build up. The tubes which have been developed are called "dark trace" tubes because the electron
beam writes a coloured line which looks dark on the
white background, in contrast with normal phosphors
where the area electron irradiated is bright. Dark trace tubes have the added advantage that they can be viewed as easily in bright as in dim light, which is important for use in ships and aircraft.

The first tubes, which were developed during World

War II, used KCl and were called Skiatrons. They were not in the end very successful, partly because insufficient optical contrast was available and partly because the KCl degraded during use. Because of this new materials have been developed, notably complex aluminosilicates containing halides ions called sodalites ($Na_6Al_6Si_6O_{24} \cdot 2NaBr$, see Phillips and Kiss[11] ; Faughnan and Shidlovsky[12]). The production of F centres in sodalites is phenomenologically very like that in alkali halides, although the production mechanisms must be different in detail. F centres are definitely produced in two stages (figure 14) and, as in alkali halides, the first stage is a consequence of impurities and rather fast (100 ev/F centre).

Early efforts to use sodalites in dark trace displays concentrated on the fast stage F centres, with the advantages of high writing speed and easy optical erasure in mind. They failed primarily because of the interference of slow stage F centres, which built up in time and which could not be erased otpically. As a result workers at RCA now favour the use of fairly "pure" sodalite, where the fast stage colouration is suppressed and slow stage F centres can be produced more efficiently. Even in this mode sodalite is much better than KCl in that a greater density of F centres is possible, giving much better optical contrast. The images then have to be erased thermally (Heyman et al[13]) but this can be achieved in just a few seconds. A dark trace tube display is shown in figure 15.

In looking at radiation damage processes and radiation induced defects applied to information storage we have seen that the potential rewards of success are very high, but nowhere has a real breakthrough yet been achieved. In dosimetry, on the other hand, we have an application where many of the effects which have been discussed at this meeting are very important and widely used, though the scale of the resulting business is nowhere near as high as information storage would be.

534

REFERENCES

1. Towsend and Kelly, "Colour centres an imperfections in insulators and semiconductors", Sussex University Press, 1973

2. Pooley and Sibley, "Colour centres and Radiation Damage" AERE-R.7347

3. Bishop et al., App.Phys.Lett. 20, 504 (1972)

4. Hobbs et al., Proc.Roys.Soc. A 332, 167 (1973)

5. Staebler and Kiss, App.Phys.Lett., 14, 93 (1969)

6. Duncan, RCA Review, 33, 248 (1972)

7. Bron et al., U.S. Patent 3, 466-616 (1965)

8. Schneider, Appl.Opt., 6, 2197 (1967)

9. Collins and Schne der, Phys.Stat.Solidi 51, 769 (1972)

10. Schneider, Appl.Optics, 10, 980 (1971)

11. Phillips and Kiss, Proc.IEEE, 56, 1072 (1968)

12. Faughnan and Shidlovsky, RCA Review, 33, 273 (1972)